TRUNCATED AND CENSORED SAMPLES

STATISTICS: Textbooks and Monographs

A Series Edited by

D. B. Owen, Coordinating Editor
*Department of Statistics
Southern Methodist University
Dallas, Texas*

R. G. Cornell, Associate Editor
for Biostatistics
University of Michigan

W. J. Kennedy, Associate Editor
for Statistical Computing
Iowa State University

A. M. Kshirsagar, Associate Editor
for Multivariate Analysis and
Experimental Design
University of Michigan

E. G. Schilling, Associate Editor
for Statistical Quality Control
Rochester Institute of Technology

Vol. 1: The Generalized Jackknife Statistic, *H. L. Gray and W. R. Schucany*
Vol. 2: Multivariate Analysis, *Anant M. Kshirsagar*
Vol. 3: Statistics and Society, *Walter T. Federer* (out of print)
Vol. 4: Multivariate Analysis: A Selected and Abstracted Bibliography, 1957-1972, *Kocherlakota Subrahmaniam and Kathleen Subrahmaniam* (out of print)
Vol. 5: Design of Experiments: A Realistic Approach, *Virgil L. Anderson and Robert A. McLean*
Vol. 6: Statistical and Mathematical Aspects of Pollution Problems, *John W. Pratt*
Vol. 7: Introduction to Probability and Statistics (in two parts), Part I: Probability; Part II: Statistics, *Narayan C. Giri*
Vol. 8: Statistical Theory of the Analysis of Experimental Designs, *J. Ogawa*
Vol. 9: Statistical Techniques in Simulation (in two parts), *Jack P. C. Kleijnen*
Vol. 10: Data Quality Control and Editing, *Joseph I. Naus* (out of print)
Vol. 11: Cost of Living Index Numbers: Practice, Precision, and Theory, *Kali S. Banerjee*
Vol. 12: Weighing Designs: For Chemistry, Medicine, Economics, Operations Research, Statistics, *Kali S. Banerjee*
Vol. 13: The Search for Oil: Some Statistical Methods and Techniques, *edited by D. B. Owen*
Vol. 14: Sample Size Choice: Charts for Experiments with Linear Models, *Robert E. Odeh and Martin Fox*
Vol. 15: Statistical Methods for Engineers and Scientists, *Robert M. Bethea, Benjamin S. Duran, and Thomas L. Boullion*
Vol. 16: Statistical Quality Control Methods, *Irving W. Burr*
Vol. 17: On the History of Statistics and Probability, *edited by D. B. Owen*
Vol. 18: Econometrics, *Peter Schmidt*
Vol. 19: Sufficient Statistics: Selected Contributions, *Vasant S. Huzurbazar (edited by Anant M. Kshirsagar)*
Vol. 20: Handbook of Statistical Distributions, *Jagdish K. Patel, C. H. Kapadia, and D. B. Owen*
Vol. 21: Case Studies in Sample Design, *A. C. Rosander*
Vol. 22: Pocket Book of Statistical Tables, *compiled by R. E. Odeh, D. B. Owen, Z. W. Birnbaum, and L. Fisher*

Vol. 23: The Information in Contingency Tables, *D. V. Gokhale and Solomon Kullback*
Vol. 24: Statistical Analysis of Reliability and Life-Testing Models: Theory and Methods, *Lee J. Bain*
Vol. 25: Elementary Statistical Quality Control, *Irving W. Burr*
Vol. 26: An Introduction to Probability and Statistics Using BASIC, *Richard A. Groeneveld*
Vol. 27: Basic Applied Statistics, *B. L. Raktoe and J. J. Hubert*
Vol. 28: A Primer in Probability, *Kathleen Subrahmaniam*
Vol. 29: Random Processes: A First Look, *R. Syski*
Vol. 30: Regression Methods: A Tool for Data Analysis, *Rudolf J. Freund and Paul D. Minton*
Vol. 31: Randomization Tests, *Eugene S. Edgington*
Vol. 32: Tables for Normal Tolerance Limits, Sampling Plans, and Screening, *Robert E. Odeh and D. B. Owen*
Vol. 33: Statistical Computing, *William J. Kennedy, Jr. and James E. Gentle*
Vol. 34: Regression Analysis and Its Application: A Data-Oriented Approach, *Richard F. Gunst and Robert L. Mason*
Vol. 35: Scientific Strategies to Save Your Life, *I. D. J. Bross*
Vol. 36: Statistics in the Pharmaceutical Industry, *edited by C. Ralph Buncher and Jia-Yeong Tsay*
Vol. 37: Sampling from a Finite Population, *J. Hajek*
Vol. 38: Statistical Modeling Techniques, *S. S. Shapiro*
Vol. 39: Statistical Theory and Inference in Research, *T. A. Bancroft and C.-P. Han*
Vol. 40: Handbook of the Normal Distribution, *Jagdish K. Patel and Campbell B. Read*
Vol. 41: Recent Advances in Regression Methods, *Hrishikesh D. Vinod and Aman Ullah*
Vol. 42: Acceptance Sampling in Quality Control, *Edward G. Schilling*
Vol. 43: The Randomized Clinical Trial and Therapeutic Decisions, *edited by Niels Tygstrup, John M. Lachin, and Erik Juhl*
Vol. 44: Regression Analysis of Survival Data in Cancer Chemotherapy, *Walter H. Carter, Jr., Galen L. Wampler, and Donald M. Stablein*
Vol. 45: A Course in Linear Models, *Anant M. Kshirsagar*
Vol. 46: Clinical Trials: Issues and Approaches, *edited by Stanley H. Shapiro and Thomas H. Louis*
Vol. 47: Statistical Analysis of DNA Sequence Data, *edited by B. S. Weir*
Vol. 48: Nonlinear Regression Modeling: A Unified Practical Approach, *David A. Ratkowsky*
Vol. 49: Attribute Sampling Plans, Tables of Tests and Confidence Limits for Proportions, *Robert E. Odeh and D. B. Owen*
Vol. 50: Experimental Design, Statistical Models, and Genetic Statistics, *edited by Klaus Hinkelmann*
Vol. 51: Statistical Methods for Cancer Studies, *edited by Richard G. Cornell*
Vol. 52: Practical Statistical Sampling for Auditors, *Arthur J. Wilburn*
Vol. 53: Statistical Signal Processing, *edited by Edward J. Wegman and James G. Smith*
Vol. 54: Self-Organizing Methods in Modeling: GMDH Type Algorithms, *edited by Stanley J. Farlow*
Vol. 55: Applied Factorial and Fractional Designs, *Robert A. McLean and Virgil L. Anderson*
Vol. 56: Design of Experiments: Ranking and Selection, *edited by Thomas J. Santner and Ajit C. Tamhane*
Vol. 57: Statistical Methods for Engineers and Scientists. Second Edition, Revised and Expanded, *Robert M. Bethea, Benjamin S. Duran, and Thomas L. Boullion*
Vol. 58: Ensemble Modeling: Inference from Small-Scale Properties to Large-Scale Systems, *Alan E. Gelfand and Crayton C. Walker*

Vol. 59: Computer Modeling for Business and Industry, *Bruce L. Bowerman and Richard T. O'Connell*
Vol. 60: Bayesian Analysis of Linear Models, *Lyle D. Broemeling*
Vol. 61: Methodological Issues for Health Care Surveys, *Brenda Cox and Steven Cohen*
Vol. 62: Applied Regression Analysis and Experimental Design, *Richard J. Brook and Gregory C. Arnold*
Vol. 63: Statpal: A Statistical Package for Microcomputers – PC-DOS Version for the IBM PC and Compatibles, *Bruce J. Chalmer and David G. Whitmore*
Vol. 64: Statpal: A Statistical Package for Microcomputers – Apple Version for the II, II+, and IIe, *David G. Whitmore and Bruce J. Chalmer*
Vol. 65: Nonparametric Statistical Inference, Second Edition, Revised and Expanded, *Jean Dickinson Gibbons*
Vol. 66: Design and Analysis of Experiments, *Roger G. Petersen*
Vol. 67: Statistical Methods for Pharmaceutical Research Planning, *Sten W. Bergman and John C. Gittins*
Vol. 68: Goodness-of-Fit Techniques, *edited by Ralph B. D'Agostino and Michael A. Stephens*
Vol. 69: Statistical Methods in Discrimination Litigation, *edited by D. H. Kaye and Mikel Aickin*
Vol. 70: Truncated and Censored Samples from Normal Populations, *Helmut Schneider*
Vol. 71: Robust Inference, *M. L. Tiku, W. Y. Tan, and N. Balakrishnan*
Vol. 72: Statistical Image Processing and Graphics, *edited by Edward J. Wegman and Douglas J. DePriest*
Vol. 73: Assignment Methods in Combinatorial Data Analysis, *Lawrence J. Hubert*
Vol. 74: Econometrics and Structural Change, *Lyle D. Broemeling and Hiroki Tsurumi*
Vol. 75: Multivariate Interpretation of Clinical Laboratory Data, *Adelin Albert and Eugene K. Harris*
Vol. 76: Statistical Tools for Simulation Practitioners, *Jack P. C. Kleijnen*
Vol. 77: Randomization Tests, Second Edition, *Eugene S. Edgington*
Vol. 78: A Folio of Distributions: A Collection of Theoretical Quantile-Quantile Plots, *Edward B. Fowlkes*
Vol. 79: Applied Categorical Data Analysis, *Daniel H. Freeman, Jr.*
Vol. 80: Seemingly Unrelated Regression Equations Models: Estimation and Inference, *Virendra K. Srivastava and David E. A. Giles*
Vol. 81: Response Surfaces: Designs and Analyses, *Andre I. Khuri and John A. Cornell*
Vol. 82: Nonlinear Parameter Estimation: An Integrated System in BASIC, *John C. Nash and Mary Walker-Smith*
Vol. 83: Cancer Modeling, *edited by James R. Thompson and Barry W. Brown*
Vol. 84: Mixture Models: Inference and Applications to Clustering, *Geoffrey J. McLachlan and Kaye E. Basford*
Vol. 85: Randomized Response: Theory and Techniques, *Arijit Chaudhuri and Rahul Mukerjee*
Vol. 86: Biopharmaceutical Statistics for Drug Development, *edited by Karl E. Peace*
Vol. 87: Parts per Million Values for Estimating Quality Levels, *Robert E. Odeh and D. B. Owen*
Vol. 88: Lognormal Distributions: Theory and Applications, *edited by Edwin L. Crow and Kunio Shimizu*
Vol. 89: Properties of Estimators for the Gamma Distribution, *K. O. Bowman and L. R. Shenton*
Vol. 90: Spline Smoothing and Nonparametric Regression, *Randall L. Eubank*
Vol. 91: Linear Least Squares Computations, *R. W. Farebrother*
Vol. 92: Exploring Statistics, *Damaraju Raghavarao*

Vol. 93: Applied Time Series Analysis for Business and Economic Forecasting, *Sufi M. Nazem*
Vol. 94: Bayesian Analysis of Time Series and Dynamic Models, *edited by James C. Spall*
Vol. 95: The Inverse Gaussian Distribution: Theory, Methodology, and Applications, *Raj S. Chhikara and J. Leroy Folks*
Vol. 96: Parameter Estimation in Reliability and Life Span Models, *A. Clifford Cohen and Betty Jones Whitten*
Vol. 97: Pooled Cross-Sectional and Time Series Data Analysis, *Terry E. Dielman*
Vol. 98: Random Processes: A First Look, Second Edition, Revised and Expanded, *R. Syski*
Vol. 99: Generalized Poisson Distributions: Properties and Applications, *P.C. Consul*
Vol. 100: Nonlinear L_p-Norm Estimation, *René Gonin and Arthur H. Money*
Vol. 101: Model Discrimination for Nonlinear Regression Models, *Dale S. Borowiak*
Vol. 102: Applied Regression Analysis in Econometrics, *Howard E. Doran*
Vol. 103: Continued Fractions in Statistical Applications, *K.O. Bowman and L.R. Shenton*
Vol. 104: Statistical Methodology in the Pharmaceutical Sciences, *Donald A. Berry*
Vol. 105: Experimental Design in Biotechnology, *Perry D. Haaland*
Vol. 106: Statistical Issues in Drug Research and Development, *edited by Karl E. Peace*
Vol. 107: Handbook of Nonlinear Regression Models, *David A. Ratkowsky*
Vol. 108: Robust Regression: Analysis and Applications, *edited by Kenneth D. Lawrence and Jeffrey L. Arthur*
Vol. 109: Statistical Design and Analysis of Industrial Experiments, *edited by Subir Ghosh*
Vol. 110: U-Statistics: Theory and Practice, *A. J. Lee*
Vol. 111: A Primer in Probability, Second Edition, Revised and Expanded, *Kathleen Subrahmaniam*
Vol. 112: Data Quality Control: Theory and Pragmatics, *edited by Gunar E. Liepins and V. R. R. Uppuluri*
Vol. 113: Engineering Quality by Design: Interpreting the Taguchi Approach, *Thomas B. Barker*
Vol. 114: Survivorship Analysis for Clinical Studies, *Eugene K. Harris and Adelin Albert*
Vol. 115: Statistical Analysis of Reliability and Life-Testing Models, Second Edition, *Lee J. Bain and Max Engelhardt*
Vol. 116: Stochastic Models of Carcinogenesis, *Wai-Yuan Tan*
Vol. 117: Statistics and Society: Data Collection and Interpretation, Second Edition, Revised and Expanded, *Walter T. Federer*
Vol. 118: Handbook of Sequential Analysis, *B. K. Ghosh and P. K. Sen*
Vol. 119: Truncated and Censored Samples: Theory and Applications, *A. Clifford Cohen*

ADDITIONAL VOLUMES IN PREPARATION

TRUNCATED AND CENSORED SAMPLES

Theory and Applications

A. CLIFFORD COHEN
University of Georgia
Athens, Georgia

Marcel Dekker, Inc.　　　New York • Basel • Hong Kong

Library of Congress Cataloging--in--Publication Data

Cohen, A. Clifford
   Truncated and censored samples: theory and applications/A. Clifford Cohen.
      p.   cm. -- -- (Statistics, textbooks and monographs)
   Includes bibliographical references and index.
   ISBN 0-8247-8447-2
   1. Sampling (Statistics) 2. Estimation theory. I. Title II. Series.
   QA276.6.C52   1991
   519.5'2-- --dc20   91-12083
                                                           CIP

This book is printed on acid-free paper.

Copyright © 1991 by MARCEL DEKKER, INC. All Rights Reserved

Neither this book nor any part may be reproduced or transmitted in any form or by any means, electronic or mechanical, including photocopying, microfilming, and recording, or by any information storage and retrieval system, without permission in writing from the publisher.

MARCEL DEKKER, INC.
270 Madison Avenue, New York, New York 10016

Current printing (last digit):
10 9 8 7 6 5 4 3 2 1

PRINTED IN THE UNITED STATES OF AMERICA

To Dorothy

Preface

Most of the current statistical literature on sampling concerns unrestricted samples. In most real-life situations, however, researchers are likely to find that their samples are either truncated or censored. In some instances, truncation or censoring is of a minor degree, and might justifiably be neglected. In others, sampling restrictions are severe and must be appropriately considered in any valid analysis of the resulting sample data. For the past 50 years, I have been concerned with the development of methodology for the analysis of such restricted sample data. The resulting publications, dealing primarily with parameter estimation from truncated and censored samples, have appeared in various professional journals, both domestic and foreign. Preparation of the manuscript for this book has provided an opportunity for the consolidation and incorporation of these scattered results into a single volume, where they might be more readily available for use by research workers in various fields of applied science.

The methodology presented in this volume is applicable wherever restricted (truncated and/or censored) samples are encountered. Important applications are to be found in business, economics, manufacturing, engineering, quality control, medical and biological sciences, management sciences, social sciences, and all areas of the physical sciences. This methodology is of particular importance in the area of research and development. Although primary emphasis has been given to my own published results, related research of numerous other writers has been recognized and incorporated into this book.

Attention is given to distributions of both continuous and discrete random variables. In the continuous category, consideration is given to the normal, lognormal, Weibull, gamma, inverse Gaussian, Rayleigh, Pareto, exponential, extreme value, and Pearson distributions. In the discrete category, the Poisson, binomial, negative binomial, and hypergeometric distributions are considered.

Maximum likelihood, moment, and modified estimators are employed for the estimation of distribution parameters. The modified estimators employ the first-order statistic in the estimation of threshold parameters in the skewed distributions. A technique that involves an iterative procedure for the completion of otherwise incomplete samples so that complete sample methods may then be employed in their subsequent analysis is also presented. Numerous tables and graphs are included as aids to facilitate and simplify computational procedures.

This volume is offered as both a college text and a reference source. However, it is primarily intended as a handbook for practitioners who need simple and efficient methods for the analysis of incomplete sample data.

The encouragement and patient guidance of Dr. Don Owen, coordinating editor for this series of monographs, and of Sandra Beberman, Maria Allegra, Brian Black, Julie Caldwell, and other staff members of Marcel Dekker, Inc., is gratefully acknowledged. Thanks are extended to Dr. Lynne Billard, Professor and former Head, and to Dr. Robert L. Taylor, Head, Department of Statistics of the University of Georgia for encouragement and support as this book was being written. Appreciation is expressed to Dawn Tolbert, Gayle Roderiguez, Molly Rema, and Connie Doss for typing and word processing.

Special thanks and appreciation are extended to Dr. Betty Jones Whitten, a coauthor of many of my previous publications, for encouragement, for numerous helpful comments, and for computing and editorial assistance.

Thanks are extended to the American Society for Quality Control, the American Statistical Association, the Institute of Mathematical Statistics, the Biometrika Trustees, and Marcel Dekker, Inc., for permission to include various tables and other previously published material in this volume.

A. Clifford Cohen

Contents

PREFACE	v
LIST OF ILLUSTRATIONS	xi
LIST OF TABLES	xiii

1. INTRODUCTION 1
    1.1 Preliminary Considerations 1
    1.2 A Historical Account 2
    1.3 Sample Types 3
    1.4 Estimators 5
    1.5 Likelihood Functions 6

2. SINGLY TRUNCATED AND SINGLY CENSORED SAMPLES FROM THE NORMAL DISTRIBUTION 8
    2.1 Preliminary Remarks 8
    2.2 Some Fundamentals 8
    2.3 Moment Estimators for Singly Truncated Samples 10
    2.4 Three-Moment Estimators for Singly Truncated Normal Distribution Parameters 13

	2.5 Maximum Likelihood Estimators for Singly Censored Samples	18
	2.6 Sampling Errors of Estimates	25
	2.7 Illustrative Examples	26
3.	MULTIRESTRICTED SAMPLES FROM THE NORMAL DISTRIBUTION	31
	3.1 Introduction	31
	3.2 Doubly Truncated Samples	31
	3.3 Doubly Censored Samples	46
	3.4 Progressively Censored Samples	50
	3.5 Some Additional Sample Types	59
	3.6 Final Comment	65
4.	LINEAR ESTIMATORS	66
	4.1 Introduction	66
	4.2 Calculation of Estimates	67
	4.3 Derivations	67
	4.4 Alternative Estimators	70
	4.5 Illustrative Examples	81
5.	TRUNCATED AND CENSORED SAMPLES FROM THE WEIBULL DISTRIBUTION	85
	5.1 Introduction	85
	5.2 Distribution Characteristics	85
	5.3 Singly Censored Samples	86
	5.4 Progressively Censored Samples	88
	5.5 The Three-Parameter Distribution	90
	5.6 Modified Maximum Likelihood Estimators (MMLE)	92
	5.7 Errors of Estimates	93
	5.8 An Illustrative Example	94
6.	TRUNCATED AND CENSORED SAMPLES FROM THE LOGNORMAL DISTRIBUTION	96
	6.1 Introduction	96
	6.2 Some Fundamentals	96
	6.3 Global Maximum Likelihood Estimation	97
	6.4 Local Maximum Likelihood and Modified Maximum Likelihood Estimators for Censored Samples	98
	6.5 Computational Procedures	101

Contents ix

	6.6	Errors of Estimates	104
	6.7	An Illustrative Example	104

7. TRUNCATED AND CENSORED SAMPLES FROM THE
   INVERSE GAUSSIAN AND THE GAMMA DISTRIBUTIONS   106

   7.1 The Inverse Gaussian Distribution   106
   7.2 The Gamma Distribution   113
   7.3 A Pseudo-Complete Sample Technique for Parameter
       Estimation from Censored Samples   121
   7.4 An Illustrative Example   123

8. TRUNCATED AND CENSORED SAMPLES FROM
   THE EXPONENTIAL AND THE EXTREME VALUE
   DISTRIBUTIONS   128

   8.1 The Exponential Distribution   128
   8.2 The Extreme Value Distribution   139

9. TRUNCATED AND CENSORED SAMPLES FROM THE
   RAYLEIGH DISTRIBUTION   146

   9.1 Introduction   146
   9.2 Some Special Cases   147
   9.3 Parameter Estimation   151
   9.4 Reliability of Estimates   156
   9.5 Illustrative Examples   159
   9.6 Some Concluding Remarks   164

10. TRUNCATED AND CENSORED SAMPLES FROM THE
    PARETO DISTRIBUTION   165

    10.1 Introduction   165
    10.2 Some Fundamentals   165
    10.3 Parameter Estimation   168
    10.4 Estimate Reliability   171
    10.5 An Illustrative Example   172

11. HIGHER-MOMENT ESTIMATES OF PEARSON
    DISTRIBUTION PARAMETERS FROM
    TRUNCATED SAMPLES   174

    11.1 Introduction   174
    11.2 The Pearson Distributions   174

	11.3 Recursion Formula for Moments of a Truncated Distribution	177
	11.4 Parameter Estimation from Doubly Truncated Samples	178
	11.5 Determining the Distribution Type	180
	11.6 Singly Truncated Samples	180
	11.7 Type III and Normal Distributions	181
	11.8 A Numerical Example	182
12.	TRUNCATED AND CENSORED SAMPLES FROM BIVARIATE AND MULTIVARIATE NORMAL DISTRIBUTIONS	185
	12.1 Introduction	185
	12.2 Estimation in the Bivariate Normal Distribution	186
	12.3 Reliability of Estimates	190
	12.4 An Illustrative Example	191
	12.5 Parameter Estimation in the Multivariate Normal Distribution	193
13.	TRUNCATED AND CENSORED SAMPLES FROM DISCRETE DISTRIBUTIONS	199
	13.1 Introduction	199
	13.2 The Poisson Distribution	199
	13.3 The Negative Binomial Distribution	208
	13.4 The Binomial Distribution	212
	13.5 The Hypergeometric Distribution	215
14.	TRUNCATED ATTRIBUTE SAMPLING AND RELATED TOPICS	220
	14.1 Truncated Attribute Acceptance Sampling	220
	14.2 Estimation from Misclassified Inspection Data	224
	14.3 Inflated Zero Distributions	236
	APPENDIX: TABLES OF CUMULATIVE STANDARD DISTRIBUTION FUNCTIONS	245
	GLOSSARY	279
	BIBLIOGRAPHY	289
	INDEX	305

List of Illustrations

2.1	Graph of Auxiliary Estimation Function $\theta(\alpha)$ for Singly Truncated Samples from the Normal Distribution	14
2.2	Efficiency Curves of Three-Moment Estimates from Singly Truncated Samples from the Normal Distribution	18
3.1	Graphs of Estimating Functions for Doubly Truncated Samples from the Normal Distribution	42
3.2	Probit Regression Line	60
5.1	The Weibull Coefficient of Variation and Its Square as Functions of the Shape Parameter	89
7.1	Graphs of α_3 as a Function of z_1 and n in Complete Samples from the Inverse Gaussian Distribution	124
7.2	Graphs of α_3 as a Function of z_1 and n in Complete Samples from the Gamma Distribution	125
8.1	Probability Density Function of the Exponential Distribution	130
8.2	Cumulative Probability Function of the Exponential Distribution	131
8.3	A Cumulative Hazard Plot of Generator Fan Data	138

8.4	Probability Density Functions of Extreme Value Distributions	142
9.1	Graphs of Estimating Functions for Truncated Samples from Two- and Three-Dimensional Rayleigh Distributions	159
10.1	Probability Density Function of the Pareto Distribution: $\alpha = 4.1$	167
10.2	Cumulative Probability Function of the Pareto Distribution: $\alpha = 4.1$	168
11.1	β_1–β_2 Curves for the Pearson System of Distribution Functions	179

List of Tables

2.1	Auxiliary Estimation Function $\theta(\alpha)$ for Singly Truncated Samples from the Normal Distribution	12
2.2	Asymptotic Variances and Efficiencies of Three-Moment Estimates from Singly Truncated Samples from the Normal Distribution	17
2.3	Auxiliary Estimation Function $\lambda(h, \alpha)$ for Singly Censored Samples from the Normal Distribution	21
2.4	Variance and Covariance Factors for Singly Truncated and Singly Censored Samples from the Normal Distribution	27
3.1	Estimating Functions $H_1(\xi_1, \xi_2)$ and $H_2(\xi_1, \xi_2)$ for Doubly Truncated Samples from the Normal Distribution	34
3.2	The Functions Q, ξQ, Q', λ, η for Progressively Censored Samples from the Normal Distribution	52
3.3	Life Distribution of Certain Biological Specimens in a Stress Environment	58
4.1	Coefficients of Linear Estimates of the Mean and Standard Deviation for Censored Samples from the Normal Distribution	72

4.2	Variances and Covariances of Linear Estimates of the Mean and Standard Deviation for Censored Samples from the Normal Distribution	80
4.3	Variances and Relative Efficiencies of Gupta's Alternative Linear Estimates of the Mean and Standard Deviation for Censored Samples from the Normal Distribution	83
6.1	Expected Values of the First-Order Statistic in Samples from the Standard Normal Distribution	101
7.1	A Random Sample from an Inverse Gaussian Distribution	126
7.2	Summary of Estimates of Inverse Gaussian Parameters	126
7.3	Successive Iterations of Censored Observations for Pseudo-Complete Samples from the Inverse Gaussian Distribution	127
8.1	A Progressively Censored Sample Consisting of Life-Span Observations of 70 Generator Fans	137
9.1	Truncated Sample Estimating Function $J_2(z)$ for Rayleigh Distribution	157
9.2	Truncated Sample Estimating Function $J_3(z)$ for Rayleigh Distribution	158
9.3	Censored Sample Estimation Function $H_3(h,z)$ for Rayleigh Distribution	160
10.1	Some Characteristics of the Pareto Distribution	169
11.1	Weight Distribution of 1000 Women Students	183
12.1	Summary of Estimates and Their Variances for Example 12.1	193
12.2	Summary of Sample Data for Example 12.2	195
12.3	Summary of Estimates for Example 12.2	196
14.1	Characteristics of Curtailed Sampling Plans	225
A.1	Cumulative Distribution Function of the Standardized Weibull Distribution	246
A.2	Cumulative Distribution Function of the Standardized Lognormal Distribution	254
A.3	Cumulative Distribution Function of the Standardized Inverse Gaussian Distribution	262
A.4	Cumulative Distribution Function of the Standardized Gamma Distribution	270

TRUNCATED AND CENSORED SAMPLES

1
Introduction

1.1 PRELIMINARY CONSIDERATIONS

Samples obtained when selection and/or observation is restricted over some portion of the sample space are, depending on the nature of the restriction, designated as either truncated or censored. Truncated samples are those from which certain population values are entirely excluded. It is perhaps more accurate to state that truncation occurs to populations, and samples described as being truncated are in fact samples from truncated populations. Censored samples are those in which sample specimens with measurements that lie in the restricted areas of the sample space may be identified and thus counted, but are not otherwise measured. In some of the earlier references, censored samples were described as truncated with a known number of missing (unmeasured) observations. According to Hald (1949), J. E. Kerrich first suggested use of the designation "censored" for these samples. In practical applications, truncated samples arise from various experimental situations in which sample selection is possible over only a partial range of the variable. Examples of this type occur frequently in manufacturing when samples are selected from production that has previously been screened to remove items above and/or below specification values. Censored samples often result from life testing and reaction time experiments where it is common practice to terminate observation prior to failure or reaction of all sample specimens.

1.2 A HISTORICAL ACCOUNT

Truncated samples were first encountered quite early in the development of modern statistics by Sir Francis Galton (1897) in connection with an analysis of registered speeds of American trotting horses. Sample data were extracted from *Wallace's Year Book*, Vols. 8–12 (1892–1896), a publication of the American Trotting Association. Recorded data consisted of running times of horses that qualified for registration by trotting around a one-mile course in not more than 2 minutes and 30 seconds while harnessed to a two-wheeled cart carrying a weight of not less than 150 pounds including the driver. No records were kept for the slower, unsuccessful trotters, and their number thus remained unknown. In today's terminology, Galton's samples would be described as singly truncated on the right at a known point.

Galton assumed his distributions to be normal, and he used sample modes as estimates of population means. He followed the simple expedient of plotting frequency polygons and locating the required values by inspection. With modes equated to medians, he located sample quartiles and used semi-interquartile ranges to estimate population standard deviations. Sample sizes varied from 982 to 1324 observations each. Agreement between observed and expected frequencies, with expected frequencies computed on the basis of estimates obtained as described above, was reasonably satisfactory for Galton's purposes.

Dissatisfaction with estimates based on modes and quartiles led Karl Pearson (1902) to propose a procedure for estimating normal distribution parameters from truncated samples by fitting parabolas to logarithms of the sample frequencies. This procedure was then employed to recalculate estimates from Galton's samples. Pearson's estimates, however, differed only slightly from those originally calculated by Galton. Pearson's interest in the truncation problem continued to motivate him, and in connection with a study of multiple normal correlation, he and Alice Lee (1908) employed the method of moments to estimate the mean and standard deviation of a normal distribution from a singly left truncated sample. Special tables, necessary for the practical application of these estimators, were provided for samples that were limited to observations in the right tail of the distribution. Subsequently, Alice Lee (1915) expanded these tables to provide for estimation from samples that included observations over the major segment of the population range.

No further published work on estimation from truncated samples appeared until R. A. Fisher (1931) derived estimators based on singly truncated samples for normal distribution parameters by employing the method of maximum likelihood, which he had introduced only 10 years earlier. Samples considered by Fisher were the same as those studied by Pearson and Lee. Furthermore, Fisher's maximum likelihood estimators were found to be identical to the Pearson and Lee moment estimators for the case of singly truncated normal samples. Fisher also derived asymptotic variances and covariances for his estimators. For the

Introduction

practical application of his results, Fisher presented tables of functions equivalent to the Pearson–Lee functions. Entries were given to more significant digits than in the Pearson–Lee tables, but the interval of the argument, 0.1, remained unchanged.

Stevens (1937), in an appendix to a paper by Bliss, derived maximum likelihood equations for estimating normal distribution parameters from samples of the types that are now described as Type I singly and doubly censored samples. The number of unmeasured (censored) observations in Stevens' samples was known, whereas this information was not available in the truncated samples considered earlier by Galton, Pearson, Lee, and Fisher. Stevens' equations were left in a form that proved rather difficult to solve, and an iterative procedure he suggested was a bit troublesome in practice. Bliss used Stevens' formulas to calculate tables that simplified the calculation of estimates to some extent. In addition to his estimating equations, Stevens also derived asymptotic variances and covariances for his estimates.

My own interest in the truncation problem dates back to 1940, when Professor Cecil Craig at the University of Michigan suggested that I select a topic in this area for my doctoral dissertation. The completed dissertation (1941) dealt with moment estimation in Pearson frequency distributions from singly and doubly truncated samples. For the next 6 years, further academic pursuits were interrupted by military service during World War II. As a result of this interruption, publication of my first paper on truncated distributions was delayed until 1949. Subsequent publications concerning various aspects of truncated and censored samples are Cohen (1950, 1951, 1955, 1957, 1959, 1961, 1963, 1965, 1966, 1969, 1973, 1975, 1976), Cohen and Norgaard (1977), Cohen and Whitten (1980, 1981, 1982, 1988), Cooley and Cohen (1970), and Whitten, Cohen, and Sundaraiyer (1988). Other writers who have made valuable contributions in this area include Hald (1949), Finney (1949), Keyfitz (1938), Birnbaum (1952, 1953), Birnbaum and Andrews (1949), Halperin (1952), Cochran (1946), Sampford (1952), Moore (1954), Grundy (1952), Quensel (1945) Des Raj (1952, 1953), Gupta (1952), Campbell (1945), Rider (1953), Plackett (1953), David and Johnson (1952), Francis (1946), Epstein and Sobel (1953), Epstein (1960), Saw (1958, 1959, 1961a, 1961b), Schneider (1986), and numerous others. Mendenhall (1958) published an extensive bibliography of papers concerning truncated and censored samples and related topics. Although the list of references included here is a long one, it is almost certain that omissions have occurred, and apologies are extended to anyone whose contributions have been overlooked.

1.3 SAMPLE TYPES

Samples to be considered in this book include those that are singly right or singly left truncated, singly right or singly left censored, doubly truncated, doubly censored, centrally truncated, centrally censored, and progressively censored.

Truncated samples are classified according to whether points of truncation (terminals) are known or unknown. When these points are unknown, they become additional parameters to be estimated from sample data. Primary interest in this book is focused on cases where terminals are known, and unless otherwise specified, these are the cases under consideration.

Censored samples are classified as Type I or Type II. In both types, we let N designate the total sample size and n the number of complete (fully measured) observations. The difference, $N - n = c$, designates the number of censored observations. In Type I samples, terminals are known (fixed) whereas n and c are observed values of random variables. In Type II samples, N, n, and c are fixed values, whereas the terminals are random (i.e., order statistics). Samples to be considered in subsequent chapters are selected from specified distributions. Sample types are more fully described in the following paragraphs.

Left singly truncated samples: For each of n observations, $x \geq T$, where T is a fixed (known) point of truncation.

Right singly truncated samples: For each of n observations, $x \leq T$, where T is a fixed (known) point of truncation.

Left singly censored samples: These samples consist of a total of N observations of which n are fully measured while $c = N - n$ are censored. For each of the censored observations, it is known only that $x < T$, whereas for each of the measured observations, $x \leq T$. In Type I samples T is a fixed (known) point of censoring. In Type II samples, $T = x_{(c+1):N}$, that is, the $(c+1)$st order statistic in a sample of size N.

Right singly censored samples: These samples also consist of a total of N observations of which n are fully measured while $c = N - n$ are censored. For each of the censored observations, it is known only that $T < x$, whereas for each of the measured observations, $x \leq T$. In Type I samples T is a fixed (known) constant. In Type II samples $T = x_{n:N}$.

Doubly truncated samples: For each of $N = n$ observations, $T_1 \leq x \leq T_2$, where T_1 and T_2 are fixed (known) points of truncation.

Doubly censored samples: These samples consist of a total of N observations of which n are fully measured and $N - n$ are censored. Each of the fully measured observations lies in the interval $T_1 \leq x \leq T_2$. There are c_1 censored observations for which it is known only that $x < T_1$, and C_2 censored observations for which it is known only that $x > T_2$. The total sample size is $N = n + c_1 + c_2$. In Type I samples T_1 and T_2 are fixed (known) constants. In Type II samples T_1 and T_2 are the order statistics $x_{c_1+1:N}$ and $x_{c_1+n:N}$, respectively.

Progressively censored samples: These samples also consist of a total of N observations. They are right censored at points $T_1 < T_2 < \ldots T_j \ldots < T_k$. At $x = T_j$, c_j observations are censored and for these it is known only that $x > T_j$. The total number of censored observations is $\Sigma_1^k c_j$, and the number

Introduction

of fully measured observations is $n = N - \Sigma_1^k c_j$. For Type I samples the T_j are fixed (known) constants. For Type II samples they are order statistics in samples of size N.

Both truncated and censored samples are further classified according to the type of distribution from which they were selected. In subsequent chapters we shall examine samples from normal, lognormal, inverse Gaussian, Weibull, gamma, Pearson, Rayleigh, extreme value, exponential, Pareto, Poisson, binomial, negative binomial, and hypergeometric distributions. Bivariate and multivariate normal distributions are also considered.

1.4 ESTIMATORS

Moment and maximum likelihood estimators are the principal estimators for calculating estimates of distribution parameters from truncated and censored samples. However, linear estimators also play an important role in the calculation of unbiased estimates from small samples. Moment-estimating equations are obtained by equating distribution moments to corresponding sample moments. With certain exceptions, maximum likelihood–estimating equations are obtained by equating to zero the first partial derivatives of the loglikelihood function with respect to the parameters. Various regularity conditions limit the applicability of maximum likelihood estimators in Weibull, gamma, and other skewed distributions that involve a threshold parameter. In those cases, modified estimators employing the first-order statistic offer advantages over moment and maximum likelihood estimators.

A major advantage of maximum likelihood estimators is the ease with which the variance–covariance matrix of estimates can be obtained. This matrix is the inverse of the Fisher information matrix with elements that are negatives of expected values of second partial derivatives of the loglikelihood function.

An alternative procedure, described in Chapter 7, for estimating parameters from singly right censored samples is due to Whitten, Cohen, and Sundaraiyer (1988). An iterative procedure is employed to estimate values of censored observations in order to create pseudocomplete samples. Complete sample estimators are then available for estimating the distribution parameters.

Linear unbiased estimators, which were mentioned earlier, are considered in Chapter 4.

The hazard plot technique, which was developed by Nelson (1968, 1969, 1972, 1982), provides a simple graphical procedure for both choosing an appropriate model (i.e., distribution) and approximating estimates of distribution parameters from progressively censored samples. This technique is described in Chapter 8.

1.5 LIKELIHOOD FUNCTIONS

Let $f(x; \theta_2, \theta_2, \ldots, \theta_r)$ and $F(x; \theta_1, \theta_2, \ldots, \theta_r)$ designate the pdf and cdf of an unrestricted (i.e., complete) distribution with parameters $\theta_1, \theta_2, \ldots, \theta_r$. Likelihood functions of truncated and censored samples to be considered in subsequent chapters are given below. To simplify notation, the pdf and the cdf are abbreviated to $f(x)$ and $F(x)$ without explicitly indicating the parameters to be estimated.

Singly left truncated at $x = T$:

$$L(\ ) = \frac{1}{1 - F(T)} \prod_{i=1}^{n} f(x_i), \qquad T \leq x_i. \tag{1.5.1}$$

Singly right truncated at $x = T$:

$$L(\ ) = \frac{1}{F(T)} \prod_{i=1}^{n} f(x_i), \qquad x_i \leq T. \tag{1.5.2}$$

Singly left censored at $x = T$:

$$L(\ ) = K[F(T)]^c \prod_{i=1}^{n} f(x_i), \qquad T \leq x_i. \tag{1.5.3}$$

Singly right censored at $x = T$:

$$L(\ ) = K[1 - F(T)]^c \prod_{i=1}^{n} f(x_i), \qquad x_i \leq T. \tag{1.5.4}$$

Doubly truncated at T_1 and T_2:

$$L(\ ) = \frac{1}{F(T_2) - F(T_1)} \prod_{i=1}^{n} f(x_i), \qquad T_1 \leq x_i \leq T_2. \tag{1.5.5}$$

Doubly censored at T_1 and T_2:

$$L(\ ) = K[F(T_1)]_{c_1}[1 - F(T_2)]_{c_2} \prod_{i=1}^{n} f(x_i), \qquad T_1 \leq x_i \leq T_2. \tag{1.5.6}$$

Progressively censored at T_j; $j = 1, 2, \ldots, k$

$$L(\ ) = K \prod_{j=1}^{k} [1 - F(T_j)]^{c_j} \prod_{i=1}^{n} f(x_i), \qquad \gamma \leq x_i \leq \infty. \tag{1.5.7}$$

In the preceding likelihood functions, K denotes ordering constants that do not depend on the parameters. The total sample size is denoted by N and the

Introduction

number of complete observations by n. Accordingly, the number of censored observations is $N - n$, and for truncated samples $N = n$. The number of observations censored at T_j is designated by c_j, but for only a single stage of censoring the subscript is omitted.

2
Singly Truncated and Singly Censored Samples from the Normal Distribution

2.1 PRELIMINARY REMARKS

In this chapter we derive and illustrate the practical application of estimators for the normal distribution mean and standard deviation from singly truncated and singly censored samples. Since the normal distribution is symmetrical about its mean, truncation and likewise censoring at a point D units to the left of the mean is equivalent to truncation or censoring at a corresponding point D units to the right of the mean. Hence, we need only to consider these restrictions for one side. For historical reasons we choose to base derivations on left truncated and left censored samples. The resulting estimators are then applicable to both left and right truncated and censored samples.

Derivations presented here follow the general outline of derivations of maximum likelihood estimators given earlier by Cohen (1950, 1959). However, since the method of moments and the method of maximum likelihood lead to identical estimators for normal distribution parameters, either method might be employed. To illustrate both methods, the method of moments will be used for truncated samples and the method of maximum likelihood will be employed for censored samples.

2.2 SOME FUNDAMENTALS

Let X designate a random variable that is normally distributed with pdf and cdf as follows:

Singly Truncated and Censored Samples

$$f(x; \mu, \sigma) = \frac{1}{\sigma\sqrt{2\pi}} \exp\left[-\frac{1}{2}\left(\frac{x-\mu}{\sigma}\right)^2\right], \quad -\infty < x < \infty, \tag{2.2.1}$$

$$F(x; \mu, \sigma) = \int_{-\infty}^{x} f(w; \mu, \sigma)\, dw.$$

In standard units, where

$$Z = \frac{X - \mu}{\sigma}, \tag{2.2.2}$$

The pdf and the cdf of the standard normal distribution become

$$\phi(z; 0, 1) = \frac{1}{\sqrt{2\pi}} e^{-z^2/2}, \quad -\infty < z < \infty, \tag{2.2.3}$$

$$\Phi(z; 0, 1) = \int_{-\infty}^{z} \phi(t; 0, 1)\, dt.$$

Let us now consider the truncated distribution that results when the normal distribution with pdf (2.2.1) is truncated on the left at $x = T$. The pdf of the resulting truncated distribution becomes

$$f_T(x; \mu, \sigma) = \frac{1}{\sigma\sqrt{2\pi}[1 - F(T)]} \exp\left[-\frac{1}{2}\left(\frac{x-\mu}{\sigma}\right)^2\right], \quad T \leq x < \infty,$$

$$= 0 \quad \text{elsewhere}. \tag{2.2.4}$$

The corresponding standardized pdf can be written as

$$\phi_T(z; 0, 1) = \frac{\phi(z; 0, 1)}{1 - \Phi(\xi)}, \quad \xi < z < \infty,$$

$$= 0 \quad \text{elsewhere}, \tag{2.2.5}$$

where ξ is the standardized point of truncation.

$$\xi = \frac{T - \mu}{\sigma}. \tag{2.2.6}$$

The kth moment of the truncated distribution about the point of truncation is

$$\bar{\mu}'_k = \frac{1}{\sigma\sqrt{2\pi}[1 - F(T)]} \int_T^\infty (x - T)^k \exp\left[-\frac{1}{2}\left(\frac{x-\mu}{\sigma}\right)^2\right] dx. \tag{2.2.7}$$

We make the standardizing transformation, $z = (x - \mu)/\sigma$ in (2.2.7), and thus $\bar{\mu}'_k(\xi)$ as a function of ξ may be expressed as

$$\bar{\mu}'_k(\xi) = \frac{\sigma^k}{[1 - \Phi(\xi)]} \int_\xi^\infty (t - \xi)^k \phi(t)\, dt, \qquad (2.2.8)$$

where $\phi(\ )$ and $\Phi(\ )$ are defined by (2.2.3), and of course $\Phi(\xi) = F(T)$.

Now let the kth moment about zero of the truncated standard normal distribution in units of the original complete distribution be defined as

$$\bar{\alpha}_k(\xi) = \frac{1}{[1 - \Phi(\xi)]} \int_\xi^\infty t^k \phi(t)\, dt. \qquad (2.2.9)$$

Let $k = 1$ and then 2 in (2.2.9) and integrate by parts to obtain

$$\bar{\alpha}_1(\xi) = \frac{\phi(\xi)}{[1 - \Phi(\xi)]}, \qquad \bar{\alpha}_2(\xi) = 1 + \frac{\xi\phi(\xi)}{1 - \Phi(\xi)}. \qquad (2.2.10)$$

In agreement with previously adopted notation, we let

$$Q(\xi) = \frac{\phi(\xi)}{1 - \Phi(\xi)}, \qquad (2.2.11)$$

which is recognized as the hazard function of the standard normal distribution.

When (2.2.11) is substituted into (2.2.10), we can write

$$\bar{\alpha}_1(\xi) = Q(\xi) \qquad \text{and} \qquad \bar{\alpha}_2(\xi) = 1 + \xi Q(\xi). \qquad (2.2.12)$$

Now let $k = 0, 1, 2$ in turn in (2.2.8), expand the binomial, and integrate to obtain

$$\bar{\mu}'_0(\xi) = 1,$$
$$\bar{\mu}'_1(\xi) = \sigma(Q - \xi), \qquad (2.2.13)$$
$$\bar{\mu}'_2(\xi) = \sigma^2[1 - \xi(Q - \xi)],$$

where $Q(\xi)$ has been abbreviated to Q. In arriving at the results of (2.2.13), integration of (2.2.8) resulted in expressions involving $\bar{\alpha}_k(\xi)$ as defined by (2.2.9), and these were subsequently evaluated from (2.2.12) to obtain the expressions given in (2.2.13).

The variance of X_T and the expected value of $(X_T - T)$, where X_T designates the truncated random variable, follow as

$$V(X_T) = \mu'_2 - (\mu'_1)^2 = \sigma^2[1 - Q(Q - \xi)], \qquad (2.2.14)$$
$$E(X_T - T) = \mu'_1 = \sigma(Q - \xi).$$

2.3 MOMENT ESTIMATORS FOR SINGLY TRUNCATED SAMPLES

Moment estimators are obtained by equating the mean $\bar{x}$ and the variance s^2 of the truncated sample to the mean and variance of the truncated normal population. Thus we have

Singly Truncated and Censored Samples

$$\bar{x} = T + \bar{\mu}'_1 \quad \text{and} \quad s^2 = \bar{\mu}'_2 - (\bar{\mu}'_1)^2 \tag{2.3.1}$$

When the expressions in (2.2.14) are substituted into (2.3.1), the estimating equations become

$$s^2 = \sigma^2[1 - Q(Q - \xi)], \tag{2.3.2}$$
$$(\bar{x} - T) = \sigma(Q - \xi).$$

The estimator for μ follows from (2.2.6) as

$$\mu^* = T - \sigma^* \xi^*. \tag{2.3.3}$$

The two equations of (2.3.2) can be solved simultaneously for estimates σ^* and ξ^*. The estimate μ^* then follows from (2.3.3).

When σ^2 is eliminated from the two equations of (2.3.2), we obtain

$$\frac{s^2}{(\bar{x} - T)^2} = \frac{1 - Q(Q - \xi)}{(Q - \xi)^2} = \alpha(\xi), \tag{2.3.4}$$

and from the second equation of (2.3.2), we write

$$\sigma = \frac{\bar{x} - T}{Q - \xi}. \tag{2.3.5}$$

Equation (2.3.4) can be solved for ξ^*, and σ^* then follows from (2.3.5). With ξ^* and σ^* thus calculated, μ^* follows from (2.3.3).

To derive the simpler estimators of Cohen (1959), we return to equations (2.3.2) and rewrite the first of these as

$$\sigma^2 = s^2 + \sigma^2 Q(Q - \xi). \tag{2.3.6}$$

We then substitute the expression for σ given by (2.3.5) into (2.3.6) and write

$$\sigma^2 = s^2 + \left(\frac{Q}{Q - \xi}\right)(\bar{x} - T)^2. \tag{2.3.7}$$

We define

$$\theta(\xi) = \frac{Q(\xi)}{Q(\xi) - \xi}, \tag{2.3.8}$$

and the estimator (2.3.7) becomes

$$(\sigma^2)^* = s^2 + \theta^* (\bar{x} - T)^2, \tag{2.3.9}$$

where ξ^* is the solution of (2.3.4) and $\theta^* = \theta(\xi^*)$.

To obtain a corresponding estimator for μ, we begin with (2.2.6) and write

$$\mu = T - \sigma \xi. \tag{2.3.10}$$

From the second equation of (2.3.2), we write

$$\sigma \xi = \sigma Q - (\bar{x} - T), \tag{2.3.11}$$

and substitute into (2.3.10) to obtain

$$\mu = \bar{x} - \sigma Q.$$

Now replace σ with the expression given in (2.3.5), and we find

$$\mu = \bar{x} - \left(\frac{Q}{Q - \xi}\right)(\bar{x} - T), \qquad (2.3.12)$$

and finally

$$\mu^* = \bar{x} - \theta^*(\bar{x} - T), \qquad (2.3.13)$$

where, as in (2.3.9), ξ^* is the solution of (2.3.4) and $\theta^* = \theta(\xi^*)$.

To facilitate the practical application of estimators (2.3.9) and (2.3.13), Table 2.1 of the auxiliary function $\theta(\alpha)$ is given. Figure 2.1 is a graph of this function.

Table 2.1 Auxiliary Estimation Function $\theta(\alpha)$ for Singly Truncated Samples from the Normal Distribution: $\hat{\alpha} = s^2/(\bar{x} - T)^2$

α	.000	.001	.002	.003	.004	.005	.006	.007	.008	.009
.050	.000004	.000005	.000006	.000007	.000009	.000011	.000013	.000015	.000017	.000020
.060	.000024	.000027	.000031	.000036	.000041	.000047	.000053	.000060	.000067	.000075
.070	.000084	.000094	.000104	.000116	.000128	.000141	.000155	.000171	.000187	.000204
.080	.000223	.000242	.000263	.000285	.000309	.000334	.000360	.000388	.000417	.000448
.090	.000481	.000515	.000550	.000588	.000627	.000668	.000711	.000756	.000802	.000851
.100	.000902	.000954	.001009	.001066	.001125	.001187	.001250	.001316	.001384	.001455
.110	.001528	.001604	.001682	.001762	.001845	.001931	.002019	.002110	.002204	.002300
.120	.002400	.002502	.002607	.002715	.002826	.002939	.003056	.003176	.003299	.003425
.130	.003554	.003687	.003822	.003961	.004103	.004249	.004398	.004550	.004705	.004865
.140	.005027	.005193	.005363	.005536	.005713	.005893	.006078	.006265	.006457	.006652
.150	.006852	.007055	.007262	.007472	.007687	.007906	.008129	.008355	.008586	.008821
.160	.009060	.009303	.009551	.009802	.010058	.010318	.010583	.010852	.011125	.011402
.170	.011684	.011971	.012262	.012557	.012857	.013162	.013471	.013785	.014103	.014426
.180	.014754	.015087	.015425	.015767	.016114	.016467	.016824	.017186	.017553	.017925
.190	.018302	.018684	.019071	.019463	.019861	.020264	.020672	.021085	.021503	.021927
.200	.022356	.022791	.023231	.023677	.024128	.024584	.025046	.025514	.025987	.026466
.210	.026950	.027440	.027936	.028438	.028946	.029459	.029978	.030503	.031035	.031572
.220	.032115	.032664	.033219	.033780	.034347	.034921	.035501	.036087	.036679	.037278
.230	.037882	.038494	.039111	.039735	.040366	.041003	.041647	.042297	.042954	.043617
.240	.044287	.044964	.045648	.046338	.047035	.047739	.048450	.049168	.049893	.050625
.250	.051364	.052110	.052863	.053623	.054390	.055165	.055947	.056736	.057533	.058337
.260	.059148	.059967	.060794	.061627	.062469	.063318	.064175	.065039	.065911	.066791
.270	.067679	.068575	.069478	.070390	.071309	.072236	.073172	.074115	.075067	.076027
.280	.076995	.077972	.078956	.079950	.080951	.081961	.082979	.084006	.085042	.086086
.290	.087139	.088200	.089271	.090350	.091438	.092534	.093640	.094755	.095879	.097012
.300	.098153	.099305	.100465	.101634	.102813	.104002	.105199	.106406	.107623	.108849
.310	.110085	.111331	.112586	.113851	.115125	.116410	.117704	.119009	.120323	.121648
.320	.122983	.124327	.125682	.127048	.128423	.129809	.131206	.132613	.134030	.135459
.330	.136897	.138347	.139807	.141278	.142760	.144253	.145757	.147272	.148798	.150335
.340	.151884	.153444	.155015	.156597	.158191	.159797	.161414	.163043	.164683	.166336
.350	.168000	.169676	.171364	.173064	.174776	.176500	.178237	.179986	.181747	.183521
.360	.185307	.187106	.188917	.190741	.192578	.194427	.196290	.198165	.200054	.201955
.370	.203870	.205798	.207740	.209694	.211663	.213644	.215640	.217649	.219672	.221709
.380	.223759	.225824	.227903	.229996	.232103	.234224	.236360	.238510	.240675	.242854
.390	.245048	.247257	.249481	.251720	.253974	.256242	.258527	.260826	.263141	.265471

Source: From Cohen and Whitten (1988), Table 8.1, pp. 134–135, by courtesy of Marcel Dekker, Inc.

Table 2.1 *Continued*

α	.000	.001	.002	.003	.004	.005	.006	.007	.008	.009
.400	.267817	.270178	.272555	.274948	.277357	.279782	.282222	.284679	.287153	.289642
.410	.292148	.294671	.297210	.299766	.302339	.304929	.307535	.310159	.312800	.315459
.420	.318134	.320828	.323539	.326267	.329014	.331778	.334560	.337361	.340179	.343016
.430	.345872	.348746	.351638	.354550	.357480	.360429	.363397	.366385	.369392	.372418
.440	.375464	.378530	.381615	.384720	.387845	.390990	.394156	.397342	.400548	.403776
.450	.407023	.410292	.413582	.416892	.420224	.423578	.426953	.430349	.433768	.437208
.460	.440670	.444154	.447661	.451190	.454742	.458316	.461913	.465533	.469177	.472843
.470	.476533	.480247	.483984	.487745	.491530	.495339	.499173	.503031	.506913	.510820
.480	.514753	.518710	.522692	.526700	.530733	.534793	.538878	.542988	.547126	.551289
.490	.555479	.559696	.563940	.568210	.572508	.576833	.581186	.585566	.589975	.594411
.500	.598876	.603369	.607891	.612442	.617022	.621631	.626269	.630937	.635635	.640362
.510	.645120	.649909	.654727	.659577	.664458	.669369	.674312	.679287	.684294	.689332
.520	.694403	.699507	.704643	.709811	.715013	.720249	.725518	.730820	.736157	.741528
.530	.746934	.752374	.757849	.763359	.768905	.774487	.780104	.785758	.791448	.797175
.540	.802938	.808739	.814578	.820454	.826368	.832320	.838311	.844340	.850409	.856517
.550	.862665	.868852	.875080	.881348	.887657	.894007	.900399	.906832	.913307	.919824
.560	.926384	.932986	.939632	.946321	.953054	.959831	.966653	.973519	.980431	.987388
.570	.994391	1.001439	1.008535	1.015677	1.022866	1.030103	1.037387	1.044720	1.052101	1.059531
.580	1.067011	1.074540	1.082119	1.089749	1.097429	1.105161	1.112944	1.120779	1.128667	1.136607
.590	1.144601	1.152648	1.160749	1.168905	1.177115	1.185381	1.193703	1.202080	1.210514	1.219006
.600	1.227554	1.236161	1.244826	1.253550	1.262333	1.271176	1.280080	1.289044	1.298069	1.307156
.610	1.316305	1.325517	1.334793	1.344132	1.353535	1.363003	1.372536	1.382136	1.391801	1.401534
.620	1.411334	1.421202	1.431139	1.441145	1.451221	1.461367	1.471585	1.481873	1.492234	1.502668
.630	1.513175	1.523756	1.534411	1.545142	1.555949	1.566832	1.577792	1.588831	1.599948	1.611144
.640	1.622420	1.633777	1.645214	1.656734	1.668337	1.680023	1.691794	1.703649	1.715590	1.727617
.650	1.739732	1.751935	1.764226	1.776607	1.789079	1.801641	1.814296	1.827044	1.839885	1.852821
.660	1.865852	1.878980	1.892205	1.905527	1.918949	1.932471	1.946094	1.959818	1.973646	1.987576
.670	2.001612	2.015753	2.030001	2.044357	2.058821	2.073395	2.088080	2.102877	2.117786	2.132810
.680	2.147949	2.163204	2.178576	2.194067	2.209678	2.225409	2.241263	2.257239	2.273341	2.289567
.690	2.305921	2.322404	2.339015	2.355758	2.372632	2.389641	2.406784	2.424063	2.441480	2.459036
.700	2.476732	2.494570	2.512552	2.530678	2.548951	2.567372	2.585942	2.604664	2.623537	2.642566
.710	2.661750	2.681091	2.700592	2.720254	2.740078	2.760067	2.780222	2.800545	2.821038	2.841703
.720	2.862541	2.883555	2.904746	2.926116	2.947668	2.969404	2.991325	3.013434	3.035732	3.058222
.730	3.080906	3.103787	3.126866	3.150146	3.173629	3.197317	3.221213	3.245320	3.269639	3.294173
.740	3.318926	3.343898	3.369094	3.394515	3.420165	3.446045	3.472160	3.498511	3.525102	3.551935

As defined by (2.3.8), $\theta(\xi)$ is a function of ξ. However, $\alpha(\xi)$, as defined by (2.3.4), is also a function of ξ. Since the estimate $\alpha(\xi^*) = s^2/(\bar{x} - T)^2$ is readily available from sample data, table entries of θ have been calculated with α as the argument rather than ξ. Accordingly, we need only set $\alpha^* = s^2/(\bar{x} - T)^2$, enter the table with this value, and interpolate to obtain θ^*. Estimates μ^* and σ^{2*} then follow from (2.3.9) and (2.3.13). Readers are again reminded that these moment estimators are identical to corresponding maximum likelihood estimators and they are applicable for both left and right singly truncated samples.

As an illustration of these calculations in a practical application, an example from Cohen (1959) is included in Section 2.7 as Example 2.7.4.

2.4 THREE-MOMENT ESTIMATORS FOR SINGLY TRUNCATED NORMAL DISTRIBUTION PARAMETERS

Explicit "higher moment" estimators of parameters of the Pearson distributions that can be calculated from moments of truncated samples without resort to special

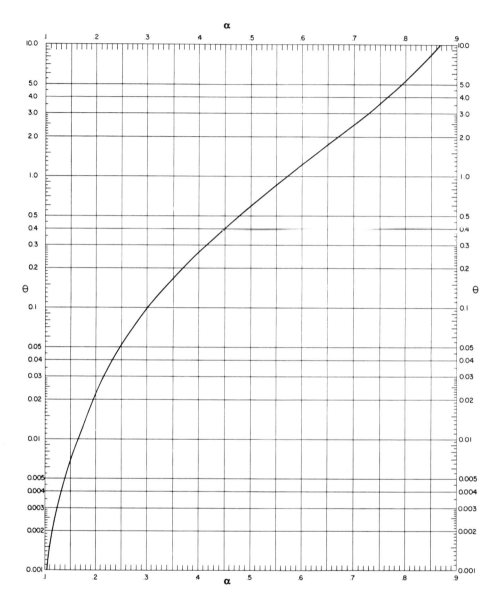

Figure 2.1 Graph of auxiliary estimation function $\theta(\alpha)$ for singly truncated samples from the normal distribution: $\hat{\alpha} = s^2/(\bar{x} - T)^2$.
Source: From Cohen (1959), Figure 1, p. 226, with permission of the Technometrics Management Committee.

Singly Truncated and Censored Samples

tables were derived by Cohen (1941). These included as a special case the three-moment estimators of normal distribution parameters for singly truncated samples. The three-moment estimators were considered in more detail by Cohen (1951). In addition to the first two moments as employed in Section 2.3, these estimators also employ the third sample moment.

The first two nonstandard moments of a singly left truncated sample about the left terminus were given by (2.2.13). In addition to these moments, we now need the third moment about the left terminal. The standard moment $\bar{\alpha}_3(\xi)$ follows from (2.2.9) and (2.2.11) as

$$\bar{\alpha}_3(\xi) = (2 + \xi^2)Q(\xi), \tag{2.4.1}$$

and $\bar{\mu}_3'(\xi)$ follows from (2.2.8) as

$$\bar{\mu}_3'(\xi) = \sigma^2 [(2 + \xi^2)Q - \xi(3 + \xi^2)], \tag{2.4.2}$$

where $Q(\xi)$ is defined by (2.2.11).

A few algebraic manipulations enable us to reduce the equations of (2.2.13) plus (2.4.2) to

$$\bar{\mu}_1' = \sigma(Q - \xi),$$
$$\bar{\mu}_2' = \sigma^2 - \sigma\xi\bar{\mu}_1', \tag{2.4.3}$$
$$\bar{\mu}_3' = 2\bar{\mu}_1'\sigma^2 - \bar{\mu}_2'.$$

Let $a = -\sigma\xi$ and let v_k' designate the kth sample moment about the terminus; that is,

$$v_k' = \sum_{i=1}^{n} \frac{(x_i - T)^k}{n}. \tag{2.4.4}$$

Equate $\bar{\mu}_k' = v_k'$, $k = 1, 2, 3$, and substitute into the last two equations of (2.4.3) to obtain

$$\sigma^2 + v_1'a = v_2',$$
$$2v_1'\sigma^2 + v_2'a = v_3'. \tag{2.4.5}$$

Estimates σ^{2*} and a^* can now be obtained as the simultaneous solution of the preceding pair of linear equations. Thereby we have

$$(\sigma^2)^* = \frac{(v_2')^2 - v_1'v_3'}{v_2' - 2(v_1')^2}, \qquad a^* = \frac{v_3' - 2v_1'v_2'}{v_2' - 2(v_1')^2}. \tag{2.4.6}$$

It subsequently follows that

$$\mu^* = T + a^*. \tag{2.4.7}$$

In order to apply these estimators, we make the following calculations from our sample data:

$$v_1' = \bar{x} - T,$$

$$v_2' = s^2 + (\bar{x} - T)^2, \qquad s^2 = \sum_{i=1}^{n} \frac{(x_i - \bar{x})^2}{n},$$

$$v_3' = \sum_{i=1}^{n} \frac{(x_i - \bar{x})^3}{n}.$$

Although it was assumed in the preceding argument that truncation occurred in the left tail of the distribution, the estimators (2.4.6) and (2.4.7) are likewise applicable when truncation is in the right tail, in which case the odd moments are negative because of the choice of origin.

2.4.1 Asymptotic Variances and Efficiencies of the Three-Moment Estimators

The delta method was employed to calculate asymptotic variances and efficiencies of the three-moment estimators which are tabulated in Table 2.2. The efficiencies are displayed graphically in Figure 2.2.

Here, n is the sample size, $V_\xi(\ )$ is the asymptotic variance, and $\text{Eff}_\xi(\ )$ is the asymptotic efficiency of the estimator $(\ )$, where ξ is the truncation point in standard units of the complete distribution. The two rows labeled $-\infty$ and ∞ give asymptotic expressions with respect to ξ for large values of ξ. Maximum likelihood estimates are designated as $\hat{a}$ and $\hat{\sigma}^2$, respectively,

2.4.2 An Illustrative Example

To illustrate the practical application of the three-moment estimators, we consider an example that was originally given by Cohen (1951).

Example 2.4.1. A complete sample of 40 observations was selected from Mahalanobis' tables (1934) of random observations from a normal population (0, 1). The sample was singly left truncated at $T = -1.000$ with the result that the retained sample consisted of $n = 32$ observations. Sample moments about the left terminal are $v_1' = 1.221906$, $v_2' = 2.245115$, and $v_3' = 4.933973$. Estimates μ^* and σ^* are calculated by substitution in (2.4.6) and (2.4.7) as

$$\mu^* = 1.000 + \frac{4.933973 - 2(1.221906)(2.245115)}{2.245115 - 2(1.221906)^2} = -0.254,$$

$$(\sigma^2) = \frac{(2.245115)^2 - (1.221906)(4.933973)}{2.245115 - 2(1.221906)^2} = 1.333763,$$

and

$$\sigma^* = \sqrt{1.333763} = 1.155.$$

Singly Truncated and Censored Samples

Table 2.2 Asymptotic Variances and Efficiencies of Three-Moment Estimates from Singly Truncated Samples from the Normal Distribution

ξ	$\frac{N}{\sigma^2}V_\xi(a^*)$	$\frac{N}{\sigma^2}V_\xi(\hat{a})$	$\text{Eff}_\xi(a^*)$ Per cent	$\frac{N}{\sigma^4}V_\xi(\sigma^{2*})$	$\frac{N}{\sigma^4}V_\xi(\hat{\sigma}^2)$	$\text{Eff}_\xi(\sigma^{2*})$ Per cent
$-\infty$	$1+O\left(\frac{1}{\xi^2}\right)$	$1+O\left(\frac{1}{\xi^2}\right)$	100	$2+O\left(\frac{1}{\xi^2}\right)$	$2+O(\xi e^{-\xi^2/2})$	100
-7	1.003	1.000	99.7	2.128	2.000	94
-6	1.005	1.000	99.5	2.176	2.000	92
-5	1.010	1.000	99	2.260	2.000	89
-4	1.027	1.001	97	2.426	2.009	83
-3	1.090	1.015	93	2.825	2.145	76
-2.5	1.190	1.057	89	3.246	2.408	74
-2	1.443	1.204	83	3.983	2.973	75
-1.5	2.138	1.688	79	5.283	4.074	77
-1	4.133	3.256	79	7.564	6.110	81
$-.5$	9.845	8.115	82	11.52	9.760	85
0	25.54	22.19	87	18.26	16.13	88
1	163.4	152.4	93	47.60	44.55	94
2	836.5	807.6	97	119.3	115.2	97
3	3387	3325	98	274.3	269.3	98
∞	$\xi^6+O(\xi^4)$	$\xi^6+O(\xi^4)$	100	$\xi^4+O(\xi^2)$	$\xi^4+O(\xi^2)$	100

Source: From Cohen (1951b), Table 2, p. 42, with permission of the Institute of Statistical Mathematics.

These estimates are to be compared with maximum likelihood estimates $\hat{\mu} = -0.534$ and $\hat{\sigma} = 1.295$, which were calculated as described in Section 2.

2.4.3 Some Comments

Although sampling errors of the three-moment estimators are somewhat greater than those of corresponding maximum likelihood estimators, the three-moment estimators might be particularly useful in providing first approximations when maximum likelihood tables are not available. In a limited investigation Cohen (1951) found reasonably close agreement between three-moment and maximum likelihood estimates when as much as 16% of the population was eliminated by truncation. Of course, in most instances the maximum likelihood estimators would be preferred.

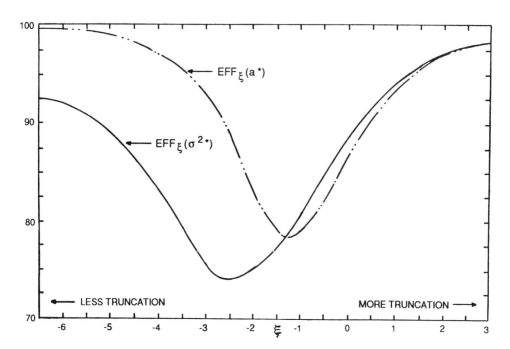

Figure 2.2 Efficiency curves of three-moment estimates from singly truncated samples from the normal distribution. Reprinted from Cohen (1951b), Fig. 1, p. 43, with permission of the Institute of Statistical Mathematics.

2.5 Maximum Likelihood Estimators for Singly Censored Samples

In Section 2.3 the method of moments was used to derive estimators for singly truncated samples from a normal distribution, and, as previously noted, these same estimators could have been derived by the method of maximum likelihood. In this section we use the method of maximum likelihood to derive corresponding estimators for singly censored samples from a normal distribution. These derivations were originally given by Cohen (1959). We consider a left singly censored sample as described in Section 1.3 of Chapter 1, with a likelihood function as given by (1.5.3) where the pdf of the normal distribution is given by (2.2.1). The loglikelihood function can be written as

$$\ln L(\ ) = c \ln \Phi(\xi) - n \ln \sigma - \frac{1}{2\sigma^2} \sum_{i=1}^{n} (x_i - \mu)^2 + \text{const.} \quad (2.5.1)$$

Estimating equations are obtained by equating to zero the partial derivatives of $\ln L$ with respect to the parameters μ and σ. We then obtain

Singly Truncated and Censored Samples

$$\frac{\partial \ln L}{\partial \mu} = -\frac{c}{\sigma}\frac{\phi(\xi)}{\Phi(\xi)} + \frac{1}{\sigma^2}\sum_{i=1}^{n}(x_i - \mu) = 0, \qquad (2.5.2)$$

$$\frac{\partial \ln L}{\partial \sigma} = -\frac{c}{\sigma}\frac{\xi\phi(\xi)}{\Phi(\xi)} - \frac{n}{\sigma} + \frac{1}{\sigma^3}\sum_{i=1}^{n}(x_i - \mu)^2 = 0.$$

In order to achieve a more compact notation we introduce the following definitions.

$$h = \frac{c}{N} \quad \text{and thus} \quad \frac{c}{n} = \frac{h}{1-h}, \qquad (2.5.3)$$

$$\Omega(h, \xi) = \left(\frac{h}{1-h}\right)\frac{\phi(\xi)}{\Phi(\xi)}. \qquad (2.5.4)$$

In subsequent usage, $\Omega(h, \xi)$ will sometimes be abbreviated to $\Omega(\xi)$, and on some occasions to Ω.

We substitute (2.5.4) into the two equations of (2.5.2) and after a few algebraic simplifications, we obtain

$$\bar{x} - \mu = \sigma\Omega(\xi), \qquad (2.5.5)$$
$$s^2 + (\bar{x} - \mu)^2 = \sigma^2[1 + \xi\Omega(\xi)],$$

where

$$\bar{x} = \sum_{i=1}^{n}\frac{x_i}{n} \quad \text{and} \quad s^2 = \sum_{i=1}^{n}\frac{(x_i - \bar{x})^2}{n}.$$

After further simplification, the two equations of (2.5.5) become

$$s^2 = \sigma^2[1 - \Omega(\Omega - \xi)], \qquad (2.5.6)$$
$$\bar{x} - T = \sigma(\Omega - \xi),$$

where from (2.2.6), $\mu = T - \sigma\xi$.

Note that equations (2.5.6) are completely analogous to equations (2.3.2) for singly truncated samples, the only difference being that here $Q(\xi)$ has been replaced by $\Omega(h,\xi)$.

When σ^2 is eliminated between the two equations of (2.5.6), we obtain

$$\frac{s^2}{(\bar{x}-T)^2} = \frac{1 - \Omega(\Omega - \xi)}{(\Omega - \xi)^2} = \alpha(h, \xi), \qquad (2.5.7)$$

and from the second equation of (2.5.6)

$$\sigma = \frac{\bar{x} - T}{\Omega - \xi}. \qquad (2.5.8)$$

Maximum likelihood estimates $\hat{\sigma}$ and $\hat{\xi}$ must, of course, satisfy equations (2.5.6) and likewise (2.5.7) and (2.5.8). Thus, $\bar{\alpha} = s^2/(\bar{x} - T)^2$ becomes the argument of primary interest here, just as it was in the case of singly truncated samples. Once more, readers are reminded that MLE, for singly truncated samples are identical to moment estimators for these samples. Corresponding equations for singly censored samples differ from those for singly truncated samples only in the substitution of $\Omega(\xi)$ for $Q(\xi)$.

The same algebraic simplifications that led to equations (2.3.7) and (2.3.12) for truncated samples, now enable us to write

$$\sigma^2 = s^2 + \left[\frac{\Omega}{\Omega - \xi}\right](\bar{x} - T)^2,$$
$$\mu = \bar{x} - \left[\frac{\Omega}{\Omega - \xi}\right](\bar{x} - T). \tag{2.5.9}$$

We define

$$\lambda(h, \xi) = \lambda(h, \alpha) = \left(\frac{\Omega}{\Omega - \xi}\right), \tag{2.5.10}$$

where $\alpha(h, \xi)$ is defined by (2.5.7). Thus, $\lambda(h, \alpha)$, as defined here, corresponds to $\theta(\alpha)$, which occurs in the case of truncated samples. As final estimators, (2.5.9) become

$$\hat{\sigma}^2 = s^2 + \lambda(h, \hat{\alpha})(\bar{x} - T)^2,$$
$$\hat{\mu} = \bar{x} - \lambda(h, \hat{\alpha})(\bar{x} - T), \tag{2.5.11}$$

where $\hat{\alpha} = s^2/(\bar{x} - T)^2$ and $h = c/N$. In complete (uncensored) samples, $h = 0$. In that case $\lambda(0, \alpha) = 0$ for all values of α, and thus the estimators (2.5.11) are applicable for complete as well as for censored samples.

Tables of the auxiliary estimating function $\lambda(h, \alpha)$ are included here as Table 2.3 to be used as aids in the calculation of estimates from sample data. In practical applications it is necessary only that we calculate $\bar{x}$ and s^2 from the n complete observations, and with $h = c/N$ and $\hat{\alpha} = s^2/(\bar{x} - T)^2$, we obtain $\lambda = \lambda(h, \hat{\alpha})$ by interpolation in Table 2.3. Estimates $\hat{\sigma}^2$ and $\hat{\mu}$ then follow from (2.5.11).

Estimators (2.5.11) are applicable for both Type I and Type II samples. The only difference to be reckoned with concerns the terminal T. In Type I samples, T is a predetermined constant, whereas c and n are observed values of random variables. In Type II samples, c and n are predetermined constants, whereas T is the observed value of a random variable. More specifically, T, in this case, is the observed value of an order statistic. In singly left censored samples, $T = x_{(c+1):N}$ whereas in singly right censored samples, $T = x_{n:N}$. In both types the total sample size is $N = n + c$.

Table 2.3 Auxiliary Estimation Function $\lambda(h,\alpha)$ for Singly Censored Samples from the Normal Distribution: $\hat{\alpha} = s^2/(\bar{x} - T)^2$

h α	.01	.02	.03	.04	.05	.06	.07	.08	.09	.10	.15
.00	.01010	.02040	.03090	.04161	.05251	.06363	.07495	.08649	.09824	.11020	.17342
.01	.01020	.02059	.03118	.04197	.05297	.06417	.07557	.08719	.09902	.11106	.17465
.02	.01029	.02077	.03145	.04233	.05341	.06469	.07618	.08787	.09978	.11190	.17586
.03	.01038	.02095	.03172	.04268	.05384	.06520	.07677	.08854	.10052	.11272	.17704
.04	.01047	.02113	.03197	.04302	.05426	.06570	.07734	.08919	.10125	.11352	.17821
.05	.01055	.02129	.03223	.04335	.05467	.06619	.07791	.08983	.10197	.11431	.17935
.06	.01064	.02146	.03247	.04367	.05507	.06667	.07846	.09046	.10267	.11508	.18047
.07	.01072	.02162	.03271	.04399	.05546	.06713	.07900	.09107	.10335	.11584	.18157
.08	.01080	.02178	.03294	.04430	.05585	.06759	.07953	.09168	.10403	.11659	.18266
.09	.01087	.02193	.03317	.04460	.05623	.06804	.08006	.09227	.10469	.11732	.18373
.10	.01095	.02208	.03340	.04490	.05660	.06848	.08057	.09285	.10534	.11804	.18479
.11	.01102	.02223	.03362	.04519	.05696	.06892	.08107	.09343	.10598	.11875	.18583
.12	.01110	.02238	.03384	.04548	.05732	.06934	.08157	.09399	.10661	.11944	.18685
.13	.01117	.02252	.03405	.04577	.05767	.06976	.08205	.09454	.10723	.12013	.18786
.14	.01124	.02266	.03426	.04604	.05802	.07018	.08254	.09509	.10785	.12081	.18886
.15	.01131	.02280	.03447	.04632	.05836	.07059	.08301	.09563	.10845	.12148	.18985
.16	.01138	.02293	.03467	.04659	.05869	.07099	.08310	.09616	.10905	.12214	.19082
.17	.01145	.02307	.03487	.04685	.05902	.07138	.08394	.09668	.10963	.12270	.19178
.18	.01151	.02320	.03507	.04712	.05935	.07177	.08439	.09720	.11021	.12343	.19273
.19	.01158	.02333	.03526	.04737	.05967	.07216	.08484	.09771	.11079	.12407	.19367
.20	.01164	.02346	.03545	.04763	.05999	.07254	.08528	.09822	.11135	.12469	.19460
.21	.01171	.02359	.03564	.04788	.06030	.07291	.08572	.09871	.11191	.12531	.19552
.22	.01177	.02371	.03583	.04813	.06061	.07329	.08615	.09921	.11246	.12592	.19643
.23	.01183	.02383	.03601	.04838	.06092	.07365	.08657	.09969	.11301	.12653	.19733
.24	.01189	.02396	.03620	.04862	.06122	.07401	.08700	.10017	.11355	.12713	.19822
.25	.01195	.02408	.03638	.04886	.06152	.07437	.08741	.10065	.11408	.12772	.19910
.26	.01201	.02420	.03656	.04909	.06182	.07473	.08783	.10112	.11461	.12831	.19997
.27	.01207	.02431	.03673	.04933	.06211	.07508	.08823	.10158	.11513	.12889	.20083
.28	.01213	.02443	.03691	.04956	.06240	.07542	.08864	.10205	.11565	.12946	.20169
.29	.01219	.02454	.03708	.04979	.06269	.07577	.08904	.10250	.11616	.13003	.20254
.30	.01224	.02466	.03725	.05002	.06297	.07611	.08943	.10295	.11667	.13059	.20338
.31	.01230	.02477	.03742	.05024	.06325	.07644	.08982	.10340	.11717	.13115	.20421
.32	.01236	.02488	.03758	.05047	.06353	.07678	.09021	.10384	.11767	.13170	.20503
.33	.01241	.02499	.03775	.05069	.06380	.07711	.09060	.10428	.11816	.13225	.20585
.34	.01247	.02510	.03791	.05090	.06408	.07743	.09098	.10472	.11865	.13279	.20666
.35	.01252	.02521	.03808	.05112	.06435	.07776	.09136	.10515	.11914	.13333	.20747
.36	.01257	.02532	.03824	.05133	.06461	.07808	.09173	.10557	.11962	.13386	.20826
.37	.01263	.02542	.03840	.05155	.06488	.07839	.09210	.10600	.12009	.13439	.20906
.38	.01268	.02553	.03855	.05176	.06514	.07871	.09247	.10642	.12057	.13491	.20984
.39	.01273	.02563	.03871	.05197	.06540	.07902	.09283	.10683	.12103	.13543	.21062
.40	.01278	.02574	.03887	.05217	.06566	.07933	.09319	.10725	.12150	.13595	.21139
.41	.01284	.02584	.03902	.05238	.06592	.07964	.09355	.10766	.12196	.13646	.21216
.42	.01289	.02594	.03917	.05258	.06617	.07994	.09391	.10806	.12242	.13697	.21292
.43	.01294	.02604	.03932	.05278	.06642	.08025	.09426	.10847	.12287	.13747	.21368
.44	.01299	.02614	.03947	.05298	.06667	.08055	.09461	.10887	.12332	.13797	.21443
.45	.01304	.02624	.03962	.05318	.06692	.08085	.09496	.10926	.12377	.13847	.21517
.46	.01309	.02634	.03977	.05338	.06717	.08114	.09530	.10966	.12421	.13896	.21591
.47	.01313	.02644	.03992	.05357	.06741	.08143	.09565	.11005	.12465	.13945	.21665
.48	.01318	.02654	.04006	.05377	.06765	.08173	.09598	.11044	.12509	.13994	.21738
.49	.01323	.02663	.04021	.05396	.06790	.08201	.09632	.11082	.12552	.14042	.21810

Source: Adapted from Cohen and Whitten (1988), Table 8.2, pp. 139–144, by courtesy of Marcel Dekker, Inc.

Table 2.3 *Continued*

α \ h	.01	.02	.03	.04	.05	.06	.07	.08	.09	.10	.15
.50	.01328	.02673	.04035	.05415	.06813	.08230	.09666	.11121	.12595	.14090	.21882
.51	.01333	.02682	.04049	.05434	.06837	.08259	.09699	.11159	.12638	.14138	.21954
.52	.01337	.02692	.04064	.05453	.06861	.08287	.09732	.11196	.12681	.14185	.22025
.53	.01342	.02701	.04078	.05472	.06884	.08315	.09765	.11234	.12723	.14232	.22095
.54	.01347	.02710	.04092	.05490	.06907	.08343	.09797	.11271	.12765	.14278	.22166
.55	.01351	.02720	.04105	.05509	.06931	.08371	.09830	.11308	.12806	.14325	.22235
.56	.01356	.02729	.04119	.05527	.06954	.08398	.09862	.11345	.12848	.14371	.22305
.57	.01360	.02738	.04133	.05546	.06976	.08426	.09894	.11382	.12889	.14417	.22374
.58	.01365	.02747	.04146	.05564	.06999	.08453	.09926	.11418	.12930	.14462	.22442
.59	.01369	.02756	.04160	.05582	.07022	.08480	.09957	.11454	.12970	.14507	.22510
.60	.01374	.02765	.04173	.05600	.07044	.08507	.09989	.11490	.13011	.14552	.22578
.61	.01378	.02774	.04187	.05617	.07066	.08534	.10020	.11526	.13051	.14597	.22645
.62	.01383	.02783	.04200	.05635	.07088	.08560	.10051	.11561	.13091	.14641	.22712
.63	.01387	.02791	.04213	.05653	.07110	.08586	.10082	.11596	.13131	.14685	.22779
.64	.01391	.02800	.04226	.05670	.07132	.08613	.10112	.11631	.13170	.14729	.22845
.65	.01396	.02809	.04239	.05687	.07154	.08639	.10143	.11666	.13209	.14773	.22910
.66	.01400	.02817	.04252	.05705	.07175	.08665	.10173	.11701	.13248	.14816	.22976
.67	.01404	.02826	.04265	.05722	.07197	.08690	.10203	.11735	.13287	.14859	.23041
.68	.01409	.02834	.04278	.05739	.07218	.08716	.10233	.11769	.13326	.14902	.23106
.69	.01413	.02843	.04290	.05756	.07239	.08742	.10263	.11804	.13364	.14945	.23170
.70	.01417	.02851	.04303	.05773	.07260	.08767	.10292	.11837	.13402	.14987	.23234
.71	.01421	.02860	.04316	.05789	.07281	.08792	.10322	.11871	.13440	.15030	.23298
.72	.01425	.02868	.04328	.05806	.07302	.08817	.10351	.11905	.13478	.15072	.23361
.73	.01430	.02876	.04341	.05823	.07323	.08842	.10380	.11938	.13515	.15113	.23425
.74	.01434	.02885	.04353	.05839	.07344	.08867	.10409	.11971	.13553	.15155	.23487
.75	.01438	.02893	.04365	.05856	.07364	.08892	.10438	.12004	.13590	.15196	.23550
.76	.01442	.02901	.04377	.05872	.07385	.08916	.10467	.12037	.13627	.15237	.23612
.77	.01446	.02909	.04390	.05888	.07405	.08941	.10495	.12070	.13664	.15278	.23674
.78	.01450	.02917	.04402	.05904	.07425	.08965	.10524	.12102	.13700	.15319	.23735
.79	.01454	.02925	.04414	.05920	.07445	.08989	.10552	.12134	.13737	.15360	.23797
.80	.01458	.02933	.04426	.05936	.07465	.09013	.10580	.12167	.13773	.15400	.23858
.81	.01462	.02941	.04438	.05952	.07485	.09037	.10608	.12199	.13809	.15440	.23918
.82	.01466	.02949	.04450	.05968	.07505	.09061	.10636	.12231	.13845	.15480	.23979
.83	.01470	.02957	.04461	.05984	.07525	.09085	.10664	.12262	.13881	.15520	.24039
.84	.01474	.02965	.04473	.06000	.07545	.09108	.10691	.12294	.13916	.15559	.24099
.85	.01478	.02972	.04485	.06015	.07564	.09132	.10719	.12325	.13952	.15599	.24158
.90	.01497	.03011	.04542	.06092	.07661	.09248	.10854	.12480	.14126	.15793	.24452
.95	.01515	.03048	.04599	.06168	.07755	.09361	.10987	.12632	.14297	.15983	.24740
1.00	.01534	.03085	.04654	.06241	.07847	.09472	.11116	.12780	.14465	.16170	.25022
1.50	.01699	.03417	.05153	.06908	.08682	.10476	.12290	.14125	.15981	.17858	.27585
2.00	.01842	.03703	.05583	.07483	.09403	.11343	.13304	.15287	.17291	.19318	.29806
2.50	.01969	.03958	.05967	.07996	.10046	.12117	.14210	.16325	.18463	.20624	.31794
3.00	.02085	.04191	.06317	.08464	.10633	.12823	.15037	.17273	.19532	.21816	.33611
3.50	.02192	.04406	.06641	.08897	.11176	.13477	.15802	.18150	.20522	.22919	.35294
4.00	.02293	.04607	.06943	.09302	.11684	.14088	.16517	.18970	.21448	.23951	.36870
5.00	.02477	.04977	.07499	.10046	.12616	.15211	.17832	.20478	.23150	.25849	.39766
6.00	.02644	.05312	.08004	.10721	.13464	.16232	.19026	.21847	.24696	.27573	.42400
7.00	.02798	.05622	.08470	.11345	.14245	.17173	.20128	.23111	.26123	.29165	.44832
8.00	.02942	.05910	.08905	.11926	.14975	.18052	.21157	.24291	.27455	.30650	.47103
10.00	.03206	.06440	.09701	.12992	.16312	.19662	.23042	.26454	.29897	.33373	.51265

Singly Truncated and Censored Samples

h \ α	.20	.25	.30	.35	.40	.45	.50	.60	.70	.80	.90
.00	.24268	.31862	.40210	.49414	.59607	.70957	.83684	1.14536	1.56148	2.17591	3.28261
.01	.24426	.32054	.40434	.49670	.59894	.71275	.84033	1.14947	1.56625	2.18139	3.28898
.02	.24581	.32243	.40655	.49923	.60178	.71590	.84378	1.15355	1.57098	2.18685	3.29532
.03	.24734	.32429	.40873	.50172	.60459	.71901	.84720	1.15759	1.57568	2.19227	3.30163
.04	.24885	.32612	.41089	.50419	.60736	.72210	.85060	1.16161	1.58035	2.19767	3.30792
.05	.25033	.32793	.41301	.50663	.61011	.72515	.85396	1.16559	1.58499	2.20304	3.31419
.06	.25179	.32972	.41511	.50904	.61283	.72817	.85729	1.16955	1.58960	2.20838	3.32043
.07	.25322	.33147	.41719	.51142	.61552	.73117	.86059	1.17347	1.59419	2.21369	3.32665
.08	.25464	.33321	.41924	.51378	.61818	.73414	.86386	1.17737	1.59874	2.21898	3.33284
.09	.25604	.33493	.42126	.51611	.62082	.73708	.86711	1.18124	1.60327	2.22424	3.33901
.10	.25741	.33662	.42326	.51842	.62343	.73999	.87033	1.18508	1.60777	2.22948	3.34516
.11	.25877	.33829	.42525	.52071	.62602	.74288	.87352	1.18890	1.61225	2.23469	3.35128
.12	.26012	.33995	.42720	.52297	.62858	.74575	.87669	1.19269	1.61669	2.23987	3.35739
.13	.26144	.34158	.42914	.52521	.63112	.74859	.87983	1.19645	1.62112	2.24503	3.36346
.14	.26275	.34320	.43106	.52743	.63364	.75140	.88295	1.20019	1.62552	2.25017	3.36952
.15	.26405	.34480	.43296	.52962	.63613	.75420	.88605	1.20390	1.62989	2.25528	3.37556
.16	.26533	.34638	.43484	.53180	.63860	.75697	.88912	1.20759	1.63424	2.26037	3.38157
.17	.26660	.34794	.43670	.53396	.64106	.75972	.89217	1.21126	1.63856	2.26543	3.38756
.18	.26785	.34949	.43855	.53610	.64349	.76245	.89519	1.21490	1.64287	2.27048	3.39353
.19	.26909	.35103	.44038	.53822	.64590	.76515	.89820	1.21852	1.64714	2.27550	3.39948
.20	.27031	.35255	.44219	.54032	.64829	.76784	.90118	1.22212	1.65140	2.28049	3.40541
.21	.27152	.35405	.44398	.54240	.65067	.77051	.90415	1.22570	1.65563	2.28547	3.41132
.22	.27273	.35554	.44576	.54447	.65302	.77315	.90709	1.22925	1.65985	2.29042	3.41721
.23	.27391	.35702	.44752	.54652	.65536	.77578	.91001	1.23279	1.66404	2.29536	3.42307
.24	.27509	.35848	.44927	.54855	.65768	.77839	.91292	1.23630	1.66821	2.30027	3.42892
.25	.27626	.35993	.45100	.55057	.65998	.78098	.91580	1.23979	1.67235	2.30516	3.43475
.26	.27741	.36137	.45272	.55257	.66227	.78356	.91867	1.24327	1.67648	2.31003	3.44056
.27	.27856	.36279	.45443	.55455	.66454	.78611	.92152	1.24672	1.68059	2.31488	3.44635
.28	.27969	.36421	.45612	.55653	.66679	.78865	.92434	1.25015	1.68467	2.31970	3.45212
.29	.28082	.36561	.45780	.55848	.66903	.79117	.92716	1.25357	1.68874	2.32451	3.45787
.30	.28193	.36700	.45946	.56042	.67125	.79368	.92995	1.25696	1.69279	2.32930	3.46360
.31	.28304	.36838	.46112	.56235	.67346	.79617	.93273	1.26034	1.69682	2.33407	3.46931
.32	.28414	.36975	.46276	.56427	.67565	.79864	.93549	1.26370	1.70082	2.33882	3.47501
.33	.28522	.37110	.46438	.56617	.67783	.80110	.93823	1.26704	1.70481	2.34355	3.48068
.34	.28630	.37245	.46600	.56806	.67999	.80354	.94096	1.27036	1.70879	2.34827	3.48634
.35	.28737	.37379	.46761	.56993	.68214	.80597	.94367	1.27367	1.71274	2.35296	3.49198
.36	.28844	.37511	.46920	.57179	.68427	.80838	.94637	1.27696	1.71668	2.35764	3.49761
.37	.28949	.37643	.47078	.57364	.68640	.81078	.94905	1.28023	1.72059	2.36230	3.50321
.38	.29053	.37774	.47235	.57548	.68851	.81316	.95172	1.28349	1.72449	2.36694	3.50880
.39	.29157	.37904	.47391	.57731	.69060	.81553	.95437	1.28673	1.72838	2.37156	3.51437
.40	.29260	.38033	.47547	.57912	.69268	.81789	.95700	1.28995	1.73224	2.37616	3.51993
.41	.29363	.38161	.47701	.58093	.69475	.82023	.95963	1.29316	1.73609	2.38075	3.52546
.42	.29464	.38288	.47854	.58272	.69681	.82256	.96223	1.29636	1.73993	2.38532	3.53098
.43	.29565	.38414	.48006	.58450	.69886	.82488	.96483	1.29953	1.74374	2.38988	3.53649
.44	.29665	.38540	.48157	.58627	.70089	.82719	.96741	1.30270	1.74754	2.39441	3.54197
.45	.29765	.38665	.48307	.58803	.70292	.82948	.96998	1.30584	1.75133	2.39893	3.54744
.46	.29864	.38788	.48456	.58978	.70493	.83176	.97253	1.30898	1.75510	2.40344	3.55290
.47	.29962	.38912	.48605	.59152	.70693	.83402	.97507	1.31209	1.75885	2.40793	3.55834
.48	.30059	.39034	.48752	.59325	.70692	.83628	.97760	1.31520	1.76259	2.41240	3.56376
.49	.30156	.39156	.48899	.59497	.71090	.83852	.98012	1.31829	1.76631	2.41685	3.56917

Table 2.3 *Continued*

α \ h	.20	.25	.30	.35	.40	.45	.50	.60	.70	.80	.90
.50	.30253	.39276	.49044	.59668	.71286	.84075	.98262	1.32136	1.77002	2.42129	3.57456
.51	.30348	.39396	.49189	.59838	.71482	.84297	.98511	1.32443	1.77371	2.42572	3.57993
.52	.30443	.39516	.49333	.60007	.71677	.84518	.98759	1.32748	1.77739	2.43013	3.58529
.53	.30538	.39635	.49476	.60175	.71870	.84738	.99006	1.33051	1.78106	2.43452	3.59064
.54	.30632	.39753	.49619	.60343	.72063	.84957	.99251	1.33353	1.78470	2.43890	3.59597
.55	.30725	.39870	.49760	.60509	.72255	.85174	.99495	1.33654	1.78834	2.44327	3.60128
.56	.30818	.39987	.49901	.60674	.72445	.85391	.99739	1.33954	1.79196	2.44762	3.60658
.57	.30910	.40103	.50041	.60839	.72635	.85606	.99981	1.34252	1.79557	2.45195	3.61187
.58	.31002	.40218	.50181	.61003	.72824	.85821	1.00222	1.34550	1.79916	2.45628	3.61714
.59	.31093	.40333	.50319	.61166	.73012	.86034	1.00462	1.34845	1.80274	2.46058	3.62240
.60	.31184	.40447	.50457	.61328	.73199	.86247	1.00700	1.35140	1.80631	2.46487	3.62764
.61	.31274	.40560	.50594	.61489	.73385	.86458	1.00938	1.35434	1.80987	2.46915	3.63287
.62	.31364	.40673	.50731	.61650	.73570	.86669	1.01175	1.35726	1.81341	2.47342	3.63808
.63	.31453	.40785	.50867	.61810	.73754	.86878	1.01411	1.36017	1.81694	2.47767	3.64328
.64	.31542	.40897	.51002	.61968	.73938	.87087	1.01645	1.36307	1.82045	2.48191	3.64847
.65	.31630	.41008	.51136	.62127	.74120	.87295	1.01879	1.36596	1.82395	2.48613	3.65364
.66	.31718	.41119	.51270	.62284	.74302	.87502	1.02112	1.36884	1.82744	2.49034	3.65880
.67	.31805	.41229	.51403	.62441	.74483	.87707	1.02343	1.37170	1.83092	2.49454	3.66394
.68	.31892	.41338	.51535	.62597	.74663	.87912	1.02574	1.37456	1.83439	2.49872	3.66907
.69	.31979	.41447	.51667	.62752	.74842	.88116	1.02804	1.37740	1.83784	2.50289	3.67419
.70	.32065	.41555	.51798	.62907	.75021	.88320	1.03032	1.38023	1.84128	2.50705	3.67930
.71	.32150	.41663	.51929	.63060	.75199	.88522	1.03260	1.38305	1.84471	2.51120	3.68439
.72	.32236	.41771	.52059	.63213	.75376	.88723	1.03487	1.38587	1.84813	2.51533	3.68947
.73	.32320	.41877	.52188	.63366	.75552	.88924	1.03713	1.38867	1.85154	2.51945	3.69453
.74	.32405	.41984	.52317	.63518	.75727	.89124	1.03938	1.39146	1.85493	2.52356	3.69958
.75	.32489	.42090	.52445	.63669	.75902	.89323	1.04162	1.39424	1.85832	2.52766	3.70462
.76	.32572	.42195	.52573	.63819	.76076	.89521	1.04386	1.39701	1.86169	2.53174	3.70965
.77	.32655	.42300	.52700	.63969	.76249	.89718	1.04608	1.39977	1.86505	2.53581	3.71467
.78	.32738	.42404	.52826	.64118	.76422	.89915	1.04830	1.40252	1.86840	2.53987	3.71967
.79	.32821	.42508	.52952	.64267	.76593	.90111	1.05051	1.40526	1.87174	2.54392	3.72466
.80	.32903	.42612	.53078	.64415	.76764	.90306	1.05270	1.40799	1.87507	2.54796	3.72964
.81	.32984	.42715	.53203	.64562	.76935	.90500	1.05490	1.41071	1.87839	2.55198	3.73460
.82	.33065	.42817	.53327	.64709	.77105	.90694	1.05708	1.41342	1.88170	2.55599	3.73956
.83	.33146	.42919	.53451	.64855	.77274	.90887	1.05925	1.41613	1.88499	2.56000	3.74450
.84	.33227	.43021	.53574	.65000	.77442	.91079	1.06142	1.41882	1.88828	2.56399	3.74943
.85	.33307	.43122	.53697	.65145	.77610	.91270	1.06358	1.42150	1.89155	2.56797	3.75434
.90	.33703	.43622	.54303	.65861	.78439	.92216	1.07425	1.43479	1.90778	2.58770	3.77876
.95	.34091	.44112	.54898	.66564	.79252	.93145	1.08474	1.44786	1.92377	2.60716	3.80289
1.00	.34471	.44592	.55481	.67253	.80051	.94058	1.09506	1.46072	1.93952	2.62636	3.82674
1.50	.37929	.48973	.60812	.73566	.87383	1.02451	1.19009	1.57980	2.08610	2.80620	4.05201
2.00	.40934	.52788	.65466	.79092	.93818	1.09836	1.27394	1.68551	2.21714	2.96832	4.25750
2.50	.43629	.56213	.69651	.84068	.99621	1.16508	1.34983	1.78156	2.33672	3.11711	4.44765
3.00	.46092	.59348	.73485	.88631	1.04949	1.22641	1.41967	1.87018	2.44742	3.25541	4.62546
3.50	.48377	.62257	.77044	.92871	1.09903	1.28348	1.48472	1.95288	2.55096	3.38516	4.79305
4.00	.50516	.64982	.80381	.96848	1.14552	1.33707	1.54584	2.03071	2.64858	3.50779	4.95202
5.00	.54450	.69998	.86526	1.04176	1.23125	1.43595	1.65870	2.17465	2.82949	3.73564	5.24865
6.00	.58030	.74563	.92122	1.10853	1.30941	1.52616	1.76174	2.30628	2.99526	3.94498	5.52232
7.00	.61336	.78782	.97294	1.17027	1.38172	1.60966	1.85716	2.42831	3.14916	4.13970	5.77765
8.00	.64424	.82722	1.02127	1.22798	1.44932	1.68774	1.94643	2.54257	3.29342	4.32249	6.01791
10.00	.70086	.89949	1.10995	1.33390	1.57344	1.83118	2.11047	2.75274	3.55906	4.65963	6.46220

Singly Truncated and Censored Samples

2.6 SAMPLING ERRORS OF ESTIMATES

The asymptotic variance–covariance matrix of the maximum likelihood estimates $(\hat{\mu}, \hat{\sigma})$ is obtained by inverting the Fisher information matrix with elements that are negatives of expected values of the second-order partial derivatives of the loglikelihood function with respect to the parameters. Accordingly, as given by Cohen (1961), we obtain

$$V(\hat{\mu}) = \frac{\sigma^2}{N}\mu_{11}, \quad V(\hat{\sigma}) = \frac{\sigma^2}{N}\mu_{22},$$
$$\text{Cov}(\hat{\mu},\hat{\sigma}) = \frac{\sigma^2}{N}\mu_{12}, \quad \rho_{\hat{\mu},\hat{\sigma}} = \frac{\mu_{12}}{\sqrt{\mu_{11}\mu_{22}}}, \quad (2.6.1)$$

where μ_{11}, μ_{12}, and μ_{22} are functions of ϕ_{11}, ϕ_{12}, and ϕ_{22}, which are elements of the information matrix.

For both left and right singly truncated samples, with $N = n$, we have

$$\mu_{11} = \frac{\phi_{22}}{\phi_{11}\phi_{22} - \phi_{12}^2}, \quad \mu_{22} = \frac{\phi_{11}}{\phi_{11}\phi_{22} - \phi_{12}^2}, \quad \mu_{12} = \frac{-\phi_{12}}{\phi_{11}\phi_{22} - \phi_{12}^2}, \quad (2.6.2)$$

and

$$\phi_{11}(\eta) = 1 - Q(\eta)[Q(\eta) - \eta],$$
$$\phi_{12}(\eta) = Q(\eta)[1 - \eta(Q(\eta) - \eta)], \quad (2.6.3)$$
$$\phi_{22}(\eta) = 2 + \eta\phi_{12}(\eta).$$

For singly left truncated samples, $\eta = \xi$. For singly right truncated samples, $\eta = -\xi$, and we delete the negative sign from μ_{12} in equation (2.6.2).

For Type I singly censored samples, with $E(n) = N[1 - \Phi(\eta)]$,

$$\mu_{11} = \frac{1}{1 - \Phi(\eta)}\left(\frac{\phi_{22}}{\phi_{11}\phi_{22} - \phi_{12}^2}\right),$$
$$\mu_{12} = \frac{1}{1 - \Phi(\eta)}\left(\frac{-\phi_{12}}{\phi_{11}\phi_{22} - \phi_{12}^2}\right), \quad (2.6.4)$$
$$\mu_{22} = \frac{1}{1 - \Phi(\eta)}\left(\frac{\phi_{11}}{\phi_{11}\phi_{22} - \phi_{12}^2}\right),$$

and

$$\phi_{11}(\eta) = 1 + Q(\eta)[Q(-\eta) + \eta],$$
$$\phi_{12}(\eta) = Q(\eta)[1 + \eta(Q(-\eta) + \eta)], \quad (2.6.5)$$
$$\phi_{22}(\eta) = 2 + \eta\phi_{12}(\eta).$$

For Type I left censored samples, $\eta = \xi$, whereas for Type I right censored samples $\eta = -\xi$, and we delete the negative sign from μ_{12} in equation (2.6.4).

For Type II censored samples, $E(n) = n$, and

$$\mu_{11} = \left(\frac{N}{n}\right) \frac{\phi_{22}}{\phi_{11}\phi_{22} - \phi_{12}^2},$$

$$\mu_{22} = \left(\frac{N}{n}\right) \frac{\phi_{11}}{\phi_{11}\phi_{22} - \phi_{12}^2}, \qquad (2.6.6)$$

$$\mu_{12} = \left(\frac{N}{n}\right) \frac{-\phi_{12}}{\phi_{11}\phi_{22} - \phi_{12}^2},$$

where

$$\phi_{11}(h, \eta) = 1 + \Omega(h, \eta)[Q(-\eta) + \eta],$$

$$\phi_{12}(h, \eta) = \Omega(h, \eta)[1 + \eta\{Q(-\eta) + \eta\}], \qquad (2.6.7)$$

$$\phi_{22}(h, \eta) = 2 + \eta\phi_{12}(h, \eta).$$

For Type II left censored samples, $\eta = \xi$. For Type II right censored samples, $\eta = -\xi$, and we delete the negative sign from μ_{12} in equation (2.6.6.).

It is to be noted that as $N \to \infty$, the ϕ_{ij} for Type II censored samples approach the ϕ_{ij} for Type I censored samples. Likewise as $N \to \infty$, $\lim(n/N) \to [1 - \Phi(\xi)]$ for left censored samples, and $\lim(n/N) \to \Phi(\xi)$ for right censored samples. Thus, limiting values of estimate variances and covariance for both types of censored samples approach equality.

To evaluate the μ_{ij} of (2.6.1), it is of course necessary to evaluate the applicable ϕ_{ij} of (2.6.2), (2.6.4), or (2.6.6), and this requires that we calculate $\hat{\xi} = (T - \hat{\mu})/\hat{\sigma}$. Table 2.4 is a computational aid that eliminates most of these tedious computations. The applicable μ_{ij} can be obtained by interpolation in this table and substituted into (2.6.1) to provide asymptotic variances and covariances of $\hat{\mu}$ and $\hat{\sigma}$.

2.7 ILLUSTRATIVE EXAMPLES

To illustrate the practical application of estimators derived in this chapter, several examples have been chosen from various sources.

Example 2.7.1. Gupta (1952) gave results of a life test on 10 laboratory mice following inoculation with a uniform culture of human tuberculosis. The test was terminated with the death of the seventh specimen. Thus, the sample in this case

Singly Truncated and Censored Samples

Table 2.4 Variance and Covariance Factors for Singly Truncated and Singly Censored Samples from the Normal Distribution

	For Truncated Samples				For Censored Samples				Percent Rest.	
η	μ_{11}	μ_{12}	μ_{22}	ρ	μ_{11}	μ_{12}	μ_{22}	ρ		η
-4.0	1.00054	-.001143	.502287	-.001613	1.00000	-.000006	.500030	-.000001	0.00	-4.0
-3.5	1.00313	-.005922	.510366	-.008277	1.00001	-.000052	.500208	-.000074	0.02	-3.5
-3.0	1.01460	-.024153	.536283	-.032744	1.00010	-.000335	.501180	-.000473	0.13	-3.0
-2.5	1.05738	-.081051	.602029	-.101586	1.00056	-.001712	.505280	-.002407	0.62	-2.5
-2.4	1.07437	-.101368	.622786	-.123924	1.00078	-.002312	.506935	-.003247	0.82	-2.4
-2.3	1.09604	-.126136	.646862	-.149803	1.00107	-.003099	.509030	-.004341	1.07	-2.3
-2.2	1.12365	-.156229	.674663	-.179434	1.00147	-.004121	.511658	-.005757	1.39	-2.2
-2.1	1.15880	-.192688	.706637	-.212937	1.00200	-.005438	.514926	-.007571	1.79	-2.1
-2.0	1.20350	-.236743	.743283	-.250310	1.00270	-.007123	.518960	-.009875	2.28	-2.0
-1.9	1.26030	-.289860	.785158	-.291398	1.00363	-.009266	.523899	-.012778	2.87	-1.9
-1.8	1.33246	-.353771	.832880	-.335818	1.00485	-.011971	.529899	-.016405	3.59	-1.8
-1.7	1.42405	-.430531	.887141	-.383041	1.00645	-.015368	.537141	-.020901	4.46	-1.7
-1.6	1.54024	-.522564	.948713	-.432293	1.00852	-.019610	.545827	-.026431	5.48	-1.6
-1.5	1.68750	-.632733	1.01846	-.482644	1.01120	-.024884	.556186	-.033181	6.68	-1.5
-1.4	1.87398	-.764405	1.09734	-.533054	1.01467	-.031410	.568471	-.041358	8.08	-1.4
-1.3	2.10982	-.921533	1.18642	-.582464	1.01914	-.039460	.582981	-.051193	9.68	-1.3
-1.2	2.40764	-1.10874	1.28690	-.629889	1.02488	-.049355	.600046	-.062937	11.51	-1.2
-1.1	2.78311	-1.33145	1.40009	-.674498	1.03224	-.061491	.620049	-.076861	13.57	-1.1
-1.0	3.25557	-1.59594	1.52746	-.715676	1.04168	-.076345	.643438	-.093252	15.87	-1.0
-0.9	3.84879	-1.90952	1.67064	-.753044	1.05376	-.094501	.670724	-.112407	18.41	-0.9
-0.8	4.59189	-2.28066	1.83140	-.786452	1.06923	-.116674	.702513	-.134620	21.19	-0.8
-0.7	5.52036	-2.71911	2.01172	-.815942	1.08904	-.143744	.739515	-.160175	24.20	-0.7
-0.6	6.67730	-3.23612	2.21376	-.841703	1.11442	-.176798	.782574	-.189317	27.43	-0.6
-0.5	8.11482	-3.84458	2.43990	-.864019	1.14696	-.217183	.832691	-.222233	30.85	-0.5
-0.4	9.89562	-4.55921	2.69271	-.883229	1.18876	-.266527	.891077	-.259011	34.46	-0.4
-0.3	12.0949	-5.39683	2.97504	-.899688	1.24252	-.327080	.959181	-.299607	38.21	-0.3
-0.2	14.8023	-6.37653	3.28997	-.913744	1.31180	-.401326	1.03877	-.343800	42.07	-0.2
-0.1	18.1244	-7.51996	3.64083	-.925727	1.40127	-.492641	1.13198	-.391156	46.02	-0.1
0.0	22.1875	-8.85155	4.03126	-.935932	1.51709	-.605233	1.24145	-.441013	50.00	0.0
0.1	27.1403	-10.3988	4.46517	-.944623	1.66743	-.744459	1.37042	-.492483	53.98	0.1
0.2	33.1573	-12.1927	4.94678	-.952028	1.86310	-.917651	1.52288	-.544498	57.93	0.2
0.3	40.4428	-14.2679	5.48068	-.958345	2.11857	-1.13214	1.70381	-.595891	61.79	0.3
0.4	49.2342	-16.6628	6.08719	-.963742	2.45318	-1.40071	1.91942	-.645504	65.54	0.4
0.5	59.8081	-19.4208	6.72512	-.968361	2.89293	-1.73757	2.17751	-.692299	69.15	0.5
0.6	72.4834	-22.5896	7.44658	-.972322	3.47293	-2.16185	2.48793	-.735459	72.57	0.6
0.7	87.6276	-26.2220	8.24204	-.975727	4.24075	-2.69858	2.86318	-.774443	75.80	0.7
0.8	105.66	-30.376	9.1178	-.97866	5.2612	-3.3807	3.3192	-.80899	78.81	0.8
0.9	127.07	-35.117	10.081	-.98119	6.6229	-4.2517	3.8765	-.83912	81.59	0.9
1.0	152.40	-40.515	11.138	-.98338	8.4477	-5.3696	4.5614	-.86502	84.13	1.0
1.1	182.29	-46.650	12.298	-.98529	10.903	-6.8116	5.4082	-.88703	86.43	1.1
1.2	217.42	-53.601	13.567	-.98694	14.224	-8.6818	6.4616	-.90557	88.49	1.2
1.3	258.61	-61.465	14.954	-.98838	18.735	-11.121	7.7804	-.92109	90.32	1.3
1.4	306.78	-70.347	16.471	-.98964	24.892	-14.319	9.4423	-.93401	91.92	1.4
1.5	362.91	-80.350	18.124	-.99074	33.339	-18.539	11.550	-.94473	93.32	1.5
1.6	428.11	-91.586	19.922	-.99171	44.986	-24.139	14.243	-.95361	94.52	1.6
1.7	503.57	-104.17	21.874	-.99256	61.132	-31.616	17.706	-.96097	95.54	1.7
1.8	591.03	-118.31	24.003	-.99332	83.638	-41.664	22.193	-.96706	96.41	1.8
1.9	691.78	-134.10	26.311	-.99398	115.19	-55.252	28.046	-.97211	97.13	1.9
2.0	807.71	-151.73	28.813	-.99457	159.66	-73.750	35.740	-.97630	97.72	2.0
2.1	940.38	-171.30	31.511	-.99509	222.74	-99.100	45.930	-.97979	98.21	2.1
2.2	1091.4	-192.92	34.405	-.99555	312.73	-134.08	59.526	-.98270	98.61	2.2
2.3	1265.4	-217.17	37.575	-.99596	441.92	-182.68	77.810	-.98514	98.93	2.3
2.4	1458.6	-243.23	40.858	-.99632	628.58	-250.68	102.59	-.98718	99.18	2.4
2.5	1677.8	-271.99	44.392	-.99665	899.99	-346.53	136.44	-.98890	99.38	2.5

Note: When truncation or Type I censoring occurs on the left, entries in this table corresponding to $\eta = \xi$ are applicable. For right truncated or Type I right censored samples, read entries corresponding to $\eta = -\xi$, but delete negative signs from μ_{12} and ρ. For both Type II left censored and Type II right censored samples, read entries corresponding to Percent Restriction = $100h$, but for right censoring delete negative signs from μ_{12} and ρ.

Source: From Cohen (1961), Table 3, p. 539, with permission of the Technometrics Management Committee.

is Type II singly right censored. Gupta assumed that logarithms to the base 10 of these life-spans were distributed normally (μ, σ^2). Survival times, Y in days from inoculation to death and logarithms, X, for the seven complete observations are as follows:

y	41	44	46	54	55	58	60
$x = \log_{10} y$	1.6128	1.6435	1.6628	1.7324	1.7404	1.7634	1.7782

Sample data are summarized as $N = 10$, $n = 7$, $c = 3$, $h = 0.3$, $T = x_{7:10} = 1.7782$, $\bar{x}_7 = 1.70479$, $s^2 = 0.003548$, and $\hat{\alpha} = 0.003548/(1.70479 - 1.7782)^2 = 0.6584$. We interpolate in Table 2.3 to obtain $\hat{\lambda} = 0.51249$. By substitution in (2.5.11), we calculate

$$\hat{\mu} = 1.70479 - (0.51249)(1.70479 - 1.7782) = 1.7424,$$

$$\hat{\sigma}^2 = 0.003548 + (0.51249)(1.70479 - 1.7782)^2 = 0.0063098,$$

$$\hat{\sigma} = 0.07943.$$

To calculate approximate asymptotic confidence intervals on μ and σ, we enter Table 2.4 with $100h = 30\%$ censoring, and interpolate to obtain $\mu_{11} = 1.13887$, $\mu_{12} = 0.207146$, $\mu_{22} = 0.820235$, and $\rho_{\hat{\mu},\hat{\sigma}} = 0.21405$. We substitute these values into (2.6.1) and calculate

$$V(\hat{\mu}) = (0.0063098/10)(1.13887) = 0.0007186,$$

$$V(\hat{\sigma}) = (0.0063098/10)(0.820235) = 0.0005176,$$

$$\text{Cov }(\hat{\mu},\hat{\sigma}) = (0.0063098/10)(0.207146) = 0.0001307.$$

Accordingly, $\sigma_{\hat{\mu}} = 0.0268$, and $\sigma_{\hat{\sigma}} = 0.0228$. Approximate 95% confidence intervals calculated as [estimate $\pm$ 1.96 (standard deviation of estimate)] follow as

$$1.698 < \mu < 1.787,$$

$$0.035 < \sigma < 0.124.$$

In the original units the mean survival time in days becomes

$$50 < \text{MST} < 61.$$

Example 2.7.2. This example was also given by Gupta (1952). It pertains to a life test of 300 electric light bulbs that was terminated with the failure of the 119th bulb. The sample is thus Type II singly right censored with $N = 300$, $n = 119$, $c = 181$, and $h = 181/300 = 0.6033$. Other data summaries are $\bar{x}_{119} = 1304.822$, $s^2_{119} = 12{,}128.25$, $T = x_{119:300} = 1450.000$ and $\hat{\alpha} =$

Singly Truncated and Censored Samples

$12,128.25/(1304.822 - 1450.000)^2 = 0.5754$. We enter Table 2.2 with $h = 0.6033$ and $\hat{\alpha} = 0.5754$ and interpolate to obtain $\hat{\lambda} = 1.3591$. We substitute in (2.5.11) and calculate $\hat{\mu} = 1502.1$ and $\hat{\sigma}^2 = 40,773.5$. Thus $\hat{\sigma} = 201.9$. We enter Table 2.4 with $100h = 60.33\%$ censoring and interpolate to obtain $\mu_{11} = 2.0219$, $\mu_{12} = 1.0508$, $\mu_{22} = 1.634$, and $\rho_{\hat{\mu},\hat{\sigma}} = 0.576$. From (2.6.1), we calculate $\sigma_{\hat{\mu}} = 16.58$ and $\sigma_{\hat{\sigma}} = 14.91$. We then calculate approximate 95% confidence intervals as

$$1470 < \mu < 1535,$$
$$172.7 < \sigma < 231.1.$$

Example 2.7.3. Nelson (1982) considered a Type I singly right censored sample consisting of life spans in thousands of miles of $N = 96$ electronic locomotive controls. Observation was discontinued at $T = 135$ (fixed) with the result that $n = 37$ failures occurred prior to termination of the test, thus leaving $c = 59$ censored observations. Following is a tabulation of the life spans in units of 1000 hours of service, of the 37 controls that failed.

Thousands of Miles to Failure of 37 Locomotive Controls						
22.5	37.5	46.0	48.5	51.5	53.0	54.5
57.5	66.5	68.0	69.5	76.5	77.0	78.5
80.0	81.5	82.0	83.0	84.0	91.5	93.5
102.5	107.0	108.5	112.5	113.5	116.0	117.0
118.5	119.0	120.0	122.5	123.0	127.5	131.0
132.5	134.0					

Nelson assumed the base 10 logarithms, X, of these life spans to be normally distributed (μ, σ^2). Thus, in summary, we have $N = 96$, $n = 37$, $c = 59$, $T = \log_{10} 135 = 2.1303$, $\bar{x}_{37} = 1.921$, $s^2_{37} = 0.0307$, and $\hat{\alpha} = 0.0307/(1.921 - 2.1303)^2 = 0.701$. The fraction of censored observations $h = 59/96 = 0.6146$. We enter Table 2.3 with $h = 0.6146$ and $\hat{\alpha} = 0.701$ and interpolate to obtain $\hat{\lambda} = 1.448$. Substitution in (2.5.11) yields $\hat{\mu} = 2.224$ and $\hat{\sigma}^2 = 0.09413$. Thus, we have $\hat{\sigma} = 0.307$ and $\hat{\xi} = (2.1303 - 2.224)/0.307 = -0.305$. We enter Table 2.3 with $\xi = -0.305$ and interpolate to find $\mu_{11} = 2.1353$, $\mu_{12} = 1.1456$, $\mu_{22} = 1.7146$, and $\rho_{\hat{\mu},\hat{\sigma}} = 0.5984$. Substitution in (2.6.1) enables us to calculate $\sigma_{\hat{\mu}} = 0.0458$ and $\sigma_{\hat{\sigma}} = 0.0410$. Approximate 95% confidence intervals are

$$2.134 < \mu < 2.314,$$
$$0.227 < \sigma < 0.387.$$

In original units of miles of service we have

$$136{,}000 < \text{mean life} < 206{,}000.$$

Example 2.7.4. To ensure meeting a maximum weight specification of 12.00 ounces for a certain electronic component in an aircraft installation, all production of this component is weighed and items that fail to meet this requirement are eliminated. A random sample of 50 units was selected from the screened production. The sample was thus singly truncated on the right with $T = 12.00$, and it was assumed to be from a normal distribution. Sample data are summarized as $n = 50$, $\bar{x} = 9.35$, $s^2 = 1.1264$, and thus $\hat{\alpha} = 1.1264/(9.35 - 12.00)^2 = 0.16040$. By interpolation in Table 2.1, we obtain $\hat{\theta} = 0.00916$ (a value that might have been read to three decimals as 0.009 from the graph of Figure 2.1). Estimates $\hat{\mu}$ and $\hat{\sigma}^2$ are calculated from (2.3.9) and (2.3.12) as

$$\hat{\sigma}^2 = 1.1264 + 0.00916(9.35 - 12.00)^2 = 1.1907,$$
$$\hat{\mu} = 9.35 - 0.00916(9.35 - 12.00) = 9.37.$$

It follows that $\hat{\sigma} = 1.091$ and $\hat{\xi} = (12.00 - 9.37)/1.091 = 2.41$.

With $\xi = 2.41$, we enter Table 2.4 and interpolate to obtain $\mu_{11} = 1.07267$, $\mu_{12} = 0.099336$, $\mu_{22} = 0.620710$, and $\rho_{\hat{\mu},\hat{\sigma}} = 0.121690$. Substitution into (2.6.1) enables us to calculate $\sigma_{\hat{\mu}} = 0.185$ and $\sigma_{\hat{\sigma}} = 0.141$. Approximate 95 percent confidence intervals calculated as for previous examples are

$$9.007 < \mu < 9.733,$$
$$0.815 < \sigma < 1.367.$$

This example was originally given by Cohen (1959).

3
Multirestricted Samples from the Normal Distribution

3.1 INTRODUCTION

This chapter is based on previous results of Cohen (1950, 1957, 1963). Maximum likelihood estimators of parameters of the normal distribution are derived for samples that are doubly truncated, doubly censored, progressively censored, centrally censored, and centrally truncated. In deriving estimating equations for the various sample types, an effort has been made to follow a routine that produces equations that are analogous in algebraic form and that can thus be solved by employing common iterative procedures.

3.2 DOUBLY TRUNCATED SAMPLES

The pdf of a doubly truncated normal distribution that is truncated on the left at T_1 and on the right at T_2 can be written as

$$f(x; \mu, \sigma) = \frac{1}{[F(T_2) - F(T_1)]\sigma\sqrt{2\pi}} \exp\left[-\frac{1}{2}\left(\frac{x-\mu}{\sigma}\right)^2\right], \quad (3.2.1)$$

$$T_1 \leq x \leq T_2, \quad \text{zero elsewhere.}$$

In standard units of the complete distribution, the truncation points are

$$\xi_1 = \frac{T_1 - \mu}{\sigma} \quad \text{and} \quad \xi_2 = \frac{T_2 - \mu}{\sigma}. \quad (3.2.2)$$

It follows that

$$F(T_1) = \Phi(\xi_1) \quad \text{and} \quad F(T_2) = \Phi(\xi_2), \quad (3.2.3)$$

where as previously defined, $\Phi(\ )$ is the cdf of the standard normal distribution $(0, 1)$.

The loglikelihood function of a random sample of size n from a distribution with pdf (3.2.1) is

$$\ln L = -n \ln[\Phi(\xi_2) - \Phi(\xi_1)] - n \ln \sigma \quad (3.2.4)$$
$$- \frac{1}{2} \sum_{i=1}^{n} (x_i - \mu)^2/\sigma^2 + \text{const.}$$

Maximum likelihood estimating equations obtained by equating to zero the partial derivatives of of $\ln L$ with respect to μ and σ are

$$\frac{\partial \ln L}{\partial \mu} = -\frac{n}{\sigma}\left[\frac{\phi_1 - \phi_2}{\Phi_2 - \Phi_1}\right] + \frac{1}{\sigma^2} \sum_{i=1}^{n} (x_i - \mu) = 0,$$
$$\frac{\partial \ln L}{\partial \sigma} = -\frac{n}{\sigma^2}\left[\frac{\xi_1\phi_1 - \xi_2\phi_2}{\Phi_2 - \Phi_1}\right] - \frac{n}{\sigma} + \frac{1}{\sigma^3} \sum_{i=1}^{n} (x_i - \mu)^2 = 0, \quad (3.2.5)$$

where $\phi(\xi_1)$, $\phi(\xi_2)$, $\Phi(\xi_1)$, and $\Phi(\xi_2)$ have been abbreviated to ϕ_1, ϕ_2, Φ_1, Φ_2. To further simplify our notation, we define

$$\bar{Q}_1 = \frac{\phi_1}{\Phi_2 - \Phi_1} \quad \text{and} \quad \bar{Q}_2 = \frac{\phi_1}{\Phi_2 - \Phi_1}. \quad (3.2.6)$$

The notation introduced here is consistent with that employed in Chapter 2 in connection with singly truncated samples. Of course, $\bar{Q}_1$ and $\bar{Q}_2$ are functions of both ξ_1 and ξ_2.

When $\bar{Q}_1$ and $\bar{Q}_2$ as defined in (3.2.6) are substituted into (3.2.5), these equations can be reduced to

$$[\bar{x}_n - \mu) = \sigma[\bar{Q}_1 - \bar{Q}_2], \quad (3.2.7)$$
$$s_n^2 + (\bar{x}_n - \mu)^2 = \sigma^2[1 + \xi_1\bar{Q}_1 - \xi_2\bar{Q}_2].$$

From (3.2.2), we can write

$$\mu = T_1 - \xi_1\sigma. \quad (3.2.8)$$

We substitute (3.2.8) into (3.2.7) and simplify to obtain

$$s_n^2 = \sigma^2[1 + \xi_1\bar{Q}_1 - \xi_2\bar{Q}_2 - (\bar{Q}_1 - \bar{Q}_2)^2], \quad (3.2.9)$$
$$(\bar{x}_n - T) = \sigma[\bar{Q}_1 - \bar{Q}_2 - \xi_1].$$

From (3.2.2) with $w = T_2 - T_1$, we can write

$$\sigma = \frac{w}{\xi_2 - \xi_1}. \tag{3.2.10}$$

We substitute this expression for σ into (3.2.9) and thereby obtain as estimating equations

$$H_1(\xi_1, \xi_2) = \frac{v_1}{w},$$
$$H_2(\xi_1, \xi_2) = \frac{s_n^2}{w_2}, \tag{3.2.11}$$

where

$$H_1(\xi_1, \xi_2) = \frac{Q_1 - \bar{Q}_2 - \xi_1}{\xi_2 - \xi_1},$$
$$H_2(\xi_1, \xi_2) = \frac{1 + \xi_1 \bar{Q}_1 - \xi_2 \bar{Q}_2 - (\bar{Q}_1 - \bar{Q}_2)^2}{(\xi_2 - \xi_1)^2}, \tag{3.2.12}$$

and where

$$v_1 = \bar{x}_n - T_1. \tag{3.2.13}$$

The two estimating equations of (3.2.11) can be solved simultaneously for estimates $\hat{\xi}_1$ and $\hat{\xi}_2$. It then follows that

$$\hat{\sigma} = \frac{w}{\hat{\xi}_2 - \hat{\xi}_1} \quad \text{and} \quad \hat{\mu} = T_1 - \hat{\sigma}\hat{\xi}_2. \tag{3.2.14}$$

It is interesting to note that the two functions $H_1(\xi_1, \xi_2)$ and $H_2(\xi_1, \xi_2)$ obey the relationships

$$H_1(\xi_1, \xi_2) = 1 - H_1(-\xi_2, -\xi_1),$$
$$H_2(\xi_1, \xi_2) = H_2(-\xi_2, -\xi_1). \tag{3.2.15}$$

From the first of the preceding equations, we note that graphs of $H_1(\xi_1, \xi_2) - K = 0$ are reflections of graphs of $H_1(\xi_1, \xi_2) - (1 - K) = 0$ about the line $\xi_1 + \xi_2 = 0$. From the second of these equations, we note that $H_2(\xi_1, \xi_2)$ is symmetrical about $\xi_1 + \xi_2 = 0$.

The algebraic forms given in (3.2.12) for $H_1(\xi_1, \xi_2)$ and $H_2(\xi_1, \xi_2)$ were first suggested by G. W. Thompson, who in collaboration with M. Friedman and E. Garelis (1954) published tables of these two functions for selected values of ξ_1 and ξ_2. More extensive tables at closer intervals of the arguments are given in Cohen and Whitten (1988) and are reproduced here as Table 3.1.

Table 3.1 Estimating Functions $H_1(\xi_1, \xi_2)$ and $H_2(\xi_1, \xi_2)$ for Doubly Truncated Samples from the Normal Distribution[a]

ξ_1 \ ξ_2	0	.1	.2	.3	.4	.5	.6	.7	.8	.9
.0	.000000	.499583	.498336	.496261	.493370	.489673	.485187	.479934	.473941	.467237
	.000000	.083305	.083221	.083075	.082863	.082577	.082207	.081742	.081170	.080477
-.1	.500417	.500000	.498754	.496684	.493802	.490122	.485661	.480441	.474492	.467845
	.083305	.083222	.083083	.082883	.082618	.082280	.081859	.081344	.080724	.079987
-.2	.501664	.501246	.500000	.497934	.495060	.491394	.486954	.481766	.475857	.469262
	.083221	.083083	.082890	.082638	.082323	.081936	.081469	.080911	.080250	.079475
-.3	.503739	.503316	.502066	.500000	.497131	.493475	.489054	.483392	.478020	.471471
	.083075	.082883	.082638	.082338	.081975	.081544	.081035	.080438	.079742	.078936
-.4	.506630	.506198	.504940	.502869	.500000	.496350	.491942	.486802	.480960	.474451
	.082863	.082618	.082323	.081975	.081569	.081098	.080551	.079921	.079195	.078364
-.5	.510327	.509878	.508606	.506525	.503650	.500000	.495599	.490474	.484655	.478180
	.082577	.082280	.081936	.081544	.081098	.080589	.080010	.079351	.078601	.077750
-.6	.514813	.514339	.513046	.510946	.508058	.504401	.500000	.494882	.489080	.482630
	.082207	.081859	.081469	.081035	.080551	.080010	.079403	.078721	.077952	.077087
-.7	.520066	.519559	.518234	.516108	.513198	.509526	.505118	.500000	.494206	.487773
	.081742	.081344	.080911	.080438	.079921	.079351	.078721	.078020	.077238	.076364
-.8	.526059	.525508	.524143	.521980	.519040	.515345	.510920	.505794	.500000	.493575
	.081170	.080724	.080250	.079742	.079195	.078601	.077952	.077238	.076447	.075571
-.9	.532763	.532155	.530738	.528529	.525549	.521820	.517370	.512227	.506425	.500000
	.080477	.079987	.079475	.078936	.078364	.077750	.077087	.076364	.075571	.074697
-1.0	.540138	.539462	.537981	.535714	.532684	.528913	.524428	.519259	.513440	.507006
	.079652	.079121	.078575	.078009	.077416	.076788	.076116	.075390	.074599	.073733
-1.1	.548142	.547383	.545826	.543491	.540399	.536576	.532047	.526844	.520999	.514549
	.078683	.078115	.077540	.076951	.076343	.075705	.075030	.074306	.073522	.072669
-1.2	.556724	.555870	.554224	.551808	.548645	.544759	.540178	.534932	.529053	.522578
	.077559	.076960	.076361	.075755	.075137	.074495	.073822	.073105	.072335	.071500
-1.3	.565831	.564866	.563119	.560611	.557366	.553409	.548766	.543468	.537548	.531042
	.076273	.075649	.075031	.074415	.073792	.073152	.072486	.071783	.071031	.070220
-1.4	.575400	.574311	.572450	.569839	.566502	.562464	.557752	.552395	.546427	.539882
	.074821	.074177	.073549	.072928	.072306	.071674	.071022	.070337	.069609	.068826

34

ξ_2 \ ξ_1	1.0	1.1	1.2	1.3	1.4	1.5	1.6	1.7	1.8	1.9
.0	.459862	.451858	.443276	.434169	.424600	.414634	.404342	.393797	.383075	.372249
	.079652	.078683	.077559	.076273	.074821	.073201	.071415	.069471	.067381	.065161
−.1	.460538	.452617	.444130	.435134	.425689	.415861	.405718	.395334	.384782	.374134
	.079121	.078115	.076960	.075649	.074177	.072545	.070756	.068817	.066740	.064541
−.2	.462019	.454174	.445776	.436881	.427550	.417847	.407841	.397603	.387204	.376715
	.078575	.077540	.076361	.075031	.073549	.071912	.070126	.068198	.066140	.063967
−.3	.464286	.456509	.448192	.439389	.430161	.420571	.410688	.400581	.390319	.379972
	.078009	.076951	.075755	.074415	.072928	.071293	.069516	.067605	.065570	.063427
−.4	.467316	.459601	.451355	.442634	.433498	.424009	.414236	.404244	.394104	.383882
	.077416	.076343	.075137	.073792	.072306	.070680	.068918	.067027	.065020	.062911
−.5	.471087	.463424	.455241	.446591	.437536	.428136	.418459	.408569	.398535	.388423
	.076788	.075705	.074495	.073152	.071674	.070062	.068320	.066455	.064480	.062407
−.6	.475572	.467953	.459822	.451234	.442248	.432926	.423331	.413530	.403588	.393571
	.076116	.075030	.073822	.072486	.071022	.069428	.067711	.065877	.063937	.061905
−.7	.480741	.473156	.465068	.456532	.447605	.438347	.428824	.419098	.409235	.399298
	.075390	.074306	.073105	.071783	.070337	.068768	.067080	.065280	.063379	.061390
−.8	.486560	.479001	.470947	.462452	.453573	.444370	.434906	.425244	.415447	.405577
	.074599	.073522	.072335	.071031	.069609	.068069	.066414	.064653	.062795	.060852
−.9	.492994	.485451	.477422	.468958	.460118	.450959	.441543	.431933	.422190	.412376
	.073733	.072669	.071500	.070220	.068826	.067319	.065703	.063984	.062172	.060278
−1.0	.500000	.492466	.484452	.476011	.467200	.458075	.448698	.439129	.429431	.419661
	.072781	.071736	.070590	.069339	.067978	.066510	.064936	.063263	.061500	.059658
−1.1	.507534	.500000	.491994	.483568	.474776	.465677	.456330	.446794	.437129	.427395
	.071736	.070715	.069597	.068379	.067056	.065630	.064102	.062479	.060768	.058982
−1.2	.515548	.508006	.500000	.491581	.482805	.473720	.464394	.454882	.445244	.435536
	.070590	.069597	.068513	.067332	.066055	.064672	.063194	.061624	.059969	.058240
−1.3	.523989	.516432	.508419	.500000	.491225	.482156	.472845	.463350	.453730	.444041
	.069339	.068379	.067332	.066194	.064960	.063630	.062206	.060692	.059096	.057427
−1.4	.532800	.525224	.517198	.508773	.500000	.490934	.481631	.472147	.462540	.452864
	.067982	.067056	.066052	.064960	.063776	.062500	.061132	.059678	.058143	.056538

[a] Top entry is $H_1(\xi_1, \xi_2)$ and bottom entry is $H_2(\xi_1, \xi_2)$. $H_1(\xi_1, \xi_2) = [\hat{Q}_1 - \hat{Q}_2 - \xi_1]/(\xi_2 - \xi_1)$; $H_2(\xi_1, \xi_2) = [1 + \xi_1 \hat{Q}_2 - \xi_2 \hat{Q}_1 - (\hat{Q}_1 - \hat{Q}_2)^2]/(\xi_2 - \xi_1)^2$.

Source: Reprinted from Cohen and Whitten (1988), Table 8.3, pp. 146–151, by courtesy of Marcel Dekker, Inc.

Table 3.1 Continued

ξ_1 \ ξ_2	2.0	2.1	2.2	2.3	2.4	2.5	2.6	2.7	2.8	2.9
.0	.361395 .062829	.350582 .060409	.339876 .057924	.329338 .055400	.319020 .052864	.308968 .050340	.299220 .047850	.289803 .045415	.280740 .043053	.272043 .040778
-.1	.363461 .062239	.352833 .059856	.342313 .057414	.331958 .054939	.321818 .052456	.311939 .049987	.302354 .047554	.293091 .045177	.284171 .042872	.275606 .040651
-.2	.366206 .061698	.355743 .059353	.345386 .056957	.335193 .054531	.325211 .052100	.315482 .049686	.306039 .047308	.296910 .044986	.288113 .042734	.279660 .040566
-.3	.369608 .061194	.359290 .058891	.349079 .056541	.339027 .054165	.329181 .051786	.319583 .049425	.310263 .047101	.301248 .044832	.292555 .042632	.284196 .040513
-.4	.373646 .060717	.363456 .058458	.353370 .056155	.343441 .053829	.333714 .051502	.324227 .049194	.315012 .046923	.306092 .044705	.297487 .042555	.289206 .040483
-.5	.378298 .060255	.368218 .058042	.358241 .055787	.348418 .053512	.338791 .051237	.329398 .048981	.320271 .046762	.311431 .044595	.302897 .042493	.294678 .040466
-.6	.383540 .059796	.373556 .057630	.363671 .055426	.353937 .053202	.344394 .050979	.335080 .048775	.326024 .046607	.317249 .044490	.308772 .042435	.300602 .040453
-.7	.389349 .059328	.379444 .057212	.369638 .055058	.359977 .052887	.350504 .050717	.341254 .048566	.332256 .046449	.323531 .044380	.315097 .042372	.306964 .040435
-.8	.395695 .058839	.385857 .056774	.376114 .054673	.366514 .052556	.357097 .050439	.347898 .048340	.338945 .046274	.330259 .044255	.321856 .042294	.313747 .040400
-.9	.402550 .058317	.392766 .056305	.383075 .054259	.373523 .052197	.364150 .050134	.354989 .048089	.346069 .046075	.337410 .044105	.329028 .042190	.320933 .040340
-1.0	.409879 .057751	.400138 .055795	.390487 .053805	.380973 .051799	.371632 .049792	.362500 .047801	.353602 .045839	.344966 .043920	.336589 .042052	.328500 .040247
-1.1	.417647 .057132	.407939 .055233	.398319 .053302	.388832 .051353	.379515 .049404	.370401 .047468	.361517 .045559	.352883 .043690	.344515 .041871	.336422 .040111
-1.2	.425814 .056450	.416131 .054611	.406533 .052740	.397064 .050851	.387763 .048960	.378659 .047080	.369781 .045226	.361148 .043409	.352775 .041640	.344673 .039925
-1.3	.434338 .055698	.424672 .053922	.415089 .052112	.405632 .050285	.396338 .048453	.387239 .046632	.378360 .044834	.369721 .043070	.361338 .041350	.353220 .039683
-1.4	.443174 .054873	.433518 .053160	.423944 .051415	.414493 .049650	.405201 .047880	.396100 .046118	.387215 .044377	.378565 .042667	.370167 .040999	.362029 .039379

36

ξ_1 \ ξ_2	.0	.1	.2	.3	.4	.5	.6	.7	.8	.9
-1.5	.585366 .073201	.584139 .072545	.582153 .071912	.579429 .071293	.575991 .070680	.571864 .070062	.567074 .069428	.561653 .068768	.555630 .068069	.549041 .067319
-1.6	.595658 .071415	.594282 .070756	.592159 .070126	.589312 .069516	.585764 .068918	.581541 .068320	.576669 .067711	.571176 .067080	.565094 .066414	.558457 .065703
-1.7	.606203 .069471	.604666 .068817	.602397 .068198	.599419 .067605	.595756 .067027	.591431 .066455	.586470 .065877	.580902 .065280	.574756 .064653	.568067 .063984
-1.8	.616925 .067381	.615218 .066740	.612796 .066140	.609681 .065570	.605896 .065020	.601465 .064480	.596412 .063937	.590765 .063379	.584553 .062795	.577810 .062172
-1.9	.627751 .065161	.625866 .064541	.623285 .063967	.620028 .063427	.616118 .062911	.611577 .062407	.606429 .061905	.600702 .061390	.594423 .060852	.587624 .060278
-2.0	.638605 .062829	.636539 .062239	.633794 .061698	.630392 .061194	.626354 .060717	.621702 .060255	.616460 .059796	.610651 .059328	.604305 .058839	.597450 .058317
-2.1	.649418 .060409	.647167 .059856	.644257 .059353	.640710 .058891	.636544 .058458	.631782 .058042	.626444 .057630	.620556 .057212	.614143 .056774	.607234 .056305
-2.2	.660124 .057924	.657687 .057414	.654614 .056957	.650921 .056541	.646630 .056155	.641759 .055787	.636329 .055426	.630362 .055058	.623886 .054673	.616925 .054259
-2.3	.670662 .055400	.668042 .054939	.664807 .054531	.660973 .054165	.656559 .053829	.651582 .053512	.646063 .053202	.640023 .052887	.633486 .052556	.626477 .052197
-2.4	.680980 .052864	.678182 .052456	.674789 .052100	.670819 .051786	.666286 .051502	.661209 .051237	.655606 .050979	.649496 .050717	.642903 .050439	.635850 .050134
-2.5	.691032 .050340	.688061 .049987	.684518 .049686	.680417 .049425	.675773 .049194	.670602 .048981	.664920 .048775	.658746 .048566	.652102 .048340	.645011 .048089
-2.6	.700780 .047850	.697646 .047554	.693961 .047308	.689737 .047101	.684988 .046923	.679729 .046762	.673976 .046607	.667744 .046449	.661055 .046274	.653931 .046075
-2.7	.710197 .045415	.706909 .045177	.703090 .044986	.698752 .044832	.693908 .044705	.688569 .044595	.682751 .044490	.676469 .044380	.669741 .044255	.662590 .044105
-2.8	.719260 .043053	.715829 .042872	.711887 .042734	.707445 .042632	.702513 .042555	.697103 .042493	.691228 .042435	.684903 .042372	.678144 .042294	.670972 .042190
-2.9	.727957 .040778	.724394 .040651	.720340 .040566	.715804 .040513	.710794 .040483	.705322 .040466	.699398 .040453	.693036 .040435	.686253 .040400	.679067 .040340
-3.0	.736281 .038601	.732598 .038526	.728443 .038489	.723823 .038482	.718745 .038497	.713219 .038523	.707254 .038552	.700864 .038574	.694063 .038580	.686870 .038561

Table 3.1 Continued

ξ_1 \ ξ_2	1.0	1.1	1.2	1.3	1.4	1.5	1.6	1.7	1.8	1.9
-1.5	.541925 .066510	.534323 .065630	.526280 .064672	.517844 .063630	.509066 .062500	.500000 .061280	.490700 .059973	.481223 .058580	.471623 .057109	.461956 .055569
-1.6	.551302 .064936	.543670 .064102	.535606 .063194	.527155 .062206	.518369 .061132	.509300 .059973	.500000 .058727	.490525 .057399	.480930 .055994	.471267 .054521
-1.7	.560871 .063263	.553206 .062479	.545118 .061624	.536650 .060692	.527853 .059678	.518777 .058580	.509475 .057399	.500000 .056137	.490406 .054800	.480745 .053395
-1.8	.570569 .061500	.562871 .060768	.554756 .059969	.546270 .059096	.537460 .058143	.528377 .057109	.519070 .055994	.509594 .054800	.500000 .053532	.490339 .052197
-1.9	.580339 .059658	.572605 .058982	.564464 .058240	.555959 .057427	.547136 .056538	.538044 .055569	.528733 .054521	.519255 .053395	.509661 .052197	.500000 .050933
-2.0	.590121 .057751	.582353 .057132	.574186 .056450	.565662 .055698	.556826 .054873	.547727 .053970	.538412 .052989	.528933 .051933	.519338 .050804	.509678 .049610
-2.1	.599862 .055795	.592061 .055233	.583869 .054611	.575328 .053922	.566482 .053160	.557376 .052323	.548059 .051410	.538579 .050423	.528986 .049364	.519328 .048240
-2.2	.609513 .053805	.601681 .053302	.593467 .052740	.584911 .052112	.576056 .051415	.566946 .050643	.557628 .049797	.548151 .048878	.538561 .047888	.528907 .046833
-2.3	.619027 .051799	.611168 .051353	.602936 .050851	.594368 .050285	.585507 .049650	.576396 .048943	.567081 .048162	.557608 .047310	.548024 .046388	.538376 .045401
-2.4	.628368 .049792	.620485 .049404	.612237 .048960	.603662 .048453	.594799 .047880	.585690 .047236	.576380 .046519	.566916 .045732	.557341 .044875	.547703 .043955
-2.5	.637500 .047801	.629599 .047468	.621341 .047080	.612761 .046632	.603900 .046118	.594797 .045535	.585497 .044880	.576044 .044155	.566483 .043363	.556857 .042507
-2.6	.646398 .045839	.638483 .045559	.630219 .045226	.621640 .044834	.612785 .044377	.603693 .043852	.594407 .043257	.584969 .042592	.575424 .041861	.565816 .041066
-2.7	.655040 .043920	.647117 .043690	.638852 .043409	.630279 .043070	.621435 .042667	.612357 .042198	.603089 .041659	.593671 .041052	.584147 .040379	.574559 .039644
-2.8	.663411 .042052	.655485 .041871	.647225 .041640	.638662 .041350	.629833 .040999	.620775 .040582	.611529 .040097	.602135 .039545	.592635 .038927	.583072 .038247
-2.9	.671500 .040247	.663578 .040111	.655327 .039925	.646780 .039683	.637971 .039379	.628937 .039011	.619717 .038577	.610351 .038076	.600880 .037511	.591345 .036884
-3.0	.679303 .038509	.671389 .038415	.663153 .038271	.654627 .038073	.645843 .037815	.636837 .037493	.627647 .037106	.618313 .036653	.608874 .036136	.599371 .035559

ξ_1 \ ξ_2	2.0	2.1	2.2	2.3	2.4	2.5	2.6	2.7	2.8	2.9
-1.5	.452273 .053970	.442624 .052323	.433054 .050643	.423604 .048943	.414310 .047236	.405203 .045535	.396307 .043852	.387643 .042198	.379225 .040582	.371063 .039011
-1.6	.461588 .052989	.451410 .051410	.442372 .049797	.432919 .048162	.423620 .046519	.414503 .044880	.405593 .043257	.396911 .041659	.388471 .040097	.380283 .038577
-1.7	.471067 .051933	.461421 .050423	.451849 .048878	.442392 .047310	.433084 .045732	.423956 .044155	.415031 .042592	.406329 .041052	.397865 .039545	.389649 .038076
-1.8	.480662 .050804	.471014 .049364	.461439 .047888	.451976 .046388	.442659 .044875	.433517 .043363	.424576 .041861	.415853 .040379	.407365 .038927	.399120 .037511
-1.9	.490322 .049610	.480672 .048240	.471093 .046833	.461624 .045401	.452297 .043955	.443143 .042507	.434184 .041066	.425441 .039644	.416928 .038247	.408655 .036884
-2.0	.500000 .048359	.490349 .047059	.480767 .045721	.471291 .044357	.461956 .042977	.452789 .041593	.443815 .040214	.435052 .038851	.426515 .037510	.418214 .036200
-2.1	.509651 .047059	.500000 .045829	.490416 .044560	.480936 .043264	.471593 .041950	.462415 .040629	.453426 .039312	.444645 .038008	.436086 .036723	.427759 .035466
-2.2	.519233 .045721	.509584 .044560	.500000 .043360	.490517 .042130	.481169 .040881	.471982 .039624	.462980 .038368	.454183 .037122	.445604 .035893	.437254 .034689
-2.3	.528709 .044357	.519064 .043264	.509483 .042130	.500000 .040966	.490648 .039781	.481455 .038586	.472444 .037390	.463633 .036202	.455037 .035028	.446665 .033877
-2.4	.538044 .042977	.528407 .041950	.518831 .040881	.509352 .039781	.500000 .038659	.490804 .037525	.481785 .036388	.472964 .035256	.464353 .034136	.455964 .033036
-2.5	.547211 .041593	.537585 .040629	.528018 .039624	.518545 .038586	.509196 .037525	.500000 .036450	.490978 .035370	.482149 .034293	.473528 .033227	.465124 .032177
-2.6	.556185 .040214	.546574 .039312	.537020 .038368	.527556 .037390	.518215 .036388	.509022 .035370	.500000 .034346	.491168 .033322	.482539 .032307	.474124 .031306
-2.7	.564948 .038851	.555355 .038008	.545817 .037122	.536367 .036202	.527036 .035256	.517851 .034293	.508832 .033322	.500000 .032350	.491367 .031385	.482945 .030432
-2.8	.573485 .037510	.563914 .036723	.554396 .035893	.544963 .035028	.535647 .034136	.526472 .033227	.517461 .032307	.508633 .031385	.500000 .030467	.491574 .029560
-2.9	.581786 .036200	.572241 .035466	.562746 .034689	.553335 .033877	.544036 .033036	.534876 .032177	.525876 .031306	.517055 .030432	.508426 .029560	.500000 .028696
-3.0	.589843 .034926	.580328 .034242	.570861 .033515	.561474 .032752	.552197 .031961	.543055 .031150	.534070 .030326	.525260 .029496	.516638 .028668	.508216 .027847

Readers are reminded that ξ_1 and ξ_2 are employed here as auxiliary parameters to facilitate the estimation of μ and σ, which remain the parameters of primary interest.

Before proceeding, it is deemed appropriate to note that estimating equations (2.3.2), which were derived in Chapter 2 for singly left truncated samples, can also be obtained as a special case of the doubly truncated equations of (3.2.9). When there is no truncation on the right, then $\phi_2 = 0$, $\Phi_2 = 1$, $\bar{Q}_2 = 0$, and $\bar{Q}_1 = Q$ [as defined by (2.2.11)], where $\xi_1 = \xi$ and $\phi_1 = \phi$. When these substitutions are made in (3.2.9), the resulting equations are identical to equations (2.3.2).

3.2.1 Sampling Errors of Estimates

As noted in Chapter 2, the asymptotic variance–covariance matrix of maximum likelihood estimates $\hat{\mu}$ and $\hat{\sigma}$ can be obtained by inverting the applicable Fisher information matrix with elements that are negatives of expected values of the second-order partial derivatives of the loglikelihood function. When samples are moderately large, the second partials, evaluated with parameters equated to their sample estimates, may serve as replacements for expected values, with only a negligible loss of information. Since the expected values are more or less intractable for doubly truncated samples, this approach becomes attractive. Applicable first and second partials for doubly truncated samples are

$$\frac{\partial \ln L}{\partial \mu} = \frac{n}{\sigma^2}[\bar{x} - \mu - \sigma(\bar{Q}_1 - \bar{Q}_2)],$$

$$\frac{\partial \ln L}{\partial \sigma} = \frac{n}{\sigma^2}[\{s^2 + (\bar{x} - \mu)^2\}/\sigma - \sigma(1 + \xi_1 \bar{Q}_1 - \xi_2 \bar{Q}_2)],$$

$$\frac{\partial^2 \ln L}{\partial \mu^2} = -\frac{n}{\sigma^2}[1 - \bar{Q}'_1 + \bar{Q}'_2], \qquad (3.2.16)$$

$$\frac{\partial^2 \ln L}{\partial \mu \, \partial \sigma} = -\frac{n}{\sigma^2}\left[\frac{2(\bar{x} - \mu)}{\sigma} - \lambda_1 + \lambda_2\right],$$

$$\frac{\partial^2 \ln L}{\partial \sigma^2} = -\frac{n}{\sigma^2}\left[\frac{3[s^2 + (\bar{x} - \mu)^2]}{\sigma^2} - 1 - \eta_1 + \eta_2\right],$$

where

$$\bar{Q}'_1 = \bar{Q}_1(\bar{Q}_1 - \xi_1), \qquad \bar{Q}'_2 = -\bar{Q}_2(\bar{Q}_2 + \xi_2),$$

$$\lambda_1 = \xi_1 \bar{Q}'_1 + \bar{Q}_1, \qquad \lambda_2 = \xi_2 \bar{Q}'_2 + \bar{Q}_2,$$

$$\eta_1 = \xi_1(\lambda_1 + \bar{Q}_1), \qquad \eta_2 = \xi_2(\lambda_2 + \bar{Q}_2).$$

The approximate variance–covariance matrix can be expressed as the inverse of the information matrix as follows

Multirestricted Samples from Normal Distribution

$$V(\hat{\mu}, \hat{\sigma}) \doteq \begin{bmatrix} -\dfrac{\partial^2 \ln L}{\partial \mu^2}\bigg|_{\substack{\mu=\hat{\mu}\\\sigma=\hat{\sigma}}} & -\dfrac{\partial^2 \ln L}{\partial \mu\, \partial \sigma}\bigg|_{\substack{\mu=\hat{\mu}\\\sigma=\hat{\sigma}}} \\ -\dfrac{\partial^2 \ln L}{\partial \sigma\, \partial \mu}\bigg|_{\substack{\mu=\hat{\mu}\\\sigma=\hat{\sigma}}} & -\dfrac{\partial^2 \ln L}{\partial \sigma^2}\bigg|_{\substack{\mu=\hat{\mu}\\\sigma=\hat{\sigma}}} \end{bmatrix}. \quad (3.2.17)$$

3.2.2 Computational Procedures

Any one of various iterative methods such as that of Newton and Raphson might be employed to calculate estimates $\hat{\xi}_1$ and $\hat{\xi}_2$, but the computations are likely to be tedious and time-consuming. As a computational aid designed to simplify these computations, a chart consisting of intersecting graphs of the equations $H_1(\hat{\xi}_1, \hat{\xi}_2) = v_1/w$ and $H_2(\hat{\xi}_1, \hat{\xi}_2) = s^2/w^2$ for selected values of s^2/w^2 and v_1/w was given by Cohen (1957). This chart is presented here as Figure 3.1. With v_1/w and s^2/w^2 calculated from sample data, estimates of ξ_1 and ξ_2 can be read as coordinates of the point of intersection of applicable intersecting graphs. These estimates are sufficiently accurate for many practical applications. Improved accuracy, when required, can be obtained by interpolation in Table 3.1. The graphs can be approximated in the vicinity of their intersection by straight lines with equations of the form

$$\xi_2 = a_1 + b_1 \xi_1, \quad (3.2.18)$$
$$\xi_2 = a_2 + b_2 \xi_1.$$

We employ Table 3.1 to determine coordinates (ξ_1, ξ_2) of points that satisfy the two equations of (3.2.11). Two points in the vicinity of their common point of intersection are required for each equation. Coefficients (a_1, b_1) and (a_2, b_2) of equations (3.2.18) are determined so that these straight lines pass through the chosen points. Thus the simultaneous solution of the two equations of (3.2.18) will provide the required estimates. If the points chosen for determination of (a_1, b_1) and (a_2, b_2) are sufficiently close to the actual point of intersection of the two graphs, then estimates $\hat{\xi}_1$ and $\hat{\xi}_2$ can be calculated quite accurately as

$$\hat{\xi}_1 = \frac{a_1 - a_2}{b_2 - b_1} \quad \text{and} \quad \hat{\xi}_2 = \frac{a_1 b_2 - a_2 b_1}{b_2 - b_1}. \quad (3.2.19)$$

In the event that estimates of ξ_1 and ξ_2 calculated as described above with the aid of Figure 3.1 and Table 3.1 are not sufficiently accurate, and thus estimates of μ and σ are not sufficiently accurate, we can employ Newton's method to calculate corrections h and g such that

$$\hat{\mu} = \mu_0 + h \quad \text{and} \quad \hat{\sigma} = \sigma_0 + g, \quad (3.2.20)$$

where μ_0 and σ_0 are estimates in need of correction. The corrections can be calculated as the simultaneous solution of the two linear equations

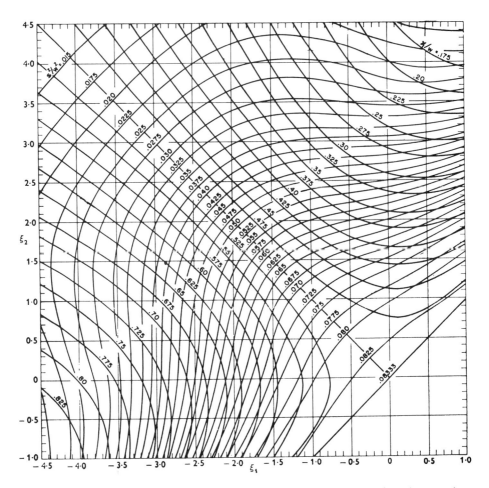

Figure 3.1 Graphs of estimating functions for doubly truncated samples from the normal distribution. (1) Locate v_1/w curve corresponding to sample value of this quantity; interpolate if necessary. (2) Follow curve located in (1) to point where it intersects with s^2/w^2 curve for corresponding sample value; if necessary, interpolate here also. (3) Coordinates of intersection determined in (2), which may be read on scales along the base and the left edge of chart, are the required values of ξ_1 and ξ_2. Reprinted from Cohen (1957), *Biometrika*, **44**, Fig. 1, p. 227, with permission of the Biometrika Trustees.

$$\begin{aligned} h\left.\frac{\partial^2 \ln L}{\partial \mu^2}\right|_{\substack{\mu=\mu_0 \\ \sigma=\sigma_0}} + g\left.\frac{\partial^2 \ln L}{\partial \mu\, \partial \sigma}\right|_{\substack{\mu=\mu_0 \\ \sigma=\sigma_0}} &= -\left.\frac{\partial \ln L}{\partial \mu}\right|_{\substack{\mu=\mu_0 \\ \sigma=\sigma_0}} \\ h\left.\frac{\partial^2 \ln L}{\partial \mu\, \partial \sigma}\right|_{\substack{\mu=\mu_0 \\ \sigma=\sigma_0}} + g\left.\frac{\partial^2 \ln L}{\partial \sigma^2}\right|_{\substack{\mu=\mu_0 \\ \sigma=\sigma_0}} &= -\left.\frac{\partial \ln L}{\partial \sigma}\right|_{\substack{\mu=\mu_0 \\ \sigma=\sigma_0}} \end{aligned} \quad (3.2.21)$$

Multirestricted Samples from Normal Distribution

The derivatives that appear in the preceding equations are given in (3.2.16).

The two equations of (3.2.21) result from Taylor's theorem when we neglect powers of h and g greater than one. As a consequence of neglecting second and higher powers of the corrections, Newton's method tends to produce rather slowly converging iterants during the first few cycles of computation unless initial approximations are in a close neighborhood of the solution. This difficulty has been recognized and discussed, for example, by Norton (1956). In an effort to overcome these objections, the method of successive substitution as described by Scarborough (1930, pp. 191–195) was employed by Cohen (1957) to deveop iterants of the form

$$\xi_1^{(i+1)} = f(\xi_1^{(i)}, \xi_2^{(i)}), \quad \text{and} \quad \xi_1^{(i+1)} = g(\xi_1^{(i)}, \xi_2^{(i)}), \quad (3.2.22)$$

where $\xi_j^{(i)}$ designates the ith iterant to ξ_j, $j = 1, 2$. Final iterants as derived by Cohen, which can be used to iterate to $\hat{\xi}_1$ and $\hat{\xi}_2$ by successive substitution, are

$$\xi_1^{(i+1)} = \frac{-\xi_2^{(i+1)} v_1/w - (\bar{Q}_1^{(i)} - \bar{Q}_2^{(i)})}{1 - v_1/w},$$

$$\xi_2^{(i+1)} = \frac{(\bar{Q}_1^{(i)} - \bar{Q}_2^{(i)}) + \{[(\bar{Q}_1^{(i)} - \bar{Q}_2^{(i)})v_1/w + \bar{Q}_2^{(i)}] - \sqrt{[(\bar{Q}_1^{(i)} - \bar{Q}_2^{(i)})v_1/w + \bar{Q}_2^{(i)}]^2 + 4s^2/w^2}\} w(v_1 - w)}{2s^2}. \quad (3.2.23)$$

With ξ_1^0 and ξ_2^0 determined from Figure 3.1 or by interpolation in Table 3.1, or even by judicious guessing, improved approximations can be obtained from the successive application of (3.2.23). In many applications these iterants result in a rapid advance toward the neighborhood of the solution during the first few cycles. Thereafter, convergence slows down as the solution is approached. This behavior is opposite to that of Newton's method, for which convergence is often slow for approximations far removed from the solution, but is more rapid in the neighborhood of the solution. The two methods thus complement each other, and in practical applications where a high degree of accuracy is required, an efficient computational procedure consists of reading initial approximations ξ_1^0 and ξ_2^0 from Figure 3.1, using two or perhaps three cycles of (3.2.23), and then obtaining final estimates with a single cycle of Newton's method. Of course, in many applications the values read from Figure 3.1 are sufficiently accurate without resorting to iteration. An advantage of Newton's method for the final cycle of iteration is that estimate variances and covariances can be obtained with very little additional effort without the necessity for further evaluations of the second partials.

3.2.3 Illustrative Examples

To illustrate and further explain the computational procedures described in Subsection 3.2.2, we offer the following examples.

Example 3.1. Cohen (1957) considered a random sample from production of a certain bushing that had been screened through go–no go gauges to eliminate items with diameters in excess of 0.6015 inches or less than 0.5985 inches. Nonconforming items were discarded and no further information was available on either their measurements or their number. For a random sample of 75 bushings selected from the screened production, $n = 75$, $\bar{x} = 0.60014933$, $T_1 = 0.5985$, $T_2 = 0.6015$, $s^2 = 0.000000371187$, $w = 0.0030$, $v_1 = \bar{x} - T_1 = 0.00164933$, $v_1/w = 0.54978$, and $s^2/w^2 = 0.041242$. Visual interpolation between the curves of Figure 3.1 gives initial approximations $\xi_1^0 = -2.50$ and $\xi_2^0 = 2.00$. With these values as approximate coordinates of the intersection of curves of the two estimating equations, we enter Table 3.1 and interpolate as necessary to obtain coordinates of points on the two curves in the vicinity of their intersection. We locate two points on the curve of $H_1(\xi_1, \xi_2) - v_1/w = 0$ with coordinates $(-2.5, 1.9734)$ and $(-2.6, 2.0666)$, and two points on the curve $H_2(\xi_1, \xi_2) - s^2/w^2 = 0$ with coordinates $(-2.5, 2.0364)$ and $(-2.6, 1.8779)$. Equations of straight lines through the first two points and through the second two points are

$$\xi_2 = 0.3566 - 0.932\xi_1,$$
$$\xi_2 = 5.9989 + 1.585\xi_1.$$

In effect, these two linear equations are replacements for the nonlinear estimating equations (3.2.11) in the vicinity of their common solution. A simultaneous solution of the two linear equations subsequently gives

$$\xi_1^{(1)} = -2.525 \quad \text{and} \quad \xi_2^{(1)} = 1.997.$$

Substitution in (3.2.14) then gives

$$\sigma_0 = 0.000\ 663\ 42 \quad \text{and} \quad \mu_0 = 0.600\ 175\ 14.$$

If the points selected for determining coefficients of the linear replacement equations are sufficiently close to the actual intersection of the original estimation curves, then estimates calculated as illustrated here will be highly accurate. However, to illustrate the procedure, we continue with one final cycle of Newton's method. With the partial derivatives of equations (3.2.16) evaluated with μ_0 and σ_0 as calculated above the correction equations (3.2.21) become

$$-150\ 706\ 576\ h + 27235297\ g = -7.076300,$$
$$27\ 235\ 297\ h - 187700371\ g = 75.019711.$$

Multirestricted Samples from Normal Distribution

The simultaneous solution of these two equations gives the corrections as

$$h = -0.000\,000\,03 \quad \text{and} \quad g = -0.000\,000\,40.$$

As final estimates, we then calculate

$$\hat{\mu} = 0.000\,175\,14 - 0.000\,000\,03 = 0.600\,175\,11,$$
$$\hat{\sigma} = 0.000\,663\,42 - 0.000\,000\,40 = 0.000\,663\,02.$$

Since the coefficients of the correction equations in h and g are approximately equal to expected values of second partial derivatives of $\ln L$, the asymptotic variance–covariance matrix of $\hat{\mu}$ and $\hat{\sigma}$ may, with very little additional effort, be approximated as

$$\begin{vmatrix} 150\,706\,576 & -27\,235\,297 \\ -27\,235\,297 & 187\,700\,371 \end{vmatrix}^{-1} = \begin{vmatrix} 6.814 \times 10^{-9} & -0.989 \times 10^{-9} \\ -0.989 \times 10^{-9} & 5.471 \times 10^{-9} \end{vmatrix}$$

Accordingly,

$$\sigma_{\hat{\mu}} = 0.000\,082\,5, \quad \sigma_{\hat{\sigma}} = 0.000\,074\,0, \quad \rho_{\hat{\mu},\hat{\sigma}} = 0.162.$$

Example 3.2. This example was selected by Cohen (1950) from tables prepared by Mahalanobis (1934) from random observations of a normal population $(0,1)$. The population was truncated on the left at $T_1 = -1.000\,000$ and on the right at $T_2 = 1.750\,000$. From this doubly truncated population 32 random observations were selected. Sample data are summarized as $n = 32$, $\bar{x} = 0.244\,625$, $s^2 = 0.556\,183\,6$, $T_1 = -1.000\,000$, $T_2 = 1.750\,000$, $w = T_2 - T_1 = 2.750\,000$, $v_1 = \bar{x} - T_1 = 1.244\,625$, $v_1/w = 0.452\,591$, $s^2/w^2 = 0.073\,545$. With these values for v_1/w and s^2/w^2, visual interpolation in Figure 3.1 yields $\xi_1^0 = -0.5$ and $\xi_2^0 = 1.3$ as initial approximations. We proceed as described for Example 3.1 and, from Table 3.1, we find $(-0.6, 1.2842)$ and $(-0.5, 1.2306)$ as coordinates of two points on the curve of the first estimating equation. We find $(-0.6, 1.2207)$ and $(-0.5, 1.2707)$ as coordinates of two points on the second estimating equation. Equations of the linear replacement equations become

$$\xi_2 = 0.9626 - 0.536\xi_1,$$
$$\xi_2 = 1.5207 + 0.500\xi_1.$$

The simultaneous solution of these equations gives

$$\hat{\xi}_1 = -0.539 \quad \text{and} \quad \hat{\xi}_2 = 1.251.$$

On substituting these values into (3.2.14), we calculate

$$\hat{\sigma} = 1.536 \quad \text{and} \quad \hat{\mu} = -0.172.$$

These estimates are considered to be sufficiently accurate for our purposes and no further efforts were made to attain additional accuracy.

3.3 DOUBLY CENSORED SAMPLES

In this section we are considering doubly censored samples from a normal distribution with pdf (2.2.1). The likelihood function as defined by (1.5.6) in this case becomes

$$L = K[\Phi(\xi_1)]_{c_1}[1 - \Phi(\xi_2)]_{c_2}(\sigma\sqrt{2\pi})^{-n} \exp\left[-\frac{1}{2\sigma^2}\sum_{i=1}^{n}(x_i - \mu)^2\right]. \quad (3.3.1)$$

The sample is censored on the left at T_1 and on the right at T_2. There are c_1 censored observations less than T_1, c_2 greater than T_2, and n complete (uncensored) observations in the interaval $T_1 \leq x \leq T_2$. The total sample size is N, where $N = c_1 + c_2 + n$. To obtain maximum likelihood estimators of μ and σ, we take logarithms of (3.3.1), differentiate with respect to these parameters, and equate to zero. The resulting estimating equations are

$$\frac{\partial \ln L}{\partial \mu} = -\frac{c_1\phi(\xi_1)}{\sigma\Phi(\xi_1)} + \frac{c_2}{\sigma}\left[\frac{\phi(\xi_2)}{1 - \Phi(\xi_2)}\right]$$

$$+ \frac{1}{\sigma^2}\sum_{i=1}^{n}(x_i - \mu) = 0, \quad (3.3.2)$$

$$\frac{\partial \ln L}{\partial \sigma} = -\frac{c_1\xi_1\phi(\xi_1)}{\sigma\Phi(\xi_1)} + \frac{c_2}{\sigma}\left[\frac{\xi_2\phi(\xi_2)}{1 - \Phi(\xi_2)}\right]$$

$$-\frac{n}{\sigma} + \frac{1}{\sigma^3}\sum_{i=1}^{n}(x_i - \mu)^2 = 0.$$

We define

$$\Omega_1 = \frac{c_1}{n}\frac{\phi(\xi_1)}{\Phi(\xi_1)} = \frac{c_1}{n}Q(-\xi_1), \quad \text{and}$$

$$\Omega_2 = \frac{c_2}{n}\left[\frac{\phi(\xi_2)}{1 - \Phi(\xi_2)}\right] = \frac{c_2}{n}Q(\xi_2), \quad (3.3.3)$$

where $Q(\xi)$ is defined by (2.2.11) and the estimating equations of (3.3.2) may be reduced to

$$(\bar{x} - \mu) = \sigma(\Omega_1 - \Omega_2), \quad (3.3.4)$$
$$s^2 + (\bar{x} - \mu)^2 = \sigma^2[1 + \xi_1\Omega_1 - \xi_2\Omega_2].$$

Note that Ω_1, as defined here, is identical to Ω as defined by (2.5.4) for singly truncated samples. Note further that the equations of (3.3.4) differ from corresponding equations (3.2.7) for doubly truncated samples only in that here, Ω_1

Multirestricted Samples from Normal Distribution

and Ω_2 have replaced $\bar{Q}_1$ and $\bar{Q}_2$. Accordingly, we follow the same steps that led to equations (3.2.11) for doubly truncated samples and thus obtain

$$\frac{\Omega_1 - \Omega_2 - \xi_1}{\xi_2 - \xi_1} = \frac{\bar{x} - T_1}{w},$$

$$\frac{1 + \xi_1\Omega_1 - \xi_2\Omega_2 - (\Omega_1 - \Omega_2)^2}{(\xi_2 - \xi_1)^2} = \frac{s^2}{w^2}. \tag{3.3.5}$$

Estimates $\hat{\xi}_1$ and $\hat{\xi}_2$ can be found as the simultaneous solution of these two equations. As in the case of doubly truncated samples, estimates of μ and σ follow as

$$\hat{\sigma} = \frac{w}{\hat{\xi}_2 - \hat{\xi}_1} \quad \text{and} \quad \hat{\mu} = T_1 - \hat{\sigma}\hat{\xi}_1. \tag{3.3.6}$$

3.3.1 Singly Censored Samples as a Special Case

Estimators derived in Chapter 2 for singly censored samples can also be obtained as a special case of those for doubly censored samples. For a singly left censored sample, it follows from (3.3.3) that $\phi_2 = 0$, $\Phi_2 = 1$, and thus $\Omega_2 = 0$. In terms of the notation employed in Chapter 2, $\Omega_1 = \Omega$, $T_1 = T$, and $\xi_1 = \xi$. Accordingly, in this case the two estimating equations of (3.3.6) become identical to equations (2.5.6), and the remaining steps in the derivation of estimators (2.5.11), as given in Chapter 2, are applicable without change.

3.3.2 Sampling Errors of Estimates

Here as was true for doubly truncated samples, when samples are moderately large, the asymptotic variance–covariance matrix $V(\hat{\mu}, \hat{\sigma})$ can be satisfactorily approximated by replacing expected values of the second partials of the log-likelihood function with their corresponding sample values. Thus the variance–covariance matrix (3.2.17) is also applicable here, but with first and second partial derivatives as follows.

$$\frac{\partial \ln L}{\partial \mu} = \frac{n}{\sigma^2}[\bar{x} - \mu - \sigma(\Omega_1 - \Omega_2)],$$

$$\frac{\partial \ln L}{\partial \sigma} = \frac{n}{\sigma^2}\left[\frac{s^2 + (\bar{x} - \mu)^2}{\sigma} - \sigma(1 + \xi_1\Omega_1 - \xi_2\Omega_2)\right],$$

$$\frac{\partial \ln L}{\partial \mu^2} = -\frac{n}{\sigma^2}[1 - \Omega_1' + \Omega_2'], \tag{3.3.7}$$

$$\frac{\partial^2 \ln L}{\partial \mu \, \partial \sigma} = -\frac{n}{\sigma^2}\left[\frac{2(\bar{x} - \mu)}{\sigma} - \lambda_1 + \lambda_2\right],$$

$$\frac{\partial^2 \ln L}{\partial \sigma^2} = -\frac{n}{\sigma^2}\left[\frac{3\{s^2 + (\bar{x} - \mu)^2\}}{\sigma^2} - 1 - \eta_1 + \eta_2\right],$$

where

$$\Omega_1' = -\Omega_1\left(\frac{n}{c_1}\Omega_1 + \xi_1\right), \qquad \Omega_2' = \Omega_2\left(\frac{n}{c_2}\Omega_2 - \xi_2\right),$$

$$\lambda_1 = \xi_1\Omega_1' + \Omega_1, \qquad \lambda_2 = \xi_2\Omega_2' + \Omega_2, \qquad (3.3.8)$$

$$\eta_1 = \xi_1(\lambda_1 + \Omega_1), \qquad \eta_2 = \xi_2(\lambda_2 + \Omega_2).$$

The partials of (3.3.7) are needed as both elements of the information matrix and coefficients in the correction equations involved in the Newton iteration procedure. Attention is invited to the similarity between equations (3.3.7) and equations (3.2.16), which apply in the case of doubly truncated samples.

3.3.3 Computational Procedures

It is of course necessary to employ iterative procedures in the solution of estimating equations (3.3.5), and it is most desirable to begin with first approximations that are reasonably close to the final solution. Since the functions Ω_1 and Ω_2 involve not only ξ_1 and ξ_2, but also c_1, c_2, and n, a chart corresponding to Figure 3.1 does not seem feasible in this case. Accordingly, we might expect to begin iteration with less accurate initial approximations than in the case of doubly truncated samples and to depend on iteration for their improvement as required. When c_1 and c_2 constitute only a small proportion of the total sample they may be neglected and initial approximations to ξ_1 and ξ_2 can be read from Figure 3.1 as though the sample were truncated rather than censored. When c_1 and c_2 are appreciable proportions of the total sample, improved approximations may be read from ordinary tables of the normal distribution cdf where

$$\frac{c_1}{N} = \Phi(\xi_1) \quad \text{and} \quad \frac{c_2}{N} = 1 - \Phi(\xi_2). \qquad (3.3.9)$$

In reasonably large samples the two sets of initial approximations suggested above should be in fairly close agreement, and either set might be used as a starting point in the iteration procedure. In some situations it might be desirable to use an average of the two. With ξ_1^0 and ξ_2^0 thus determined, the Cohen successive substitution iterants (3.2.29) might be employed to improve the initial approximations. However, for censored samples, we must replace $\bar{Q}_1$ and $\bar{Q}_2$ in (3.2.29) with Ω_1 and Ω_2, respectively. After one or perhaps two cycles of successive substitution, we might wish to switch to Newton's method for the final cycle of iteration. Newton's correction equations in this case are the same as (3.2.21) for

Multirestricted Samples from Normal Distribution

doubly truncated samples but with the partial derivatives as given by (3.3.7) for doubly censored samples.

3.3.4 An Illustrative Example

Example 3.3 In a time–mortality experiment used by Cohen (1957) as an illustrative example, the first observation was delayed for a fixed time interval with the result that c_1 experimental specimens died before observation began. The experiment was subsequently terminated with c_2 survivors. Actual survival times were recorded for n specimens that died during the period of full observation. The sample thus is doubly censored, and it was assumed that logarithms, X, of survival times in days are distributed normally (μ, σ). The sample data are summarized as follows: $N = 47$, $n = 40$, $c_1 = 2$, $c_2 = 5$, $T_1 = 1.301032$, $T_2 = 1.903090$, $w = 0.602060$, $\bar{x} = 1.620111$, $v_1 = 0.319081$, $s^2 = 0.0217392$, $v_1/w = 0.529982$, $s^2/w^2 = 0.0599741$. When we neglect the information provided by c_1 and c_2, we can use Figure 3.1 to obtain approximations $\xi_1 \doteq -1.71$ and $\xi_2 \doteq 1.36$. We subsequently read from tables of normal curve areas where $c_1/N = 2/47 = \Phi(\xi_1)$ and $c_2/N = 5/47 = 1 - \Phi(\xi_2)$ to obtain $\xi_1 \doteq 1.72$ and $\xi_1 \doteq 1.25$. The two sets of approximations are in reasonably close agreement, and we elect to begin the iterative process with initial approximations $\xi_1^0 = -1.72$ and $\xi_2^0 = 1.28$. The iterants of (3.2.23) with $\bar{Q}_1$ and $\bar{Q}_2$ replaced by Ω_1, and Ω_2 were employed and after two cycles $\xi_1 = -1.689$ and $\xi_2 = 1.282$. As a demonstration of Newton's method for this example we used these values to calculate approximations:

$$\sigma_0 = 0.602\,060/(1.282 + 1.689) = 0.202\,646,$$

$$\mu_0 = 1.301\,030 - 0.202\,646(-1.689) = 1.643\,299.$$

We substitute these values into the derivatives of (3.3.7) and the correction equations (3.2.21) become

$$-1117.362\,555\,h + 52.976\,746\,g = 0.009\,567\,35,$$

$$52.976\,746\,h + 1791.254\,507\,g = 0.234\,567\,59.$$

As a simultaneous solution, we calculate $h = -0.000\,00236$ and $g = 0.000\,130\,89$. As final estimates we calculate

$$\hat{\mu} = 1.643\,299 - 0.000\,002 = 1.643\,297,$$

$$\hat{\sigma} = 0.202\,646 + 0.000\,131 = 0.202\,777.$$

These estimates correspond to $\hat{\xi}_1 = -1.68790$ and $\hat{\xi}_2 = 1.28122$.

We use the coefficients of the correction equations to approximate the asymptotic variance–covariance matrix as

$$\begin{vmatrix} 1117.3626 & -52.9767 \\ -52.9767 & 1791.2545 \end{vmatrix}^{-1} \doteq \begin{vmatrix} 0.000\,896\,2 & -0.000\,026\,5 \\ -0.000\,026\,5 & 0.000\,559\,1 \end{vmatrix}$$

Thus we have

$$\sigma_{\hat{\mu}} \doteq 0.029\,94, \qquad \sigma_{\hat{\sigma}} \doteq 0.023\,64, \qquad \rho_{\hat{\mu},\hat{\sigma}} \doteq 0.0374.$$

3.4 PROGRESSIVELY CENSORED SAMPLES

Progressively censored samples have previously been considered by Herd (1956, 1960), Roberts (1962a, 1962b), Cohen (1963), and perhaps others. Herd referred to these samples as being multicensored, Roberts referred to them as being hypercensored. More recently, the designation of progressive censoring seems to be more or less accepted as the appropriate descriptive terminology. This section is based on the author's 1963 presentation.

In life and dosage–response studies, progressively censored samples arise, when at various stages of an experiment, some though not all of the survivors are removed from further observation. Sample specimens remaining after each state of censoring continue to be observd until failure (or specified reaction) or until a subsequent stage of censoring. When test facilities are limited and when prolonged tests are expensive, the early censoring of a substantial number of sample specimens leaves test facilities free for other tests, while specimens that remain on test until subsequent failure provide information on a limited number of the more extreme sample values. In progressive censoring as in single stage censoring, a distinction is made between Type I censoring, in which censoring times are predetermined, and Type II censoring, in which censoring occurs when the number of survivors has dropped to predetermined levels. Progressive censoring of a somewhat more complex type in which both times of censoring and the number of items censored are the result of random causes was considered by Sampford (1952) in connection with response-time studies involving animals. In this section we are primarily concerned with maximum likelihood estimation based on progressively censored samples from the normal distribution. We also consider a probit regression procedure, previously described by Cohen (1963), and a hazard plot technique described by Nelson (1969).

The loglikelihood function of a progressively censored sample from a normal distribution (μ,σ) may, in accordance with the definition given in Chapter 1, be written as

$$\ln L = -n \ln \sigma - \frac{1}{2\sigma^2} \sum_{i=1}^{n} (x_i - \mu)^2$$

$$+ \sum_{j=1}^{k} c_j \ln [1 - \Phi(\xi_j)] + \text{const.}, \quad (3.4.1)$$

where

$$\Phi(\xi_j) = \int_{-\infty}^{\xi_\alpha} \phi(t)\, dt = F(T_j) = \int_{-\infty}^{T_\alpha} f(x; \mu, \sigma)\, dx, \qquad (3.4.2)$$

and

$$\xi_j = \frac{T_j - \mu}{\sigma}.$$

We differentiate (3.4.1) with respect to μ and σ and equate to zero to obtain the maximum likelihood estimating equations:

$$\frac{\partial \ln L}{\partial \mu} = \frac{n}{\sigma}\left[\frac{\bar{x} - \mu}{\sigma} + \sum_{j=1}^{k} \frac{c_j}{n} Q_j\right] = 0,$$

$$\frac{\partial \ln L}{\partial \sigma} = \frac{n}{\sigma}\left[\frac{s^2 + (\bar{x} - \mu)^2}{\sigma^2} - 1 + \sum_{j=1}^{k} \frac{\xi_j c_j}{n} Q_j\right] = 0, \qquad (3.4.3)$$

where $Q_j = Q(\xi_j)$ has been defined by (2.2.11) and where $\bar{x}$ and s^2 are the mean and the variance of the n complete sample observations. The pair of equations (3.4.3) may be simplified to the form

$$(\bar{x} - \mu) = -\sigma \sum_{i=1}^{n} \frac{c_j}{n} Q_j,$$

$$s^2 + (\bar{x} - \mu)^2 = \sigma^2\left[1 - \sum_{j=1}^{k} \frac{\xi_j c_j}{n} Q_j\right]. \qquad (3.4.4)$$

Any one of various standard iterative procedures may be used to solve the pair of equations (3.4.4) for estimates $\hat{\mu}$ and $\hat{\sigma}$. For example, Newton's method, which was employed in connection with the solution of corresponding equations for doubly truncated and doubly censored samples, can also be used in this instance. Both equations (3.2.20) and (3.2.21) are applicable with the first partial derivatives as given in (3.4.3) and the second partials as given below.

$$\frac{\partial^2 \ln L}{\partial \mu^2} = -\frac{n}{\sigma^2}\left[1 + \sum_{j=1}^{k} \frac{c_j}{n} Q'_j\right],$$

$$\frac{\partial^2 \ln L}{\partial \mu\, \partial \sigma} = -\frac{n}{\sigma^2}\left[\frac{2(\bar{x} - \mu)}{\sigma} + \sum_{j=1}^{k} \frac{c_j}{n} \lambda_j\right], \qquad (3.4.5)$$

$$\frac{\partial^2 \ln L}{\partial \sigma^2} = -\frac{n}{\sigma^2}\left[\frac{3\{s^2 + (\bar{x} - \mu)^2\}}{\sigma^2} - 1 + \sum_{i=1}^{n} \frac{c_j}{n} \eta_j\right],$$

where

$$Q'_j = Q_j(Q_j - \xi_j),$$
$$\lambda_j = \xi_j Q'_j + Q_j, \qquad (3.4.6)$$
$$\eta_j = \xi_j(\lambda_j + Q_j).$$

The number of iterations required for a specified degree of accuracy will to a considerable extent depend on the initial approximations. Satisfactory first approximations can, in many applications, be obtained from sample percentiles. The pattern of censoring in specific samples will of course determine which percentiles are available, but, in general, a suitable first approximation to $\hat{\mu}$ may be calculated as

$$\mu_0 = \frac{P_i + P_{100-i}}{2},$$

where P_i is the ith percentile of the sample. It is desirable, though not necessary, to restrict i to the interval $25 \leq i \leq 50$.

In some circumstances, it might be desirable to employ the probit technique or the Nelson (1969) hazard plot technique, both of which are considered later in this chapter, to provide more accurate first approximations to both $\hat{\mu}$ and $\hat{\sigma}$.

As a computational aid, which might simplify evaluation of the second partials of (3.4.5), Table 3.2 of the functions Q, ξQ, Q', λ, and η has been reproduced here from Cohen (1963).

Herd (1956) employed an iterative procedure for the solution of the estimating

Table 3.2 The Functions Q, ξQ, Q', λ, and η for Progressively Censored Samples from the Normal Distribution

ξ	Q	ξQ	$Q' = Q(Q-\xi)$	$\lambda = Q + \xi Q'$	$\eta = \xi(Q + \lambda)$
−1.0	0.28760	−0.28760	0.37031	−0.083	−0.205
−0.5	0.50916	−0.25458	.51382	0.252	−0.381
0	0.79788	0	.63662	0.798	0
0.5	1.14108	0.57054	.73152	1.507	1.324
1.0	1.52514	1.52514	.80090	2.326	3.851
1.1	1.60580	1.76638	.81221	2.499	4.515
1.2	1.68755	2.02506	.82277	2.675	5.235
1.3	1.77033	2.30143	.83263	2.853	6.010
1.4	1.85406	2.59568	.84185	3.033	6.841
1.5	1.93868	2.90802	.85045	3.214	7.730
1.6	2.02413	3.23861	.85849	3.398	8.675
1.7	2.11036	3.58761	.86600	3.583	9.678
1.8	2.19731	3.95516	.87302	3.769	10.739
1.9	2.28495	4.34140	.87958	3.956	11.858
2.0	2.37322	4.74644	.88572	4.145	13.036
2.5	2.82274	7.05685	.91103	5.100	19.808
3.0	3.28310	9.84930	.92944	6.071	28.064
3.5	3.75137	13.12980	.94307	7.052	37.812

Source: Adapted from Cohen (1963), Table 1, p. 331, with permission of the Technometrics Management Committee.

equations, which in some respects is simpler than the traditional method of Newton. However, when close first approximations are available, Newton's method is quite satisfactory.

3.4.1 Estimation of Mean When σ Is Known

In certain routine applications the standard deviation σ may be known, and in this special case, we need only the first equation of (3.4.4), which we now write in the form

$$G(\mu) = \bar{x} - \mu + \sigma \sum_{j=1}^{k} \frac{c_j}{n} Q_j = 0. \quad (3.4.7)$$

With the aid of tables of Q, included here in abridged form as one column of Table 3.2, this equation can be easily solved for $\hat{\mu}$ by using standard iterative methods. In most applications the simple "trial and error" procedure might be preferred. Once we have two values μ_1 and μ_2 in a sufficiently narrow interval such that $G(\mu_1) \leq 0 \leq G(\mu_2)$, we can interpolate for the required estimate $\hat{\mu}$.

3.4.2 Estimation of σ When the Mean Is Known

In applications where μ is known in advance, σ can be estimated from the second equation of (3.4.4), which we now write as

$$H(\sigma) = s^2 + (\bar{x} - \mu)^2 - \sigma^2[1 - \sum_{j=1}^{k} \frac{\xi_j c_j}{n} Q_j]. \quad (3.4.8)$$

This equation can be solved for $\hat{\mu}$ in much the same manner as that suggested for solving (3.4.7) for $\hat{\mu}$. Certain values of ξQ are available here in Table 3.2 as computational aids. Additional values of this function can easily be calculated directly from tables of ordinates and areas of the standard normal distribution.

3.4.3 Single-Stage Censoring as a Special Case

In the special case where only a single stage of censoring occurs, $k = 1$ and the estimation equations (3.4.4) become

$$(\bar{x} - \mu) = -\sigma \left(\frac{c}{n} Q\right),$$
$$s^2 + (\bar{x} - \mu)^2 = \sigma^2 \left[1 - \frac{\xi c}{n} Q\right]. \quad (3.4.9)$$

Since only one stage of censoring is involved the subscripts used in (3.3.4) have been dropped from (3.4.9)

Although based on a right censored sample, these equations are equivalent to equations (2.5.5) for singly left censored samples. When we replace ξ with $-\xi$ in (3.4.9), and recognize that in left censored samples, $(\bar{x} - \mu)$ is negative, the two sets of estimating equations become identical.

3.4.4 Errors of Estimates

The asymptotic variance–covariance matrix of $\hat{\mu}$, $\hat{\sigma}$ is given as

$$\begin{bmatrix} -E\left(\dfrac{\partial^2 \ln L}{\partial \mu^2}\right) & -E\left(\dfrac{\partial^2 \ln L}{\partial \mu \, \partial \sigma}\right) \\ -E\left(\dfrac{\partial^2 \ln L}{\partial \mu \, \partial \sigma}\right) & -E\left(\dfrac{\partial^2 \ln L}{\partial \sigma^2}\right) \end{bmatrix}^{-1} \doteq \begin{bmatrix} V(\hat{\mu}) & \mathrm{Cov}(\hat{\mu}, \hat{\sigma}) \\ \mathrm{Cov}(\hat{\mu}, \hat{\sigma}) & V(\hat{\sigma}) \end{bmatrix}, \qquad (3.4.10)$$

where E symbolizes expectation. In practice, when Newton's method of iteration is employed, final values of the applicable second partial derivatives (3.4.5), which are available from the Newton correction equations (3.2.21), should approximately equal their expected values and may accordingly be used to approximate the variance–covariance matrix (3.4.10).

For the special case in which σ is known

$$V(\hat{\mu}) \sim \frac{\sigma^2}{n\left[1 + \sum_{j=1}^{n} \dfrac{c_j}{n} Q'_j\right]}, \qquad (3.4.11)$$

and for the special case in which μ is known

$$V(\hat{\sigma}) \sim \frac{\sigma^2}{n}\left[\frac{3\{s^2 + (\bar{x} - \mu)^2\}}{\sigma^2} - 1 + \sum_{j=1}^{k} \frac{c_j}{\mu} \eta_j\right]^{-1}, \qquad (3.4.12)$$

where Q' and η are given by (3.4.6).

3.4.5 First Approximations to $\hat{\mu}$ and $\hat{\sigma}$ By Using Probits

This subsection is concerned with the use of a probit technique to obtain first approximations to $\hat{\mu}$ and $\hat{\sigma}$ when samples from a normal distribution are progressively censored as described in Section 3.4. For convenience, we assume that sample data are grouped into k classes with boundaries $T_o, T_1, \ldots, T_k$. Let f_j designate the number of observations in the jth class, that is, in the interval $T_{j-1} \leq x \leq T_j$. As in Section 3.4, let c_j designate the number of items withdrawn (censored) at $X = T_j$. We assume that censoring can occur only at one of the class boundaries T_j. As in Section 3.4, $N = n + \sum_{j=1}^{n} c_j$, where N is the total sample size and $n = \sum_{j=1}^{k} f_j$. It is noted that some of the c_j may be zeros.

Multirestricted Samples from Normal Distribution

We let $f(x)$ designate the pdf of X, and the conditional density subject to the restriction that $x > T_j$, may be expressed as

$$f(x|x > T_j) = \frac{f(x)}{1 - F_j}. \qquad (3.4.13)$$

Expected frequencies may then be expressed as follows

$$E(f_1) = N \int_{T0}^{T1} F(x)\, dx = Np_1$$

$$E(f_j) = \left[N - \sum_{i=1}^{j-1} \frac{c_i}{1 - F_i} \right] p_j, \quad j = 2, 3, \ldots, k, \qquad (3.4.14)$$

where

$$p_j = \int_{T_{j-1}}^{T_j} f(x)\, dx. \qquad (3.4.15)$$

Accordingly, estimates of p_j may be obtained as

$$p_1^* = \frac{f_1}{N},$$

$$p_j^* = f_j \bigg/ \left[N - \sum_{i=1}^{j-1} \frac{c_i}{1 - f_i^*} \right], \quad j = 2, 3, \ldots, k \qquad (3.4.16)$$

and estimates of F_j follow as

$$F_1^* = p_1^*$$

$$F_j^* = F_{j-1}^* + p_j^*, \quad j = 2, 3, \ldots, k. \qquad (3.4.17)$$

The asterisks serve to distinguish estimates from parameter values.

We recall that the standardized points of censoring are $\xi_j = (T_j - \mu)/\sigma$, and thus $F(T_j) = \Phi(\xi_j)$. We use (3.4.17) to calculate the F_j. The ξ_j can then be obtained from a table of the standard normal cdf. The probit estimate y_j^* is given as

$$y_j^* = 5 + \xi_j^*, \qquad (3.4.18)$$

and thus the equation of the probit regression line becomes

$$y = 5 + \frac{1}{\sigma}(x - \mu). \qquad (3.4.19)$$

The $k - 1$ points with coordinates (T_j, y_j^*), $1 \leq j \leq k - 1$, should lie approximately on this line. Accordingly, in a practical application we plot these points on rectangular-cross-section graph paper and sketch in the best-fitting straight line. This might be done as a least-square fit, but in many applications a fit by eye is satisfactory. The probit estimate μ_0 can be read from the fitted line as the value of x that corresponds to the probit value $y = 5$. The probit estimate σ_0 is the reciprocal of the slope. Details of these calculations are illustrated with examples in Section 3.4.7.

3.4.6 The Cumulative Hazard Plot Technique

The cumulative hazard plot technique, proposed by Nelson (1969), provides an alternative to the probit technique for use in obtaining first approximations to parameter estimates from progressively censored samples. This procedure is applicable not only to the normal distribution, but to various other distributions as well. A description of the hazard plot procedure is given in Chapter 8, where it is illustrated as a technique for estimation in the exponential distribution. When individual (ungrouped) sample observations are available, the hazard plot approach is likely to be preferred, but when data are grouped, the probit technique might produce better results with less effort. It is recognized, of course, that estimates produced by both the probit and the hazard plot procedures should be regarded as first approximations that may require improvement from iteration methods such as that of Newton.

The hazard function can serve to characterize a distribution in lieu of the density function or the cumulative distribution function. In a life-span context, the hazard function $h(y)$ of a continuous distribution of life Y might be defined as

$$h(y) = \frac{f(y)}{1 - F(y)}, \qquad (3.4.20)$$

where $f(y)$ and $F(y)$ are respectively the pdf and the cdf of Y.

The hazard function is also referred to as the hazard rate, the instantaneous failure rate, and the force of mortality. The cumulative hazard function is defined as

$$H(y) = \int_{-\infty}^{y} h(t)\, dt, \qquad -\infty < y < \infty. \qquad (3.4.21)$$

For a more complete account of the hazard function, its characteristics, and its applications, readers are referred to Nelson (1982).

3.4.7 Illustrative Examples

In order to illustrate the calculations involved in practical applications of the probit technique, two examples given by Cohen (1963) have been selected. The first pertains to a life test conducted on an electronic component and the second to a life test on certain biological specimens in a stress environment.

Multirestricted Samples from Normal Distribution

Example 3.4. A random sample of 300 units of a certain type of electronic component were place on test, and the life-span x was recorded for each item that failed. At the end of 1650 hours, 50 of the survivors were withdrawn (censored), and the test was terminated after 1735 hours with 95 survivors. In the notation employed here, the sample is summarized as: $N = 300$, $T_0 = 0$, $T_1 = 1650$, $T_2 = 1735$, $f_1 = 120$, $f_2 = 35$, $n = 120 + 35 = 155$, $c_1 = 50$, $c_2 = 95$, $\bar{x} = 1544.8$, and $s^2 = 17{,}022$.

To obtain first approximations to $\hat{\mu}$ and $\hat{\sigma}$, we estimate F_1 and F_2 as described in Section 3.4.5 and interpolate linearly between the points (T_1, y_1^*) and (T_2, y_2^*), which determine a straight line that approximates the probit regression line. From equations (3.4.16) and (3.4.17), we calculate

$$p_1^* = F_1^* = 120/300 = 0.4000,$$
$$p_2^* = 35/[300 - 50/(1 - 0.4000)] = 0.1615,$$
$$F_2^* = 0.4000 + 0.1615 = 0.5615.$$

The linear interpolation with F_1^* and F_2^* transformed into probits y^* is summarized as follows:

X	F*	y*
1650	0.4000	4.747
1703	0.5000	5.000
1735	0.5615	5.155

Initial approximations to the required estimates are then calculated as

$$\mu_0 = 1703, \qquad \sigma_0 = \frac{1735 - 1650}{5.155 - 4.747} = 208.$$

We elect to employ Newton's method to improve these approximations. With the partials of (3.4.3) and (3.4.5) evaluated for these approximations and thus for $\xi_1^o = -0.255$ and $\xi_2^o = 0.154$, the correction equations (3.2.21) become

$$-0.00572\, h + 0.00268\, g = 0.00188,$$
$$0.00268\, h - 0.00717\, g = -0.00285.$$

The simultaneous solution of these equations yields $h = -0.17$ and $g = 0.33$. Accordingly, the corrected estimates become

$$\hat{\mu} = 1703 - 0.17 = 1702.83,$$
$$\hat{\sigma} = 208 + 0.33 = 208.33.$$

A single cycle of iteration is considered sufficient for the purposes of this illustration, but the accuracy might be further improved by a second cycle of iteration with 1702.83 and 208.33 as new approximations. The asymptotic variance–covariance matrix now follows approximately as

$$\begin{vmatrix} 0.00572 & -0.00268 \\ -0.00268 & 0.00717 \end{vmatrix}^{-1} = \begin{vmatrix} 212 & 79 \\ 79 & 169 \end{vmatrix}$$

The approximate standard errors and correlation coefficient follow as

$$\sigma_{\hat{\mu}} = 14.56, \quad \sigma_{\hat{\sigma}} = 13.00, \quad \rho_{\hat{\mu},\hat{\sigma}} = 0.417.$$

Example 3.5. For this example, 316 biological specimens were observed in a stress environment. Life-spans were recorded in days as deaths occurred. The resulting data were grouped as shown in Table 3.3. Ten surviving specimens were withdrawn (censored) after 36.5 days and ten more were withdrawn after 44.5 days. The p_j^* and the F_j^* were calculated by using equations (3.4.16) and (3.4.17). These and the corresponding probit values are also shown in Table 3.3.

Table 3.3 Life Distribution of Certain Biological Specimens in a Stress Environment

j	Boundaries D_j	Midpoints x_j	f_j	c_j	p_j^*	F_j^*	Probits y_j^*
0	24.5					.000000	
1	26.5	25.5	1	0	.003165	.003165	2.270
2	28.5	27.5	1	0	.003165	.006330	2.507
3	30.5	29.5	4	0	.012658	.018988	2.925
4	32.5	31.5	18	0	.056962	.075950	3.567
5	34.5	33.5	18	0	.056962	.132912	3.887
6	36.5	35.5	37	10	.117089	.250001	4.326
7	38.5	37.5	45	0	.148678	.398679	4.743
8	40.5	39.5	57	0	.188326	.587005	5.220
9	42.5	41.5	39	0	.128855	.715860	5.571
10	44.5	43.5	43	10	.142071	.857931	6.071
11	46.5	45.5	20	0	.086104	.944035	6.590
12	48.5	47.5	9	0	.038747	.982782	7.116
13	50.5	49.5	1	0	.004305	.987087	7.229
14	52.5	51.5	3	0	.012916	1.00000	
TOTALS			296	20			

Source: From Cohen (1963), p. 337 with permission of the Technometrics Management Committee.

Multirestricted Samples from Normal Distribution

The 13 points (T_j, y_j^*), $j = 1, \ldots, 13$, were plotted as shown in Figure 3.2, and the probit regression line was sketched by eye. Initial approximations $\mu_0 = 39.8$ and $\sigma_0 = 4.55$ are obtained from the regression line. One cycle of Newton's method gave $\mu_1 = 39.6$ and $\sigma_1 = 4.60$ as new approximations which were then used as starting points for a second cycle of the iteration process. For this second cycle the correction equations become

$$-14.591 h + 0.801 g = 0.261,$$
$$0.801 h + 28.067 g = -0.331.$$

On solving this pair of equations simultaneously, we obtain as new corrections $h = -0.017$ and $g = 0.011$, and the final estimates become $\hat{\mu} = 39.583$ and $\hat{\sigma} = 4.611$. The asymptotic variance–covariance matrix is now approximated as

$$\begin{vmatrix} 14.591 & -0.801 \\ -0.801 & 28.067 \end{vmatrix}^{-1} = \begin{vmatrix} 0.069 & 0.002 \\ 0.002 & 0.036 \end{vmatrix}$$

Standard errors and the correlation coefficient follow as

$$\sigma_{\hat{\mu}} = 0.263, \qquad \sigma_{\hat{\sigma}} = 0.190, \qquad \rho_{\hat{\mu},\hat{\sigma}} = 0.0401.$$

3.5 SOME ADDITIONAL SAMPLE TYPES

In this section we consider (1) doubly censored samples in which the total number of censored observations is known, but not the number in each tail separately, (2) centrally censored samples, and (3) centrally truncated samples. Although samples of these types may not occur frequently in practical applications, they are deemed to be of sufficient interest to justify their inclusion here.

3.5.1 Doubly Censored Samples in Which the Total Number of Censored Observations Is Known, But Not the Number in Each Tail Separately

These samples differ from the doubly censored samples considered in Section 3.3 in that $c = c_1 + c_2$ is known, but c_1 and c_2 separately remain unknown. The total sample size is $N = n + c$, where n is the number of complete observations. The likelihood function for a sample of this type, where K is an ordering constant, is

$$L = K[\Phi_1 + (1 - \Phi_2)]^c (\sigma\sqrt{2\pi})^{-n} \exp\left[-\frac{1}{2\sigma^2} \sum_{i=1}^{n} (x_i - \mu)^2\right]. \quad (3.5.1)$$

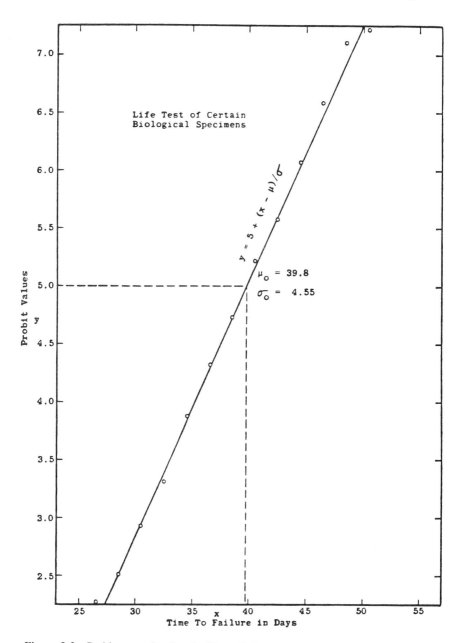

Figure 3.2 Probit regression line for Example 3.5. Reprinted from Cohen (1963), Fig. 1, p. 338, with permission of the Technometrics Management Committee.

Multirestricted Samples from Normal Distribution

The points of censoring are T_1 and T_2. Their standardized values are ξ_1 and ξ_2. We take logarithms of (3.5.1), differentiate and equate to zero to obtain the estimating equations

$$\frac{\partial \ln L}{\partial \mu} = \frac{c}{\sigma}\left[\frac{\phi_2 - \phi_1}{\Phi_1 + (1 - \Phi_2)}\right] + \frac{1}{\sigma^2}\sum_{i=1}^{n}(x_i - \mu) = 0,$$

$$\frac{\partial \ln L}{\partial \sigma} = \frac{c}{\sigma}\left[\frac{\xi_2\phi_2 - \xi_1\phi_1}{\Phi_1 + (1 - \Phi_2)}\right] - \frac{n}{\sigma} + \frac{1}{\sigma^3}\sum_{i=1}^{n}(x_i - \mu)^2 = 0. \quad (3.5.2)$$

In order to simplify the notation, we define

$$U_1(\xi_1, \xi_2) = \frac{c}{n}\left[\frac{\phi_1}{\Phi_1 + (1 - \Phi_2)}\right] \quad \text{and}$$

$$U_2(\xi_1, \xi_2) = \frac{c}{n}\left[\frac{\phi_2}{\Phi_1 + (1 - \Phi_2)}\right]. \quad (3.5.3)$$

When U_1 and U_2, as thus defined, are substituted into (3.5.2), the estimating equations are subsequently reduced to

$$\bar{x} - \mu = \sigma(U_1 - U_2), \quad (3.5.4)$$
$$s^2 + (\bar{x} - \mu)^2 = \sigma^2[1 + \xi_1 U_1 - \xi_2 U_2].$$

Note that these equations differ from those of (3.3.4) only in that U_1 and U_2 have replaced Ω_1 and Ω_2. Accordingly, by following the same steps that were employed in deriving (3.3.5), we obtain

$$\frac{U_1 - U_2 - \xi_1}{\xi_2 - \xi_1} = \frac{\bar{x} - T}{w},$$

$$\frac{1 + \xi_1 U_1 - \xi_2 U_2 - (U_1 - U_2)^2}{(\xi_2 - \xi_1)^2} = \frac{s^2}{w^2}. \quad (3.5.5)$$

These equations are completely analogous to equations (3.3.5) for the case in which c_1 and c_2 are known separately. Estimates $\hat{\xi}_1$ and $\hat{\xi}_2$ can be calculated by solving the two equations of (3.5.5) simultaneously. The method of solution is the same as that described in Section 3.3 for the solution of equations (3.3.5). With $\hat{\xi}_1$ and $\hat{\xi}_2$ thus determined, it then follows that

$$\hat{\sigma} = \frac{w}{\hat{\xi}_2 - \hat{\xi}_1} \quad \text{and} \quad \hat{\mu} = T_1 - \hat{\sigma}\hat{\xi}_1. \quad (3.5.6)$$

In this case, the second partial derivatives of the likelihood function for use with Newton's method of iteration and in the information matrix, which is then inverted to produce the variance–covariance matrix, are

$$\frac{\partial^2 \ln L}{\partial \mu^2} = -\frac{n}{\sigma^2}[1 - U'_1 + U'_2],$$

$$\frac{\partial^2 \ln L}{\partial \mu \, \partial \sigma} = -\frac{n}{\sigma^2}\left[\frac{2(\bar{x} - \mu)}{\sigma} - \lambda_1 + \lambda_2\right], \quad (3.5.7)$$

$$\frac{\partial^2 \ln L}{\partial \sigma^2} = -\frac{n}{\sigma^2}\left[\frac{3\{s^2 + (\bar{x} - \mu)^2\}}{\sigma^2} - 1 - \eta_1 + \eta_2\right],$$

where

$$U'_1 = -U_1\left(\frac{n}{c}U_1 + \xi_1\right), \quad U'_2 = U_2\left(\frac{n}{c}U_2 - \xi_2\right),$$

$$\lambda_1 = \xi_1 U'_1 + U_1, \quad \lambda_2 = \xi_2 U'_2 + U_2,$$

$$\eta_1 = \xi_1(\lambda_1 + U_1), \quad \eta_2 = \xi_2(\lambda_2 + U_2).$$

Initial approximations for use as starting points in the iteration procedure might be obtained from the chart of Figure 3.1 with the value of c neglected.

Estimating equations given in Chapter 2 for singly censored samples can be obtained as a special case of (3.4.4). Proof of this statement is left as an exercise for the reader.

3.5.2 Centrally Censored Samples

In samples of this type, observations in both tails are complete, but only a count is made for central observations such that $T_1 < x < T_2$. We assume that T_1 and T_2 are known censoring points and that c observations are censored. There are n_1 complete observations in the left tail and n_2 in the right tail. Let $n = n_1 + n_2$ and the total sample size is $N = n + c$. The likelihood function for a sample of this type in which K is an ordering constant may be expressed as

$$L = K[\Phi_2 - \Phi_1]^c (\sigma\sqrt{2\pi})^{-n} \exp{-\frac{1}{2\sigma^2}\sum_{i=1}^{n}(x_i - \mu)^2}. \quad (3.5.8)$$

To obtain estimating equations, we take logarithms of (3.5.8), differentiate, and equate to zero. We then have

$$\frac{\partial \ln L}{\partial \mu} = -\frac{c}{n}\left[\frac{\phi_2 - \phi_1}{\Phi_2 - \Phi_1}\right] + \frac{1}{\sigma^2}\sum_{i=1}^{n}(x_i - \mu) = 0,$$

$$\frac{\partial \ln L}{\partial \sigma} = -\frac{c}{n}\left[\frac{\xi_2\phi_2 - \xi_1\phi_1}{\Phi_2 - \Phi_1}\right] - \frac{n}{\sigma} + \frac{1}{\sigma^3}\sum_{i=1}^{n}(x_i - \mu)^2 = 0. \quad (3.5.9)$$

With $\bar{Q}_1$ and $\bar{Q}_2$ as defined by (3.2.6) for doubly truncated samples, we substitute these quantities into (3.5.9) and simplify to obtain

Multirestricted Samples from Normal Distribution

$$(\bar{x} - \mu) = \sigma \left(\frac{c}{n}\right)[\bar{Q}_2 - \bar{Q}_1], \tag{3.5.10}$$

$$s^2 + (\bar{x} - \mu)^2 = \sigma^2 \left[1 + \frac{c}{n}(\xi_2 \bar{Q}_2 - \xi_1 \bar{Q}_1)\right].$$

After further algebraic simplification, these equations are reduced to

$$\frac{\dfrac{c}{n}(\bar{Q}_2 - \bar{Q}_1) - \xi_1}{\xi_2 - \xi_1} = \frac{\bar{x} - T_1}{w},$$

$$\frac{1 + \dfrac{c}{n}(\xi_2 \bar{Q}_2 - \xi_1 \bar{Q}_1) - \left(\dfrac{c}{n}\right)^2 (\bar{Q}_2 - \bar{Q}_1)^2}{(\xi_2 - \xi_1)^2} = \frac{s^2}{w^2}. \tag{3.5.11}$$

These equations can be solved iteratively for estimates $\hat{\xi}_1$ and $\hat{\xi}_2$ by employing procedures similar to those outlined for the solution of equations (3.2.11) and (3.3.5). With $\hat{\xi}_1$ and $\hat{\xi}_2$ obtained as the simultaneous solution of (3.5.11), it follows that

$$\hat{\sigma} = \frac{w}{\hat{\xi}_2 - \hat{\xi}_1} \quad \text{and} \quad \hat{\mu} = T_1 - \hat{\sigma}\hat{\xi}_1. \tag{3.5.12}$$

First approximations to ξ_1 and ξ_2 can be obtained from the cdf of the standard normal distribution from the relations

$$\frac{n_1}{N} = \Phi(\xi_1) \quad \text{and} \quad \frac{n_2}{N} = 1 - \Phi(\xi_2). \tag{3.5.13}$$

Second partials of the likelihood function for use in Newton's method of iteration and in the applicable variance–covariance matrix are

$$\frac{\partial^2 \ln L}{\partial \mu^2} = -\frac{n}{\sigma^2}\left[1 + \frac{c}{n}(\bar{Q}'_1 - \bar{Q}'_2)\right],$$

$$\frac{\partial^2 \ln L}{\partial \mu \, \partial \sigma} = -\frac{n}{\sigma^2}\left[\frac{2(\bar{x} - \mu)}{\sigma} - \frac{c}{n}(\lambda_2 - \lambda_1)\right], \tag{3.5.14}$$

$$\frac{\partial^2 \ln L}{\partial \sigma^2} = -\frac{n}{\sigma^2}\left[\frac{3\{s^2 + (\bar{x} - \mu)^2\}}{\sigma^2} - 1 - \frac{c}{n}(\eta_2 - \eta_1)\right],$$

where

$$\bar{Q}'_1 = \bar{Q}_1(\bar{Q}_1 - \xi_1), \qquad \bar{Q}'_2 = -\bar{Q}_2(\bar{Q}_2 + \xi_2),$$

$$\lambda_1 = \xi_1 \bar{Q}'_1 + \bar{Q}_1, \qquad \lambda_2 = \xi_2 \bar{Q}'_2 + \bar{Q}_2, \tag{3.5.15}$$

$$\eta_1 = \xi_1(\lambda_1 + \bar{Q}_1), \qquad \eta_2 = \xi_2(\lambda_2 + \bar{Q}_2).$$

Estimating equations for singly censored samples given in Chapter 2 can also be derived from (3.5.10) as a special case, but proof of this statement is left as an exercise for the reader.

3.5.3 Centrally Truncated Samples

In samples of this type, observation is possible only in the two tails of the distribution. Observation is not possible for $T_1 < x < T_2$. The sample consists of $N = n = n_1 + n_2$ observations, where n_1 observations are less than or equal to T_1 and n_2 observations are greater than or equal to T_2. These samples differ from the centrally censored samples only in that in the truncated case, the value of c is unknown. The likelihood function of a sample of this type from a normal distribution is

$$L = [\Phi_1 + (1 - \Phi_2)]^{-n} (\sigma\sqrt{2\pi})^{-n} \exp\left[-\frac{1}{2\sigma^2} \sum_{i=1}^{n} (x_i - \mu)^2\right]. \quad (3.5.16)$$

We take logarithms of this equation, differentiate and equate to zero to obtain estimating equations

$$\frac{\partial \ln L}{\partial \mu} = \frac{n}{\sigma}\left[\frac{\phi_1 - \phi_2}{\Phi_1 + (1 - \Phi_2)}\right] + \frac{1}{\sigma^2} \sum_{i=1}^{n} (x_i - \mu) = 0,$$

$$\frac{\partial \ln L}{\partial \sigma} = \frac{n}{\sigma}\left[\frac{\xi_1\phi_1 - \xi_2\phi_2}{\Phi_1 + (1 - \Phi_2)}\right] - \frac{n}{\sigma} + \frac{1}{\sigma^3} \sum_{i=1}^{n} (x_i - \mu)^2 = 0. \quad (3.5.17)$$

In order to simplify the notation and thereby reduce the estimating equations to forms that are analogous to corresponding equations for other types of doubly restricted samples, we define

$$J_1(\xi_1,\xi_2) = \frac{\phi_1}{\Phi_1 + (1 - \Phi_2)} \quad \text{and}$$

$$J_2(\xi_1,\xi_2) = \frac{\phi_2}{\Phi_1 + (1 - \Phi_2)}. \quad (3.5.18)$$

It is to be noted that $U_1 = (c/n)J_1$ and $U_2 = (c/n)J_2$ where U_1 and U_2 were defined by (3.4.3) in connection with estimators based on doubly censored samples for which the total number of censored observations was known but not the number in each tail separately.

With J_1 and J_2 as thus defined, the estimating equations of (3.5.16) become

$$\bar{x} - \mu = \sigma(J_2 - J_1),$$
$$s^2 + (\bar{x} - \mu)^2 = \sigma^2(1 + \xi_2 J_2 - \xi_1 J_1). \quad (3.5.19)$$

After further algebraic simplification, these equations become

Multirestricted Samples from Normal Distribution

$$\frac{J_2 - J_1 - \xi_1}{\xi_2 - \xi_1} = \frac{\bar{x} - T_1}{w},$$

$$\frac{1 + \xi_2 J_2 - \xi_1 J_1 - (J_2 - J_1)^2}{(\xi_2 - \xi_1)^2} = \frac{s^2}{w^2}.$$

(3.5.20)

The simultaneous solution of these equations involves essentially the same procedures as those employed in solving corresponding estimating equations for other doubly restricted samples. With $\hat{\xi}_1$ and $\hat{\xi}_2$ thus calculated, it then follows that

$$\hat{\sigma} = \frac{w}{\hat{\xi}_2 - \hat{\xi}_1} \quad \text{and} \quad \hat{\mu} = T_1 - \hat{\sigma}\hat{\xi}_1.$$

(3.5.21)

First approximations to $\hat{\xi}_1$ and $\hat{\xi}_2$ can be obtained from the estimators given in Chapter 2 for singly truncated samples since a centrally truncated sample is in fact two separate samples, one that is singly left truncated and one that is singly right truncated. An average of the two sets of estimators obtained from these two tails of the sample should provide satisfactory first approximations for use in an iterative solution of (3.5.18).

Second partial derivatives of the likelihood function for use in the Newton method of iteration and for calculating elements of the information matrix, which is then inverted to produce the variance–covariance matrix, are

$$\frac{\partial^2 \ln L}{\partial \mu^2} = -\frac{n}{\sigma^2}[1 - J'_2 + J'_1],$$

$$\frac{\partial^2 \ln L}{\partial \mu \, \partial \sigma} = -\frac{n}{\sigma^2}\left[\frac{2(\bar{x} - \mu)}{\sigma} - \lambda_2 + \lambda_1\right],$$

(3.5.22)

$$\frac{\partial^2 \ln L}{\partial^2 \sigma^2} = \frac{n}{\sigma^2}\left\{\frac{3[s^2 + (\bar{x} - \mu)^2]}{\sigma^2} - 1 - \eta_2 + \eta_1\right\},$$

where

$$J'_1 = -J_1(J_1 + \xi_1), \qquad J'_2 = J_2(J_2 - \xi_2),$$

$$\lambda_1 = \xi_1 J'_1 + J_1, \qquad \lambda_2 = \xi_2 J'_2 + J_2,$$

(3.5.23)

$$\eta_1 = \xi_1(\lambda_1 + J_1), \qquad \eta_2 = \xi_2(\lambda_2 + J_2).$$

3.6 FINAL COMMENT

In deriving estimating equations for the various sample types considered in this chapter, an effort has been made to follow a routine that produces estimating equations that are analogous in algebraic form, and that can thus be solved by employing common iterative procedures.

4
Linear Estimators

4.1 INTRODUCTION

Although maximum likelihood estimators (MLEs) are optimal with respect to estimate variances, they are not unbiased. In large samples and even in moderate-sized samples, bias is essentially negligible. However, in small samples it becomes a major source of error. In order to eliminate bias in small samples from distributions of the type $(1/\sigma)f[(x - \mu)/\sigma]$, the method of weighted least squares, based on the Gauss–Markov least-square theorem, was developed. Using this method, best linear unbiased estimators (BLUE) of the location and scale parameters; μ and σ, which employ order statistics in a systematic manner and which have minimum variance in the class of linear unbiased estimators, can be obtained. Some of the major contributors to the development of the theory underlying these estimators have been Lloyd (1952), Teichroew (1956), Ogawa (1957), Sarhan (1954, 1955a, b), Sarhan and Greenberg (1956, 1957, 1958, 1962), David (1970), Gupta (1952), Blom (1956), and numerous others.

The parent distribution enters into the least-square method only through expected values, variances, and covariances of its order statistics. None of the higher-order moments are used, and thus some of the sample information is lost. The method of maximum likelihood uses the entire distribution without any loss of information. The efficiency of the BLUEs is therefore expected to be somewhat less than that of the MLEs. In small samples, however, this loss of efficiency

Linear Estimators

is not of significant proportions, and the elimination of bias more than compensates for any loss of efficiency.

4.2 CALCULATION OF ESTIMATES

Estimates are calculated as sums of products of the complete (uncensored) observations and appropriate coefficients. Thus, we write

$$\mu^* = \sum_{i=c_1+1}^{c_1+n} a_{1:i} x_{i:N},$$

$$\sigma^* = \sum_{i=c_1+1}^{c_1+n} a_{2:i} x_{i:N},$$

(4.2.1)

where N is the total sample size, n is the number of complete (uncensored) observations, c_1 is the number of left censored observations, and c_2 is the number of right censored observations. It follows that $N = n + c_1 + c_2$. For a complete sample, $c_1 = c_2 = 0$ and $N = n$. In a singly left censored sample, $c_1 \geq 1$ and $c_2 = 0$. In a singly right censored sample, $c_2 \geq 1$ and $c_1 = 0$. Coefficients $a_{1:i}$ and $a_{2:i}$ are obtained by ordering the uncensored sample observations according to magnitude $(x_{1:N} \leq x_{2:N} \leq \cdots \leq x_{n:N})$ and applying the method of least squares to obtain the linear combination of observations that produces unbiased parameter estimates with minimum variance. These coefficients have been calculated and tabulated by Sarhan and Greenberg (1956, 1962), for all possible conditions of censoring. Entries in the 1956 tables are given to 8 decimals for sample sizes up to 10. In the 1962 tables, entries are given to 4 decimals for sample sizes up to 20. Corresponding variances and covariances and their efficiencies (relative to complete samples) were also calculated and tabulated. The original 1956 tables for sample sizes up to 10 have been reproduced here as Tables 4.1 and 4.2. When N exceeds 20, the calculation and tabulation of the coefficients becomes complicated and time-consuming. However, for samples of sizes greater than 20, maximum likelihood estimates are most likely to be preferred.

4.3 DERIVATIONS

Let $x_{i:N}$, $i = c_1 + 1, c_1 + 2, \ldots, c_1 + n$, designate the ith observed order statistic in a sample of size N, and let

$$z_{i:N} = \frac{x_{i:N} - \mu}{\sigma}$$

(4.3.1)

designate the corresponding standardized order statistic. Let $E(Z_{i:N}) = \mu_{i:N}$ and let

$$\mathbf{V} = \sigma_{i,j:N} \qquad (4.3.2)$$

designate the variance–covariance matrix of $z_{i:N}$, where $\sigma_{i,j:N} = \text{cov}(z_{i:N}, z_{j:N})$. The elements $\mu_{i:N}$ and $\sigma_{i:N}$, of course, depend on the underlying distribution. For the uniform distribution

$$\mu_{i:N} = \frac{i}{N+1},$$

$$\sigma_{i,i:N} = \frac{i(N-i-1)}{(N+1)^2(N+2)}, \qquad (4.3.3)$$

$$\sigma_{i,j:N} = \frac{i(N-j+1)}{(N+1)^2(N+2)}.$$

For all distributions,

$$E(x_{i:N}) = \mu + \sigma \mu_{i:N} \qquad (4.3.4)$$

and

$$\text{Cov}(x_{i:N}, x_{j:N}) = \sigma^2 \sigma_{i,j:N}. \qquad (4.3.5)$$

Let

$$\mathbf{X}' = (x_{c_1+1}, \ldots, x_{c_1+n}), \quad \boldsymbol{\theta}' = (\mu, \sigma),$$
$$\mathbf{E}' = (\mu_{c_1+1}, \ldots, \mu_{c_1+n}), \qquad (4.3.6)$$

and let $\mathbf{W} = (\mathbf{1_N}, \mathbf{E})$, where $\mathbf{1_N}$ is an $N \times 1$ column of ones.

In matrix notation

$$E(\mathbf{X}) = \mathbf{W}\boldsymbol{\theta} \qquad (4.3.7)$$

and the variance–covariance matrix of $\mathbf{X}$ is

$$V(\mathbf{X}) = \sigma^2 \mathbf{V}. \qquad (4.3.8)$$

From (4.3.8), the weighted least-square solution of $\boldsymbol{\theta}'^* = (\mu^*, \sigma^*)$ becomes

$$\boldsymbol{\theta}^* = (\mathbf{W}'\mathbf{V}^{-1}\mathbf{W})^{-1} \mathbf{W}'\mathbf{V}^{-1}\mathbf{X}, \qquad (4.3.9)$$

which is the best linear unbiased estimator (BLUE) of $\boldsymbol{\theta}$ based on a Type II censored sample. From (4.3.8), it follows that

$$E(\boldsymbol{\theta}^*) = \boldsymbol{\theta}, \qquad (4.3.10)$$

Linear Estimators

and the variance–covariance matrix of $\boldsymbol{\theta}^*$ is

$$V(\boldsymbol{\theta}^*) = \sigma^2(\mathbf{W}'\mathbf{V}^{-1}\mathbf{W}) = (v_{ij}). \tag{4.3.11}$$

Let the determinant be designated as

$$\Delta_W = |\mathbf{W}'\mathbf{V}^{-1}\mathbf{W}| \tag{4.3.12}$$

and let

$$\Gamma = \left(\frac{1}{\Delta_W}\right)\mathbf{V}^{-1}(\mathbf{1}_n\mathbf{E}' - \mathbf{E}\mathbf{1}_n')\mathbf{V}^{-1}. \tag{4.3.13}$$

It then follows that

$$\mu^* = -\mathbf{E}'\Gamma\mathbf{X} = \sum_{i=c_1+1}^{c_1+n} a_{1i:N}\, x_{i:N},$$

$$\sigma^* = \mathbf{1}_n'\Gamma\mathbf{X} = \sum_{i=c_1+1}^{c_1+n} a_{2i:N}\, x_{i:N},$$

$$v_{11} = \sigma^2 \frac{(\mathbf{E}'\mathbf{V}^{-1}\mathbf{E})}{\Delta_w}, \tag{4.3.14}$$

$$v_{22} = \frac{\sigma^2\,(\mathbf{1}_n'\mathbf{V}^{-1}\mathbf{1}_n)}{\Delta_w},$$

$$v_{12} = v_{21} = \frac{-\sigma^2(\mathbf{1}_n\mathbf{V}^{-1}\mathbf{E})}{\Delta_w}.$$

For symmetric distributions, $\mu_{i:N} = -\mu_{(N-i+1):N}$ and $\sigma_{i,j:N} = \sigma_{(N-i+1),(N-j+1):N}$ ($i < j$). If the distribution is symmetric and if censoring is symmetric Type II, then $c_1 = c_2 = c$, and $\mathbf{1V}^{-1}\mathbf{E} = \mathbf{E}'\mathbf{V}^{-1}\mathbf{1}$, in which case $v_{12} = \text{Cov}(\mu^*, \sigma^*) = 0$. It follows that

$$\mu^* = \frac{\mathbf{1}'\mathbf{V}^{-1}\mathbf{X}}{\mathbf{1}'\mathbf{V}^{-1}\mathbf{1}},$$

$$\sigma^* = \frac{\mathbf{E}'\mathbf{V}^{-1}\mathbf{X}}{\mathbf{E}'\mathbf{V}^{-1}\mathbf{E}}. \tag{4.3.15}$$

These are symmetric linear functions of the ordered observations $x_{i:N}$ ($i = c_1 + 1, \ldots, c_1 + n$). For example, $a_{1i} = a_{1(N-i+1)}$ and $a_{2i} = a_{2(N-i+1)}$. Furthermore, the variances in this case are

$$V(\mu^*) = \frac{\sigma^2}{\mathbf{1}'\mathbf{V}^{-1}\mathbf{1}},$$

$$V(\sigma^*) = \frac{\sigma^2}{\mathbf{E}'\mathbf{V}^{-1}\mathbf{E}}. \tag{4.3.16}$$

Accordingly, the least-square estimators can be calculated along with their variances and covariances from expected values of variances and covariances of order statistics from random samples from distributions of the type $(1/\sigma)[f(x - \mu)/\sigma]$. The means, variances, and covariances of order statistics are available for a number of distributions. Tietjen et al. (1977) gave these results for the normal distribution. Corresponding results have been given by Barnett (1966) for the Cauchy distribution; by Govindarajulu (1966) for the double exponential; by Gupta et al. (1967) for the logistic distribution; by Tiku and Kumra (1981) for Student's t distribution; by Prescott (1974) for the gamma distribution; by Balakrishnan and Kocherlakota (1985) for the double Weibull distributions; by Balakrishnan and Malik (1985) for log-logistic distributions; by Balakrishnan (1985) for the half-logistic distribution; and by David et al. (1977) for normal distributions in the presence of outliers.

In this chapter we are primarily concerned with the normal distribution. To facilitate the calculation of BLUE estimates from normal distribution samples, Sarhan and Greenberg (1956) tabulated the coefficients a_{1i} and a_{2i} together with their variances and covariances for sample sizes up to 10. Subsequently, in 1962 these tables were extended to include samples up to 20. As a convenience to readers, the 1956 tables for sample sizes of 10 or less are reproduced here as Tables 4.1 and 4.2. When sample sizes exceed 10, but do not exceed 20, readers may seek access to the 1962 tables, or they may elect to employ alternate estimators such as those due to Gupta which follow in Section 4.4. Gupta's estimators are easy to calculate and they are at least nearly optimal in comparison to the BLUEs. The maximum likelihood estimators given in Chapters 2 and 3 are recommended for larger samples.

4.4 ALTERNATIVE ESTIMATORS

Alternative linear estimators for the mean and standard deviation of a normal population were suggested by Gupta (1952) for use when sample sizes are greater than 10, since in these cases exact estimates are tedious to calculate and since variances and covariances of the order statistics in larger samples are not readily available. Gupta's original estimators were only for singly censored samples, but they were subsequently extended to doubly censored samples. These alternative estimators can be expressed as

$$\mu^{**} = \sum_{i=c_1+1}^{c_1+n} b_{1i} x_i,$$
$$\sigma^{**} = \sum_{i=c_1+1}^{c_1+n} b_{2i} x_i,$$

(4.4.1)

Linear Estimators

where the coefficients b_{1i} and b_{2i} are

$$b_{1i} = \frac{1}{n} - \frac{\bar{\mu}_k(\mu_i - \bar{\mu}_k)}{\sum\limits_{j=c_1+1}^{c_1+n} (\mu_j - \bar{\mu}_k)^2},$$

$$b_{2i} = \frac{\mu_i - \bar{\mu}_k}{\sum\limits_{j=c_1+1}^{c_1+n} (\mu_j - \bar{\mu}_k)^2},$$

(4.4.2)

and where $\bar{\mu}_k$ is the arithmetic mean of expected values of the uncensored (complete) sample observations; that is,

$$\bar{\mu}_k = \frac{1}{n} \sum_{j=c_1+1}^{c_1+n} \mu_j.$$

(4.4.3)

Table 4.3 gives variances and relative efficiencies of the alternative estimates for samples of size 10. In a relatively recent paper, Saw (1959) proposed a nonlinear estimate for σ that is unbiased and asymptotically efficient. However, its calculation is limited to the case of single censoring.

Other simplified alternatives to the BLUEs include the "nearly best linear" estimators of location and scale parameters proposed by Blom (1956).

4.5 ILLUSTRATIVE EXAMPLES

Example 4.1. Gupta (1952) presented the following data giving the number of days to death (y) of the first 7 in a sample of 10 mice following inoculation with a uniform culture of human tuberculosis. The test was terminated with the death of the seventh sample specimen. Accordingly, the sample was singly right censored Type II. Gupta assumed that logarithms (x) (to the base 10) of these observations were normally distributed (μ, σ).

y	41	44	46	54	55	58	60
$x = \log_{10} y$	1.6128	1.6435	1.6628	1.7324	1.7404	1.7634	1.7782

For this sample, $c_1 = 0$, $c_2 = 3$, $N = 10$, and $n = 7$. In order to calculate best linear estimates, we apply equations (4.1) with coefficients a_{1i} and a_{2i} obtained from Table 4.1. Thus we have

$$\mu^* = (0.02443)(1.6128) + (0.06356)(1.6435) + (0.08178)(1.6628)$$
$$+ (0.09617)(1.7324) + (0.10886)(1.7404) + (0.12074)(1.7634)$$
$$+ (0.50447)(1.7782) = 1.746,$$

Table 4.1 Coefficients of Linear Estimates of the Mean and Standard Deviation for Censored Samples from the Normal Distribution

N	c_1	c_2	$x_{1:N}$	$x_{2:N}$	$x_{3:N}$	$x_{4:N}$	$x_{5:N}$	$x_{6:N}$	$x_{7:N}$	$x_{8:N}$	$x_{9:N}$	$x_{10:N}$
2	0	0	.50000000 −.88622693	.50000000 .88622693								
3	0	0	.33333333 −.59081795	.33333333 .00000000	.33333333 .59081795							
3	0	1	.00000000 −1.18163590	1.00000000 1.18163590								
4	0	0	.25000000 −.45394040	.25000000 −.11018073	.25000000 .11018073	.25000000 .45394040						
4	0	1	.11606577 −.69713303	.24083805 −.12681665	.64309618 .82394968							
4	0	2	−.40555169 −1.36544125	1.40555169 1.36544125								
4	1	1		.50000000 −1.68343717	.50000000 1.68343717							
5	0	0	.20000000 −.37238157	.20000000 −.13521392	.20000000 .00000000	.20000000 .13521392	.20000000 .37238157					
5	0	1	.12515679 −.51173274	.18304590 −.16678091	.21471643 .02740065	.47708089 .65111300						
5	0	2	−.06377484 −.78958387	.14982836 −.21211572	.91394649 .98169958							
5	0	3	−.74110683 −1.49712813	1.74110683 1.49712813								
5	1	1		.38929103 1.01006230	.22141794 .00000000	.38929103 1.01006230						

5	1	2	.16666667 -.31752484	.00000000 -2.02012460	1.00000000 2.02012460	.16666667 .04321165	.16666667 .13855961	.166 6667 .317 2484
6	0	0		.16666667 -.13855961	.16666667 -.04321165	.16666667 .04321165	.16666667 .13855961	
	0	1	.11828773 -.40969394	.15097353 -.16845737	.16803141 -.04061162	.18280232 .07395248	.37990501 .54481044	
	0	2	.01848367 -.55281996	.12260668 -.20913743	.17614875 -.02897078	.68276091 .79092817		
	0	3	-.21591800 -.82435700	.00485167 -.27604235	1.15106633 1.10039835			
	0	4	-1.02609712 -1.59884545	2.02609712 1.59884545				
	1	1		.31977001 -.75309024	.18022999 -.08286186	.18022999 .08286186	.31977001 .75309024	
	1	2		.15386831 -1.14382882	.17811699 -.08783818	.66802470 1.23166700		
	1	3		-.45784434 -2.27165234	1.45784434 2.27165234			
	2	2			.50000000 -2.48081297	.50000000 2.48081297		
7	0	0	.14285714 -.27781036	.14285714 -.13509780	.14285714 -.06246312	.14285714 .00000000	.14285714 .06246312	.1-285714 .1-509780	.14285714 .27781036
	0	1	.10882014 -.34400143	.12954538 -.16098444	.13997050 -.06807607	.14873929 .01143886	.15705206 .09006788	.3 587262 .4 155609	
	0	2	.04654966 -.43696302	.10721153 -.19432593	.13743095 -.07179355	.16260139 .03213315	.54615647 .67094936		
	0	3	-.07380239 -.58481466	.06771901 -.24284221	.13752310 -.07174176	.86856027 .89639863			

Table 4.1 Continued

N	c_1	c_2	$x_{1:N}$	$x_{2:N}$	$x_{3:N}$	$x_{4:N}$	$x_{5:N}$	$x_{6:N}$	$x_{7:N}$	$x_{8:N}$	$x_{9:N}$	$x_{10:N}$
7	0	4	−.34744504 −.86817366	−.01345514 −.32689877								
	0	5	−1.27331716 −1.68122579	2.27331716 1.68122579	1.36090107 1.19507242							
	1	1		.27183155 −.61077842	.15198954 −.10607146	.15235782 .00000000	.15198954 .10607146	.27183155 .61077842				
	1	2		.17480153 −.52879521	.14322942 −.12575459	.16338504 .02477706	.51858402 .92977274					
	1	3		−.05916442 −1.24627431	.12704486 −.15477199	.93211956 1.40304629						
	1	4		−.87159736 −2.47116575	1.87159736 2.47116575							
	2	2			.41569461 −1.41760741	.16861078 .00000000	.41569461 1.41760741					
	2	3			.00000000 −2.83521483	1.00000000 2.83521483						
8	0	0	.12500000 −.24758623	.12500000 −.12944776	.12500000 −.07130849	.12500000 −.02295726	.12500000 .02295726	.12500000 .07130849	.12500000 .12944776	.12500000 .24758623		
	0	1	−.09966946 −.29775817	.11366201 −.15150866	.12079601 −.07963530	.12649213 −.02000181	.13176484 .03635632	.13698715 .09505131	.27042840 .41749631			
	0	2	−.05691876 −.36375811	.09621316 −.17875554	.11531993 −.08808945	.13090112 −.01319507	.14512418 .05698091	.45552284 .58681726				
	0	3	−.01672011 −.45862177	.06765099 −.21555013	.10840409 −.09699747	.14131770 .00022386	.69934643 .77094550					
	0	4	−.15491146 −.61090114	.01760383 −.27072110	.10013416 −.10611506	1.03717346 .98779730						

0	5	−.46316724 −.90154196	−.08553509 −.38806282	1.54870633 1.27349478						
0	6	−1.49153218 −1.75016272	2.49153218 1.75016272							
1	1		.23666261 .51837013	.13147170 −.11152847	.13186568 −.03605506	.13 88668 .03 05506	.13147170 .11152847	.23666261 .51837013		
1	2		.17163950 −.66079246	.12217411 −.13189331	.13381130 −.03179351	.14 17854 .06 02434	.42819655 .76145495			
1	3		.04303423 −.88940019	.10609003 −.16049536	.14058990 −.01973924	.71923584 1.06463479				
1	4		−.25191114 −1.33367137	.07414879 −.20659974	.17776234 1.54227111					
1	5		−1.24622969 −2.63572419	2.24622969 2.63572419						
2	2			.35694901 −1.03574984	.14305099 −.06736528	.14305099 .06736528	.35694901 1.03574984			
2	3			.17417957 −1.66608726	.14291598 −.06776700	.65390445 1.65384432				
2	4			−.47614906 −3.12199414	1.47614906 3.12199414					
3	3				.50000000 −3.27837896	.50000000 3.27837896				
0	0	.1111111 −.22373410	.1111111 −.12226850	.1111111 −.07509922	.1111111 −.03596908	.1111111 .00000000	.1111111 .03596908	.1111111 .07509922	.1111111 .12326850	.1111111 .22373410
0	1	.09148453 −.26325227	.10175129 −.14211324	.10665020 −.08407936	.11060016 −.03090545	.1 419180 .06618347	.11765844 .04915546	.12118875 .09539434	.12326850 .23647484 .37570705	
0	2	.06023974 −.31280447	.08756442 −.16465459	.10056358 −.09375143	.11099937 −.03636313	.1 038626 .0 604052	.12937290 .06776492	.30087375 .52385821		

9

75

Table 4.1 Continued

N	c_1	c_2	$x_{1:N}$	$x_{2:N}$	$x_{3:N}$	$x_{4:N}$	$x_{5:N}$	$x_{6:N}$	$x_{7:N}$	$x_{8:N}$	$x_{9:N}$	$x_{10:N}$
9	0	3	.01040388 −.37968567	.06597348 −.19359128	.09230224 −.10482346	.11331998 −.03325300	.13204377 .03166420	.58595666 .67968024				
	0	4	−.07313367 −.47658635	.03155502 −.23351551	.08087391 −.11807995	.11994592 −.02556713	.84075879 .85374893					
	0	5	−.22717960 −.63301232	−.02842070 −.29411786	.06443680 −.13477100	1.19116347 1.06220121						
	0	6	−.56642662 −.93553052	−.15208218 −.40469106	1.71850879 1.34022157							
	0	7	−1.68675768 −1.80924841	2.68675768 1.80924841								
	1	1		.20970437 −.45274773	.11586689 −.11065463	.11624355 −.05323076	.11633036 .00000000	.11624355 .05323070	.11586689 .11065463	.20970437 .45274773		
	1	2		.16260603 −.55443214	.10736149 −.12906081	.11483307 −.05627598	.12137684 .01089528	.12749454 .07752140	.36632802 .65135225			
	1	3		.07989097 −.70150427	.09363515 −.15346705	.11399182 −.05777177	.13204087 .02991545	.58039119 .88279764				
	1	4		−.07681209 −.93990137	.06990479 −.18956185	.11531222 −.05576340	.89162509 1.18522662					
	1	5		−.42723635 −1.40567629	.02184604 −.25344578	1.40539031 1.65912206						
	1	6		−1.58736731 −2.77525943	2.58736731 2.77525943							
	2	2			.31343459 .83170114	.12427928 −.08848435	.12457225 .00000000	.12427928 .08848435	.31343459 .83170114			
	2	3			.20395900 −1.12219549	.11906520 −.10231994	.13296497 .02227016	.54401083 1.20224528					

9	2	4	.10000000 -.20438349	.10000000 -.11719379	-.05271887 -1.69944478							.10000000 .20438349
	2	5	.08432557 -.23641943	.09206394 -.13341377	-.09568489 -.08507858						.10000000 .11719379	.21013724 .34229613
	3	3	.06045239 -.27530680	.08044709 -.15233660	-.08978431 -.09469015							
	3	4	.02442821 -.32524451	.06355637 -.17575100	-.08178041 -.10578536							
10	0	0	-.03158434 -.39304766	.03829283 -.20633250	-.07072145 -.11917222							
	0	1	-.12396225 -.49191253	-.00163103 -.24905990	-.05485447 -.13615341							
	0	2	-.29230821 -.65203409	-.07092971 -.31497342	-.03053696 -.15928304							
	0	3	-.65962405 -.90246073	-.21376659 -.43568764	1.87339063 1.39814837							
	0	4	-1.86335148 -1.86082625	2.80335148 1.86082625								
	0	5										
	0	6										
	0	7										
	0	8										
	1	1		.18335246 -.40337309	.10363971 -.10738113	.10405602 .02013735	.10405602 .02013735	.10395180 .06163683	.10363971 .10738113	.18335246 .40337309		
	1	2		.15245229 -.48025640	.09611128 -.12350392	.10126825 -.06738392	.10126825 -.06738392	.11376585 .08265445	.32087559 .57261056			

Table 4.1 Continued

N	c_1	c_2	$x_{1:N}$	$x_{2:N}$	$x_{3:N}$	$x_{4:N}$	$x_{5:N}$	$x_{6:N}$	$x_{7:N}$	$x_{8:N}$	$x_{9:N}$	$x_{10:N}$
10	1	3		.0943044 −.5815474	.08164767 −.14396103	.09790057 −.07337756	.10054691 −.00974899	.12039990 .05136065	.49326580 .75988167			
	1	4		−.00426039 −.73588101	.06648054 −.17194771	.00383130 −.07066018	.11787039 .00307342	.72607816 .98441548				
	1	5		−.18661330 −.98311476	.03511935 −.21446716	.08919507 −.08594598	1.06220888 1.28352789					
	1	6		−.58772619 −1.46776140	−.02885426 −.29176358	1.61658046 1.75952498						
	1	7		−1.89997078 −2.89604971	2.89997078 2.89604971							
	2	2			.27977552 −.70208441	.10994078 −.09470422	.11028371 −.03101584	.11028371 −.03101584	.10994078 .09470422	.27977552 .70208441		
	2	3			.20496319 −.88982266	.10382533 −.11005067	.11220127 −.02620385	.11982080 .05494874	.45018942 .97112842			
	2	4			.06055627 −1.19522527	.09352928 −.13182552	.11776906 −.01442865	.72814539 1.34147944				
	2	5			−.26483993 −1.79471038	.07347216 −.16877724	1.19136777 1.96348761					
	2	6			−1.34060718 −3.56767731	2.34060718 3.56767731						
	3	3				.38067840 −1.28320587	.11932160 −.05586351	.11932160 .05586351	.38067840 1.28320587			
	3	4				.18711959 −1.97905791	.11984110 −.05532554	.69303931 2.03438315				
	3	5				−.48466706 −3.95105520	1.48466706 3.95105520					
	4	4					.50000000 −4.07605089	.50000000 4.07605089				

Source: Adapted from Sarhan and Greenberg (1956), Table II, pp. 434–440, with permission of the Institute of Mathematical Statistics

Linear Estimators 79

$$\sigma^* = (-0.32524)(1.6128) + (-0.17575)(1.6435) + (-0.10579)(1.6628)$$
$$+ (-0.05018)(1.7324) + (0.00056)(1.7404) + (0.04686)(1.7634)$$
$$+ (0.61066)(1.7782) = 0.091.$$

From Table 4.2, we read

$$V(\mu^*) = 0.1167\sigma^2, \qquad V(\sigma^*) = 0.0989\sigma^2, \qquad \text{Cov}(\mu^*, \sigma^*) = 0.260\sigma^2.$$

Estimates of the standard errors thus are

$$\sigma_{\mu^*} = 0.091\sqrt{0.1167} = 0.031 \qquad \text{and} \qquad \sigma_{\sigma^*} = 0.091\sqrt{0.0989} = 0.029.$$

Gupta (1952) calculated the alternative estimates for this example to be

$$\mu^{**} = 1.748, \qquad \sigma^{**} = 0.094,$$

and the standard errors of these estimates to be

$$\sigma_{\mu^{**}} = 0.033 \qquad \text{and} \qquad \sigma_{\sigma^{**}} = 0.031.$$

Maximum likelihood estimates as calculated in Chapter 2 (Example 2.7.1) are

$$\hat{\mu} = 1.7424 \qquad \text{and} \qquad \hat{\sigma} = 0.07943,$$

with standard errors

$$\sigma_{\hat{\mu}} = 0.0268 \qquad \text{and} \qquad \sigma_{\hat{\sigma}} = 0.0228.$$

Readers will note that differences between the three sets of estimates from this example are small. Standard errors of the maximum likelihood estimates are slightly less than those of the two sets of linear estimates, but the linear estimates are unbiased, whereas the maximum likelihood estimates do not enjoy this property.

Example 4.2. This example was given by Sarhan and Greenberg (1962). Students were learning to measure concentrations of strontium-90 in samples of milk. Measurements were made on a prepared test mixture that contained 9.22 micromicrocuries of strontium-90 per liter. The skill of the inexperienced observers was suspect for observations in the tails of the distribution. For this reason it was decided that from a total of 10 observations, the two smallest and the three largest observations would be censored. The remaining sample was thus doubly censored Type II with observations __, __, 8.2, 8.4, 9.1, 9.8, 9.9, __, __, __. Best linear estimates calculated with coefficients from Table 4.1 are

$$\mu^* = (0.20496)(8.2) + (0.10383)(8.4) + (0.11220)(9.1) + (0.11982)(9.8)$$
$$+ (0.45919)(9.9) = 9.29,$$

$$\sigma^* = (-0.88982)(8.2) + (-0.11005)(8.4) + (-0.02620)(9.1)$$
$$+ (0.05495)(9.8) + (0.97113)(9.9) = 1.69.$$

Table 4.2 Variances and Covariances of Linear Estimates of the Mean and Standard Deviation for Censored Samples from the Normal Distribution

N	c_1		0	1	2	3	4	5	6	7	8
2	0	$V(\mu^*)$	.50000000								
		$V(\sigma^*)$	.57079633								
		$\text{Cov}(\mu^*, \sigma^*)$	0								
3	0	$V(\mu^*)$	.33333333	.44867111							
		$V(\sigma^*)$	.27548197	.63782627							
		$\text{Cov}(\mu^*, \sigma^*)$	0	.20443088							
4	0	$V(\mu^*)$	.25000000	.28701097	.51299280						
		$V(\sigma^*)$	.18005021	.30207500	.67303155						
		$\text{Cov}(\mu^*, \sigma^*)$	0	.06720309	.35673623						
	1	$V(\mu^*)$		.29819962							
		$V(\sigma^*)$		.70572111							
		$\text{Cov}(\mu^*, \sigma^*)$		0							
5	0	$V(\mu^*)$	.20000000	.21772341	.28393206	.61122918					
		$V(\sigma^*)$	.13332124	.19476301	.31808583	.69570833					
		$\text{Cov}(\mu^*, \sigma^*)$	0	.03299936	.12335995	.47492042					
	1	$V(\mu^*)$		.22579422	.28683366						
		$V(\sigma^*)$		.32968861	.74060939						
		$\text{Cov}(\mu^*, \sigma^*)$		0	.15837417						
6	0	$V(\mu^*)$	.16666667	.17688040	.20682717	.29985334	.71864005				
		$V(\sigma^*)$	.10570264	.14277434	.20436158	.32919818	.71192848				
		$\text{Cov}(\mu^*, \sigma^*)$	0	.01945869	.06240445	.17016841	.57052121				
	1	$V(\mu^*)$		.18249856	.20698549	.32960016					
		$V(\sigma^*)$		.21017023	.34598687	.76280252					
		$\text{Cov}(\mu^*, \sigma^*)$		0	.05766917	.28373032					

n	r							
6	2	$V(\mu^*)$			.21474267	.20713842	.32479943	.82636223
		$V(\sigma^*)$			.77471679	.21136168	.33752616	.72430317
		$\text{Cov}(\mu^*, \sigma^*)$			0	.08805478	.20989335	.65033968
7	0	$V(\mu^*)$	.14285714	.14942352	.16600214	.20946061	.39543079	
		$V(\sigma^*)$	.08749856	.11233100	.14927902	.35717357	.77852463	
		$\text{Cov}(\mu^*, \sigma^*)$	0	.01276945	.03751914	.10645593	.38638223	
	1	$V(\mu^*)$		.15346296	.16680538	.21044686		
		$V(\sigma^*)$		.15269771	.22005773	.36219907		
		$\text{Cov}(\mu^*, \sigma^*)$		0	.02997909	0		
	2	$V(\mu^*)$			.17312828			
		$V(\sigma^*)$			.79619765			
		$\text{Cov}(\mu^*, \sigma^*)$			.12726434			
8	0	$V(\mu^*)$	.12500000	.12954025	.13988098	.16225105	.21375841	.35409951
		$V(\sigma^*)$	.07461121	.09242321	.11706951	.15419334	.21678723	.34408411
		$\text{Cov}(\mu^*, \sigma^*)$	0	.00899283	.02405718	.05377493	.11055552	.21421558
	1	$V(\mu^*)$		.13255044	.14091823	.16234808	.22331839	.47112520
		$V(\sigma^*)$		.11928888	.15943388	.22720134	.36548626	.79012118
		$\text{Cov}(\mu^*, \sigma^*)$		0	.01832825	.05643660	.14825957	.47276076
	2	$V(\mu^*)$			.14567896	.16269076	.23922424	.93099884
		$V(\sigma^*)$			.22999555	.37323003	.81130052	.73416689
		$\text{Cov}(\mu^*, \sigma^*)$			0	.04936270	.23240354	.71959780
	3	$V(\mu^*)$				.16818086		
		$V(\sigma^*)$				.81706103		
		$\text{Cov}(\mu^*, \sigma^*)$				0		

Table 4.2 *Continued*

N	c_1		0	1	2	3	4	5	6	7	8
							c_2				
9	0	$V(\mu^*)$	.11111111	.11441700	.12139976	.13515042	.16291237	.22407238	.38537601	1.03120529	
		$V(\sigma^*)$	.06501502	.07812444	.09604682	.12074575	.58100000	.22116450	.31943152	.74228259	
		$Cov(\mu^*,\sigma^*)$	0	.00665033	.01775128	.03618024	.06838314	.13048810	.27432806	.77806418	
	1	$V(\mu^*)$		.11673259	.12244352	.13518326	.16466906	.24349912	.55048580		
		$V(\sigma^*)$		.09758705	.12420684	.16148335	.23270051	.37199378	.79983467		
		$Cov(\mu^*,\sigma^*)$		0	.01232979	.03498177	.07983082	.18461869	.54702931		
	2	$V(\mu^*)$			.12605965	.13612535	.16566283	.28513252			
		$V(\sigma^*)$			.16624730	.23712095	.38138044	.82268191			
		$Cov(\mu^*,\sigma^*)$			0	.02670942	.06198622	.32159927			
	3	$V(\mu^*)$				.14097979	.16610128				
		$V(\sigma^*)$				.38407790	.83165694				
		$Cov(\mu^*,\sigma^*)$				0	.10603703				
10	0	$V(\mu^*)$	.10000000	.10250531	.10749284	.11666310	.13559472	.16441008	.23658560	.41735056	1.12690295
		$V(\sigma^*)$	.05759553	.06806087	.08129462	.09891696	.12572658	.16131247	.22479946	.35390665	.74912245
		$Cov(\mu^*,\sigma^*)$	0	.00512045	.01324471	.02595717	.04645253	.08157225	.14831978	.30108776	.83064069
	1	$V(\mu^*)$		.10433167	.10845084	.11682744	.13338367	.17122372	.26722758	.63041997	
		$V(\sigma^*)$		.08241682	.10135027	.12799713	.16847458	.23711263	.37726676	.80752873	
		$Cov(\mu^*,\sigma^*)$		0	.00884085	.02377304	.05004837	.10067393	.21667108	.61197840	
	2	$V(\mu^*)$			.11126628	.11795177	.13388616	.17665047	.34008123		
		$V(\sigma^*)$			.12920079	.17132071	.24257700	.38772575	.83163934		
		$Cov(\mu^*,\sigma^*)$			0	.01678448	.05047738	.12926308	.39861255		
	3	$V(\mu^*)$				.12182093	.13426086	.18655569			
		$V(\sigma^*)$				.24418850	.39205377	.84259481			
		$Cov(\mu^*,\sigma^*)$				0	.04204412	.19638665			
	4	$V(\mu^*)$					.13832514				
		$V(\sigma^*)$					.84582536				
		$Cov(\mu^*,\sigma^*)$					0				

Source: Adapted from Sarhan and Greenberg (1956), Table III, pp. 442–444, with permission of the Institute of Mathematical Statistics.

Table 4.3 Variances and Relative Efficiencies of Gupta's Alternative Linear Estimates of the Mean and Standard Deviation for Censored Samples from the Normal Distribution

N	c_1		0	1	2	3	4	5	6	7	8
							c_2				
10	0	μ^* V.	.10000000	.10308928	.10961456	.12148953	.14281760	.18261677	.26336721	.45614307	1.12690295
		R.E.	100.00	99.43	98.06	96.03	93.51	91.13	89.83	91.50	100.00
		σ^* V.	.05766121	.07022636	.08642325	.10748774	.13638670	.17888915	.24796349	.38065951	.74912245
		R.E.	99.87	96.92	94.07	92.03	90.72	90.17	90.66	92.97	100.00
	1	μ^* V.		.10534509	.11034725	.12008840	.13904835	.17855378	.27531177	.63041997	
		R.E.		99.04	98.29	97.28	96.29	95.89	97.06	100.00	
		σ^* V.		.08489780	.10545474	.13383112	.17586018	.24533968	.38411228	.80752873	
		R.E.		97.08	96.11	95.64	95.80	96.65	98.22	100.00	
	2	μ^* V.			.11330806	.12042277	.13682291	.17922164	.34008123		
		R.E.			98.20	97.95	97.85	98.57	100.00		
		σ^* V.			.13414020	.17683334	.24714875	.38944383	.83163934		
		R.E.			96.32	96.88	98.02	99.56	100.00		
	3	μ^* V.				.12376594	.13558005	.18655569			
		R.E.				98.43	99.03	100.00			
		σ^* V.				.24837948	.39221654	.84259430			
		R.E.				96.16	99.96	100.00			
	4	μ^* V.					.13832641				
		R.E.					100.00				
		σ^* V.					.84582653				
		R.E.					100.00				

Source: Adapted from Sarhan and Greenberg (1956), Table V, p. 448, with permission of the Institute of Mathematical Statistics.

From Table 4.2, we read

$$V(\mu^*) = 0.13613\,\sigma^2, \quad V(\sigma^*) = 0.23712\,\sigma^2, \quad \text{Cov}(\mu^*, \sigma^*) = 0.02671\,\sigma^2.$$

Standard errors follow as

$$\sigma_{\mu^*} = 1.69\sqrt{0.13613} = 0.624 \quad \text{and} \quad \sigma_{\sigma^*} = 1.69\sqrt{0.23712} = 0.823.$$

5
Truncated and Censored Samples from the Weibull Distribution

5.1 INTRODUCTION

To a considerable extent, this chapter is based on previous results by the author (Cohen, 1965, 1973, 1975). The Weibull distribution has in recent years assumed a position of prominence in the field of reliability and life testing where samples are often either truncated or censored. From a computational point of view, this distribution is particularly appealing, since its cumulative distribution function (cdf) can be expressed explicitly as a simple function of the random variable. Various topics associated with this distribution have been considered by numerous writers. Among these are Dubey (1963), Esary and Proschan (1963), Jaech (1964), Kao (1958, 1959), Lehman (1963), Leone et al. (1960), Lloyd and Lipow (1962), Menon (1963), Procassini and Romano (1961), Proschan (1963), Nelson (1968, 1969, 1972, 1982), Bain (1978), Cohen and Whitten (1982, 1988), Harter and Moore (1965, 1967), Mann (1968), Mann and Fertig (1975a, b), Lemon (1975), Wingo (1973), Ringer and Sprinkle (1972), Rockette et al. (1974), Koniger (1981), Zanakis (1977, 1979a,b), Zanakis and Mann (1981), Wycoff et al. (1980), and others.

5.2 DISTRIBUTION CHARACTERISTICS

The pdf of the two-parameter Weibull distribution with origin at zero is

$$f(x; \beta, \delta) = \frac{\delta}{\beta^\delta} x^{\delta-1} \exp - \left(\frac{x}{\beta}\right)^\delta, \qquad 0 \le x < \infty, \qquad (5.2.1)$$
$$= 0 \quad \text{elsewhere.}$$

and the cdf is

$$F(x; \beta, \delta) = 1 - \exp\left[-\left(\frac{x}{\beta}\right)^\delta\right], \qquad (5.2.2)$$

where β and δ are the scale and shape parameters, respectively.

The expected value (mean), variance, and other characteristics of this distribution are

$$E(X) = \beta\Gamma_1, \qquad V(X) = \beta^2[\Gamma_2 - \Gamma_1^2],$$

$$\text{Me}(X) = \beta(\ln 2)^{1/\delta}, \qquad \alpha_3(X) = \frac{\Gamma_3 - 3\Gamma_2\Gamma_1 + 2\Gamma_1^3}{[\Gamma_2 - \Gamma_1^2]^{3/2}}, \qquad (5.2.3)$$

$$\text{Mo}(X) = \beta\left[\frac{\delta - 1}{\delta}\right]^{1/\delta}, \qquad \alpha_4(X) = \frac{\Gamma_4 - 4\Gamma_3\Gamma_1 + 6\Gamma_2\Gamma_1^2 - 3\Gamma_1^4}{[\Gamma_2 - \Gamma_1^2]^2},$$

where

$$\Gamma_k = \Gamma\left(1 + \frac{k}{\delta}\right), \qquad (5.2.4)$$

and $\Gamma(\ )$ is the gamma function

$$\Gamma(z) = \int_0^\infty t^{z-1} e^{-t}\, dt. \qquad (5.2.5)$$

The coefficient of variation is

$$\text{CV}(X) = \frac{\sqrt{V(X)}}{E(X)} = \frac{\sqrt{\Gamma_2 - \Gamma_1^2}}{\Gamma_1}. \qquad (5.2.6)$$

When $\delta > 1$, this distribution is bell-shaped with a discernible mode as given by (5.2.3), but when $\delta \leq 1$, it is reverse J-shaped. When $\delta = 1$, it becomes the exponential distribution. The Weibull distribution is usually perceived to be positively skewed, but for $\delta > \delta_0 = 3.6023494257197$ (cf. Cohen, 1973), it becomes negatively skewed. For $\delta = \delta_0$, it is almost normal with $\alpha_3 = 0$ and $\alpha_4 = 2.72$.

Our primary concern in this volume is with parameter estimation when samples are truncated and/or censored. Estimation in complete unrestricted samples has been quite fully considered by the various writers whose works are referenced in Section 5.1.

5.3 SINGLY CENSORED SAMPLES

In a typical life test, N specimens are placed on test, and the elapsed time is recorded as each failure occurs. Finally, at some predetermined, fixed time T,

Samples from the Weibull Distribution

or after some predetermined, fixed number of sample specimens have failed, the test is terminated. In both cases, the data collected consist of n fully measured observations $\{x_i\}$, $i = 1, 2, \ldots, n$ plus the information that $c = N - n$ specimens survived beyond the time of censoring, T. For Type I censoring, T is fixed and n is the observed value of a random variable. For Type II censoring, n is fixed and T is the observed value of a random variable, that is, the nth-order statistic in a random sample of size N.

For both Type I and Type II censoring, the likelihood function of a random sample from the two-parameter Weibull distribution may be written as

$$L = K \left[\prod_{i=1}^{n} \left(\frac{\delta}{\beta^{\delta}}\right) x_i^{\delta-1} \right] \left[\exp - \sum_{i=1}^{n} \left(\frac{x_i}{\beta}\right)^{\delta} \right] [1 - F(T)]^c, \quad (5.3.1)$$

where K is an ordering constant that does not depend on the parameters.

We replace the scale parameter β with θ, where

$$\theta = \beta^{\delta}, \quad (5.3.2)$$

and take logarithms of (5.3.1) to obtain

$$\ln L = \text{const} + n \ln \delta - n \ln \theta - \frac{1}{\theta} \sum_{i=1}^{n} x_i^{\delta} + (\delta - 1) \sum_{i=1}^{n} \ln x_i \quad (5.3.3)$$
$$+ c \ln[1 - F(T)].$$

Since $F(x) = 1 - \exp-(x^{\delta}/\theta)$, (5.3.3) may now be written as

$$\ln L = n \ln \delta - n \ln \theta + (\delta - 1) \sum_{i=1}^{n} \ln x_i \quad (5.3.4)$$
$$- \frac{1}{\theta} \sum_{i=1}^{n} x_i^{\delta} - \frac{c}{\theta} T^{\delta} + \text{const}.$$

When $\delta > 1$, maximum likelihood estimating equations for parameters δ and θ are

$$\frac{\partial \ln L}{\partial \delta} = \frac{n}{\delta} + \sum_{i=1}^{n} \ln x_i - \frac{1}{\theta} \sum_{i=1}^{n} x_i^{\delta} \ln x_i - \frac{cT^{\delta}}{\theta} \ln T = 0,$$
$$\frac{\partial \ln L}{\partial \theta} = -\frac{n}{\theta} + \frac{1}{\theta^2} \sum_{i=1}^{n} x_i^{\delta} + \frac{cT^{\delta}}{\theta^2} = 0. \quad (5.3.5)$$

The introduction of θ as a scale parameter to replace β results in estimating equations that are much easier to solve. When θ is eliminated between the two equations of (5.3.5), we obtain the following equation in which $\hat{\delta}$ is the only unknown.

$$\left[\frac{\sum_{i=1}^{n} x_i^{\hat{\delta}} \ln x_i + cT^{\hat{\delta}} \ln T}{\sum_{i=1}^{n} x_i^{\hat{\delta}} + cT^{\hat{\delta}}} - \frac{1}{\hat{\delta}}\right] = \frac{1}{n}\sum_{i=1}^{n} \ln x_i. \qquad (5.3.6)$$

The maximum likelihood estimate $\hat{\delta}$ follows as the solution of this equation. Although Newton's method or other standard iterative procedure might be employed to obtain a solution, in many applications a simple trial-and-error approach will be preferred. Once two values δ_i and δ_j within a sufficiently narrow interval have been found such that $\delta_i \leq \hat{\delta} \leq \delta_j$, linear interpolation will yield the final estimate.

With $\hat{\delta}$ thus determined, $\hat{\theta}$ follows from the second equation of (5.3.5) as

$$\hat{\theta} = \frac{\sum_{i=1}^{n} x_i^{\hat{\delta}} + cT^{\hat{\delta}}}{n}, \qquad (5.3.7)$$

and

$$\hat{\beta} = \hat{\theta}^{1/\hat{\delta}}. \qquad (5.3.8)$$

As a computational aid which might be helpful in choosing a first approximation to $\hat{\delta}$ for use in solving (5.3.6), graphs of the Weibull coefficient of variation and its square as functions of δ have been reproduced from Cohen (1965) and are given here as Figure 5.1. In order to use these graphs, it is necessary to estimate the coefficient of variation from the sample data. For a complete (uncensored) sample, CV = $s/\bar{x}$. For a singly censored sample in which the proportion of censored observations is small, a satisfactory approximation can be calculated as

$$CV = \frac{\sqrt{\sum_{i=1}^{n}(x_i - \bar{x}_N)^2 + c(T - \bar{x}_N)^2}}{\bar{x}_N}, \qquad (5.3.9)$$

where

$$\bar{x}_N = \frac{1}{N}\left[\sum_{i=1}^{n} x_i + cT\right]. \qquad (5.3.10)$$

5.4 PROGRESSIVELY CENSORED SAMPLES

The loglikelihood function of a progressively censored sample from a two-parameter Weibull distribution with pdf (5.2.1) can be written as

Samples from the Weibull Distribution

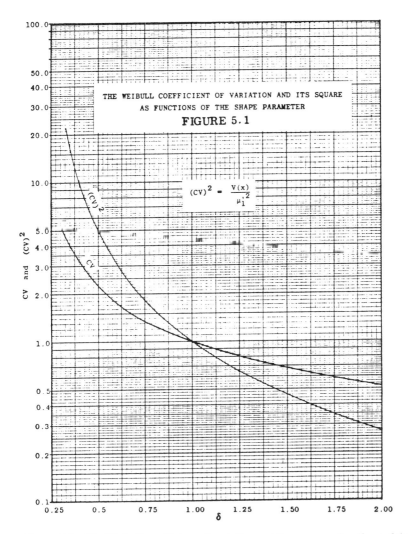

Figure 5.1 The Weibull coefficient of variation and its square as functions of the shape parameter. Reprinted from Cohen (1965a), Fig. 1, p. 586, with permission of the Technometrics Management Committee.

$$\operatorname{Ln} L = n \ln \delta - n \ln \theta - \frac{1}{\theta} \sum_{i=1}^{n} x_i^{\delta} + (\delta - 1) \sum_{i=1}^{n} \ln x_i \qquad (5.4.1)$$
$$+ \sum_{j=1}^{k} c_j \ln[1 - F(T_j)] + \text{const.},$$

where c_j is the number of observations censored at T_j.

For the bell-shaped distribution with $\delta > 1$, the maximum likelihood estimating equations for parameters θ and δ are

$$\frac{\partial \ln L}{\partial \delta} = \frac{n}{\delta} + \sum_{i=1}^{n} \ln x_i - \frac{1}{\theta} \Sigma^* x_i^\delta \ln x_i = 0,$$

$$\frac{\partial \ln L}{\partial \theta} = -\frac{n}{\theta} + \frac{1}{\theta^2} \Sigma^* x_i^\delta = 0,$$

(5.4.2)

where Σ^* signifies summation over the entire sample with the observations censored at T_j assigned the value $x_j = T_j$. Thus,

$$\Sigma^* x_i^\delta \ln x_i = \sum_{i=1}^{n} x_i^\delta \ln x_i + \sum_{j=1}^{k} c_j T_j^\delta \ln T_j,$$

$$\Sigma^* x_i^\delta = \sum_{i=1}^{n} x_i^\delta + \sum_{j=1}^{k} c_j T_j^\delta.$$

(5.4.3)

When θ is eliminated between the two equations of (5.4.2), the algebraic form of the resulting equation is analogous to that of (5.3.6), and it can be solved as described for that equation.

5.5 THE THREE-PARAMETER DISTRIBUTION

When the origin is at an unknown point γ such that $x > \gamma$, then γ becomes an additional (threshold) parameter that must be estimated from sample data. Although in many applications the origin is known to be at zero, there are numerous situations in which it is unknown. In this more general three-parameter case, the random variable X is replaced by $(X - \gamma)$. In equations (5.2.3) we now have

$$E(X) = \gamma + \beta \Gamma_1,$$

$$\text{Me}(X) = \gamma + \beta (\ln 2)^{1/\delta},$$

$$\text{Mo}(X) = \gamma + \beta \left(\frac{\delta - 1}{\delta}\right)^{1/\delta},$$

(5.5.1)

but $V(X)$, $\alpha_3(X)$, and $\alpha_4(X)$ are invariant under changes of the origin.

With the addition of the threshold parameter, the loglikelihood function of a singly right censored sample as given by (5.3.3) now becomes

$$\ln L = n \ln \delta - n \ln \theta - \frac{1}{\theta} \sum_{i=1}^{n} (x_i - \gamma)^\delta$$

$$+ (\delta - 1) \sum_{i=1}^{n} \ln(x_i - \gamma) + \frac{c(T - \gamma)^\delta}{\theta} + \text{const.}$$

(5.5.2)

Samples from the Weibull Distribution

and that of a progressively censored sample is

$$\ln L = n \ln \delta - n \ln \theta - \frac{1}{\theta} \Sigma^* (x_i - \gamma)^\delta \qquad (5.5.3)$$

$$+ (\delta - 1) \sum_{i=1}^{n} \ln(x_i - \gamma) + \text{const.}$$

where $\gamma \leq x < \infty$, and where

$$\Sigma^*(x_i - \gamma)^\delta = \sum_{i=1}^{n} (x_i - \gamma)^\delta + \sum_{j=1}^{n} c_j(T_j - \gamma)^\delta. \qquad (5.5.4)$$

Maximum likelihood estimating equations that are valid when $\delta > 1$ are

$$\frac{\partial \ln L}{\partial \delta} = \frac{n}{\delta} + \sum_{i=1}^{n} \ln(x_i - \gamma) - \frac{1}{\theta} \Sigma^*(x_i - \gamma)^\delta \ln(x_i - \gamma) = 0,$$

$$\frac{\partial \ln L}{\partial \theta} = -\frac{n}{\theta} + \frac{1}{\theta^2} \Sigma^* (x_i - \gamma)^\delta = 0, \qquad (5.5.5)$$

$$\frac{\partial \ln L}{\partial \gamma} = \frac{\delta}{\theta} \Sigma^* (x_i - \gamma)^{\delta-1} - (\delta - 1) \sum_{i=1}^{n} (x_i - \gamma)^{-1} = 0.$$

ML estimates, $\hat{\delta}$, $\hat{\theta}$, and $\hat{\gamma}$ follow as the simultaneous solution of these three equations. Although solutions exist provided $\delta > 1$, computational difficulties are likely to be encountered unless δ is somewhat greater than this boundary value. Any standard iterative procedure might be employed in the solution, but the trial-and-error procedure is likely to be preferred in most applications.

When θ is eliminated between the first two equations of (5.5.5), we obtain

$$\frac{\Sigma^*(x_i - \gamma)^\delta \ln(x_i - \gamma)}{\Sigma^*(x_i - \gamma)^\delta} - \frac{1}{\delta} = \frac{1}{n} \sum_{i=1}^{n} \ln(x_i - \gamma), \qquad (5.5.6)$$

which differs from equation (5.3.6) only in the presence of the threshold parameter γ. For any given value of γ, equation (5.5.6) then becomes a function of δ alone.

We must now solve equation (5.5.6) and the last equation of (5.5.5) for estimates $\hat{\delta}$ and $\hat{\gamma}$. Accordingly, we choose a first approximation $\gamma_1 < x_{1:N}$. We substitute this value into (5.5.6), solve for δ_1, and calculate θ_1 from the second equation of (5.5.5). We then test these approximations by substitution into the last equation of (5.5.5). If this result is zero, then $\gamma_1 = \hat{\gamma}$, $\delta_1 = \hat{\delta}$, and $\theta_1 = \hat{\theta}$, and no further calculations are necessary. Otherwise, we select a second approximation and proceed through a second cycle of calculations. We continue until we find two approximations γ_i and γ_j such that the absolute value $|\gamma_i - \gamma_j|$ is sufficiently small and such that

$$\left. \frac{\partial \ln L}{\partial \gamma} \right|_i \geq 0 \geq \left. \frac{\partial \ln L}{\partial \gamma} \right|_j.$$

Final estimates are obtained by linear interpolation between γ_i and γ_j, and between δ_i and δ_j. We calculate $\hat{\theta}$ for the second equation of (5.5.5) as

$$\hat{\theta} = \frac{1}{n}\Sigma^*(x_i - \hat{\gamma})^{\hat{\delta}}, \tag{5.5.7}$$

and

$$\hat{\beta} = \hat{\theta}^{1/\hat{\delta}}. \tag{5.5.8}$$

In selecting approximations to γ, we recall that

$$E(X_{1:N}) = \gamma + \left(\frac{\theta}{N}\right)^{1/\delta} \Gamma_1, \tag{5.5.9}$$

and of course

$$\gamma < x_{1:N}. \tag{5.5.10}$$

5.6 MODIFIED MAXIMUM LIKELIHOOD ESTIMATORS (MMLE)

Maximum likelihood estimators of parameters of the Weibull distribution are subject to regularity restrictions that limit their usefulness. They are valid only if $\delta > 1$, and for the three-parameter distribution the asymptotic variance–covariance matrix is not valid unless $\delta > 2$ ($\alpha_3 < 0.63$). In an effort to circumvent these disadvantages of the MLE, Cohen Whitten, and Ding (1984) proposed modified moment estimators which employ the first-order statistic as an estimator of the threshold parameter in the three-parameter Weibull when samples are complete. When samples are censored, it seems more appropriate to employ a similar modification of the maximum likelihood estimators, and that is the topic for consideration in this section. The estimating equations are $\partial \ln L/\partial \delta = 0$, $\partial \ln L/\partial \theta = 0$, and $E(X_{1:N}) = x_{1:N}$, and in their expanded form, these equations become

$$\frac{\partial \ln L}{\partial \delta} = \frac{n}{\delta} + \sum_{i=1}^{n} \ln(x_i - \gamma) - \frac{1}{\theta}\Sigma^*(x_i - \gamma)^{\delta} \ln(x_i - \gamma) = 0,$$

$$\frac{\partial \ln L}{\partial \theta} = -\frac{n}{\theta} + \frac{1}{\theta^2}\Sigma^*(x_i - \gamma)^{\delta} = 0, \tag{5.6.1}$$

$$E(X_{1:N}) = \gamma + \left(\frac{\theta}{N}\right)^{1/\delta} \Gamma_1 - x_{1:N} = 0.$$

Samples from the Weibull Distribution

To solve these equations simultaneously for the MMLE, we proceed as for the MLE and with the first approximation, γ_1, substituted into equation (5.5.6), we solve for δ_1. We then "test" by substituting these approximations into the third equation of (5.6.1). If this equation is satisfied, then no further calculations are required. Otherwise, we continue with additional approximations to γ until, as in the case of the MLE, we find two approximations γ_i and γ_j in a sufficiently narrow interval such that $\gamma_i \lesssim \hat{\gamma} \lesssim \gamma_j$. Final estimates are then obtained by linear interpolation, as in the case of the MLE.

5.7 ERRORS OF ESTIMATES

The asymptotic variance–covariance matrix of the MLE is essentially the same as that for complete samples. For censored samples, however, it is expedient to replace expected values of the second partials with their sample evaluations. In progressively censored samples from the three-parameter Weibull distribution, second partials of the loglikelihood function are

$$\frac{\partial^2 \ln L}{\partial \delta^2} = -\frac{n}{\delta^2} - \frac{1}{\theta} \Sigma^*(x_i - \gamma)^\delta [\ln(x_i - \gamma)]^2,$$

$$\frac{\partial^2 \ln L}{\partial \theta^2} = \frac{n}{\theta^2} - \frac{2}{\theta^3} \Sigma^*(x_i - \gamma)^\delta,$$

$$\frac{\partial^2 \ln L}{\partial \gamma^2} = -(\delta - 1) \sum_{i=1}^{n} (x_i - \gamma)^{-2}$$
$$- \frac{\delta(\delta - 1)}{\theta} \Sigma^*(x_i - \gamma)^{\delta - 2},$$

$$\frac{\partial^2 \ln L}{\partial \delta\, \partial \theta} = \frac{1}{\theta_2} \Sigma^*(x_i - \gamma)^\delta \ln(x_i - \gamma),$$

$$\frac{\partial^2 \ln L}{\partial \delta\, \partial \gamma} = -\sum_{i=1}^{n} (x_i - \gamma)^{-1} + \frac{1}{\theta} \Sigma^*(x_i - \gamma)^{\delta - 1}$$
$$+ \frac{\delta}{\theta} \Sigma^*(x_i - \gamma)^{\delta - 1} \ln(x_i - \gamma),$$

$$\frac{\partial^2 \ln L}{\partial \theta\, \partial \gamma} = -\frac{\delta}{\theta^2} \Sigma^*(x_i - \gamma)^{\delta - 1}.$$

(5.7.1)

These same results are applicable in the case of singly censored samples when we set $k = 1$.

The approximate variance–covariance matrix of the MLE from censored samples accordingly can be expressed as

$$\left| \begin{array}{ccc} -\dfrac{\partial^2 \ln L}{\partial \delta^2} \bigg|_{\hat{\gamma}}^{\hat{\delta}} & -\dfrac{\partial^2 \ln L}{\partial\,\partial\theta} \bigg|_{\hat{\gamma}}^{\hat{\delta}} & -\dfrac{\partial^2 \ln L}{\partial \delta\, \partial\gamma} \bigg|_{\hat{\gamma}}^{\hat{\delta}} \\ -\dfrac{\partial^2 \ln L}{\partial \theta\, \partial \delta} \bigg|_{\hat{\gamma}}^{\hat{\delta}} & -\dfrac{\partial^2 \ln L}{\partial \theta^2} \bigg|_{\hat{\gamma}}^{\hat{\delta}} & -\dfrac{\partial^2 \ln L}{\partial \theta\, \partial \gamma} \bigg|_{\hat{\gamma}}^{\hat{\delta}} \\ -\dfrac{\partial^2 \ln L}{\partial \gamma\, \partial \delta} \bigg|_{\hat{\gamma}}^{\hat{\delta}} & -\dfrac{\partial^2 \ln L}{\partial \gamma\, \partial \theta} \bigg|_{\hat{\gamma}}^{\hat{\delta}} & -\dfrac{\partial^2 \ln L}{\partial \gamma^2} \bigg|_{\hat{\gamma}}^{\hat{\delta}} \end{array} \right|^{-1} = \left| \begin{array}{ccc} V(\hat{\delta}) & \mathrm{Cov}(\hat{\delta},\hat{\theta}) & \mathrm{Cov}(\hat{\delta},\hat{\gamma}) \\ \mathrm{Cov}(\hat{\theta},\hat{\delta}) & V(\hat{\theta}) & \mathrm{Cov}(\hat{\theta},\hat{\gamma}) \\ \mathrm{Cov}(\hat{\gamma},\hat{\delta}) & \mathrm{Cov}(\hat{\gamma},\hat{\theta}) & V(\hat{\gamma}) \end{array} \right|.$$

(5.7.2)

Although these results have been derived for maximum likelihood estimates, various simulation studies (cf. Cohen and Whitten, 1982) have indicated that they also provide close estimates of variances and covariances of the modified estimators MMLE.

5.8 AN ILLUSTRATIVE EXAMPLE

Example 5.1. A computer-generated Type II progressively censored sample from a Weibull population with $\gamma = 100$, $\beta = 100$, and $\delta = 2$, originally given by Cohen (1975), has been selected to illustrate the practical application of estimators presented in this chapter. This sample simulates results from a life test of 100 items. When the sixth failure occurred at $T_1 = 124.63$, 10 specimens selected randomly from the survivors were withdrawn (censored). When the fortieth failure occurred at $T_2 = 174.22$, 15 randomly selected survivors were withdrawn, and the test was terminated at $T_3 = 239.35$ with seven remaining survivors. These data are summarized in the notation employed here as $N = 100$, $c_1 = 10$, $T_1 = 124.63$, $c_2 = 15$, $T_2 = 174.22$, $c_3 = 7$, $T_3 = 249.35$, $n = N - (c_1 + c_2 + c_3)$ 68, $x_{1:100} = 109.12$, $\sum_{i=1}^{68} x_i = 11{,}577.47$, and $\bar{x}_{68} = 170.257$. For the population, $E(X) = 188.623$, $\sqrt{V(X)} = 46.325$, and $\alpha_3(X) = 0.6311$. The 68 complete ordered observations are tabulated as follows:

109.12	130.53	144.09	158.31	177.19	198.11	222.11
113.37	131.98	148.83	158.92	180.57	199.23	224.83
117.73	133.14	150.23	160.13	181.99	203.27	227.27
119.56	134.52	150.79	161.31	184.02	206.55	230.88
119.82	135.73	151.88	162.09	185.43	208.76	235.14
<u>124.63</u>	136.71	153.07	165.45	187.21	210.69	237.43
125.21	137.88	154.18	166.62	189.77	213.32	246.08
126.93	138.63	154.97	168.23	191.63	215.08	<u>249.35</u>
128.25	141.11	155.26	169.98	194.88	218.43	
129.41	142.33	156.82	<u>174.22</u>	196.91	219.37	

Samples from the Weibull Distribution

Both maximum likelihood and modified maximum likelihood estimates were calculated from these data with the following results.

Parameters	MLE	MMLE	Population values
γ	106.93	102.38	100
θ	1635.05	3905.30	10000
δ	1.64	1.81	2
β	91.50	96.66	100
$E(X)$	188.80	188.32	188.62
$V(X)$	51.27	49.81	46.33
$\alpha_3(X)$	0.92	0.77	0.63

Approximate asymptotic variances and covariances obtained from (5.7.2) are

$$V(\hat{\delta}) = 0.0447, \quad V(\hat{\theta}) = 2.71 \times 10^6, \quad V(\hat{\gamma}) = 10.02, \quad \text{Cov}(\hat{\delta}, \hat{\theta})$$
$$= 345, \quad \text{Cov}(\hat{\delta}, \hat{\gamma}) = -0.4364, \quad \text{Cov}(\hat{\theta}, \hat{\gamma}) = -3558.$$

Although estimates of θ differ considerably from the population value, estimates of other parameters compare more favorably with corresponding population values. It is to be noted that the MMLEs for this example are a slight improvement over the MLE.

6
Truncated and Censored Samples from the Lognormal Distribution

6.1 INTRODUCTION

The lognormal is a "long-tailed" positively skewed distribution that is an appropriate model in life-span and reaction-time studies where data are often highly skewed. It has been studied extensively by numerous investigators. Among these are Yuan (1933), Cohen (1951, 1976, 1988), Aitchison and Brown (1957), Hill (1963), Johnson and Kotz (1970), Giesbrecht and Kempthorne (1976), Kane (1978, 1982), Cohen et al. (1985), Wingo (1975, 1976), Munro and Wixley (1970), Stedinger (1980), Rukhin (1984), Crow and Shimizu (1988), and many others. Estimation of lognormal parameters from complete samples has been effectively treated by various writers among those referenced here. In this chapter we are primarily concerned with parameter estimation in the three-parameter lognormal distribution from truncated and censored samples. Both modified maximum likelihood estimators, which employ the first-order statistic, and local maximum likelihood estimators are considered. Because of regularity problems to be discussed later, global maximum likelihood estimation is not always feasible.

6.2 SOME FUNDAMENTALS

The lognormal distribution derives its name from the relation it bears to the normal distribution. If the random variable $Y = \ln(X - \gamma)$ is normally distributed

Samples from the Lognormal Distribution

(μ, σ^2), then X is lognormally distributed (γ, μ, σ). When γ is known, it is a simple matter to make the transformation from X to Y, and subsequent analyses including parameter estimation require only the use of well-known normal distribution theory. When the threshold parameter γ is unknown, estimation procedures become more complex. The pdf of the three-parameter lognormal distribution follows from the definition as

$$f(x; \gamma, \mu, \sigma^2) = \frac{1}{\sigma\sqrt{2\pi}\,(x - \gamma)} \exp\left[-\frac{1}{2}\left[\frac{\ln(x - \gamma) - \mu}{\sigma}\right]^2\right],$$

$$\gamma \leq x < \infty, \sigma^2 > 0, \quad (6.2.1)$$

$$= 0, \quad \text{elsewhere.}$$

The cumulative distribution function may be expressed as

$$F(x; \gamma, \mu, \sigma^2) = \Phi\left[\frac{\ln(x - \gamma) - \mu}{\sigma}\right], \quad (6.2.2)$$

where $\Phi(\ )$ is the cumulative standard normal distribution function.

In the notation employed here, σ^2 and μ are the variance and mean of Y, but they become the shape and scale parameters respectively of the lognormal variate, X. On some occasions it is more expedient to employ $\beta = \exp(\mu)$ as the scale parameter and $\omega = \exp(\sigma^2)$ as the shape parameter of the lognormal distribution.

The expected value, median, mode, variance, coefficient of variation, β_1 and β_2 (Pearson's betas) were given by Yuan (1933) as

$$E(X) = \gamma + \beta\sqrt{\omega},$$

$$\text{Me}(X) = \gamma + \beta, \quad \text{Mo}(X) = \gamma + \frac{\beta}{\omega},$$

$$V(X) = \beta^2\omega(\omega - 1), \quad \text{cv}(X) = \sqrt{\omega - 1}, \quad (6.2.3)$$

$$\beta_1 = \alpha_3^2 = (\omega + 2)^2(\omega - 1),$$

$$\beta_2 = \alpha_4 = \omega^4 + 2\omega^3 + 3\omega^2 - 3,$$

where $E(\ )$ is the expected value symbol, $V(\ )$ is the variance, $\text{cv}(\ )$ is the coefficient of variation, and α_3 and α_4 are the third and fourth standard moments. The coefficient of variation is defined as

$$\text{cv}(X) = \frac{\sqrt{V(X)}}{E(X) - \gamma}. \quad (6.2.4)$$

6.3 GLOBAL MAXIMUM LIKELIHOOD ESTIMATION

For a complete (uncensored) sample consisting of observations $\{x_i\}, i = 1, \ldots, n$, from the three-parameter lognormal distribution, the likelihood function is

$$L = \prod_{i=1}^{n} f(x_i; \gamma, \mu, \sigma^2), \tag{6.3.1}$$

where $f(x; \gamma, \mu, \sigma^2)$ is given by (6.2.1). Without any loss of generality, we assume the observations to be ordered so that $x_{1:n}$ is the smallest observation in a random sample of size n. By definition, maximum likelihood estimates are those estimates that maximize the likelihood function (6.3.1). It becomes immediately obvious that $L(\ )$ becomes infinite as $\gamma \to x_{1:n}$. It would thus appear that we should take $\hat{\gamma} = x_{1:n}$ as our estimate. However, Hill (1963) demonstrated the existence of paths along which the likelihood function of any ordered sample $x_{1:n}, \ldots, x_{n:n}$ tends to infinity as (γ, μ, σ^2) approach $(x_{1:n}, -\infty, \infty)$. Accordingly, this global maximum leads to the inadmissible estimates, $\hat{\mu} = -\infty$ and $\hat{\sigma}^2 = \infty$, regardless of the sample. This result holds for both complete and censored samples.

As an alternative to global maximum likelihood estimation in this case, Cohen (1951), Harter and Moore (1966), and Cohen and Whitten (1980) equated partial derivatives of the loglikelihood function to zero and solved the resulting equations to obtain local maximum likelihood estimators (LMLE). Harter and Moore (1966) and later Calitz (1973) noted that these LMLE appear to possess most of the desirable properties ordinarily associated with the MLE. These comments apply both to complete and to censored samples.

In a further effort to avoid the problems associated with maximum likelihood estimators, Cohen and Whitten (1980) proposed the use of modified estimators that equate functions of the first-order statistic to corresponding sample values. Both modified moment and modified maximum likelihood estimators were considered. A slight preference was expressed for the modified moment estimators when samples were complete, and the modified maximum likelihood estimators were preferred when samples were censored.

As previously mentioned, here we are concerned with both local maximum likelihood and modified maximum likelihood estimators for parameters of the three-parameter lognormal distribution based on truncated and censored samples.

6.4 LOCAL MAXIMUM LIKELIHOOD AND MODIFIED MAXIMUM LIKELIHOOD ESTIMATORS FOR CENSORED SAMPLES

For a progressively censored sample from a three-parameter lognormal distribution, the loglikelihood function is

$$\ln L = -n \ln \sigma - \sum_{i=1}^{n} \ln(x_i - \gamma) - \frac{1}{2\sigma^2} \sum_{i=1}^{n} [\ln(x_i - \gamma) - \mu]^2 \tag{6.4.1}$$

Samples from the Lognormal Distribution

$$+ \sum_{d=1}^{k} c_j \ln [1 - F(T_j)] + \ln K,$$

where K is an ordering constant and

$$F(T_j) = \int_{\gamma}^{T_j} f(x; \gamma, \mu, \sigma^2)\, dx = \Phi(\xi_j) = \Phi\left[\frac{\ln(T_j - \gamma) - \mu}{\sigma}\right], \quad (6.4.2)$$

where $\Phi(\ )$ is the cdf of the standard normal distribution $(0, 1)$. In the context of a life test, the total sample size is N and the sample data consist of n full-term (uncensored) observations plus $N - n$ partial-term (censored) observations. Censoring occurs progressively in k stages at times $T_1 < \cdots < T_j < \cdots < T_k$. At time T_j, c_j sample specimens, selected randomly from the survivors at that time, are removed. It follows that

$$N = n - \sum_{j=1}^{k} c_j \quad (6.4.3)$$

Singly censored samples constitute a special case of progressively censored samples in which $k = 1$.

6.4.1 Local Maximum Likelihood Estimators

To obtain local maximum likelihood estimating equations based on progressively censored samples from the three-parameter lognormal distribution, we differentiate the loglikelihood equation of (6.4.1) with respect to the parameters and equate to zero. We accordingly obtain

$$\frac{\partial \ln L}{\partial \mu} = \frac{1}{\sigma^2} \sum_{i=1}^{n} [\ln(x_i - \gamma) - \mu] + \frac{1}{\sigma} \sum_{j=1}^{k} c_j Q_j = 0,$$

$$\frac{\partial \ln L}{\partial \sigma} = -\frac{n}{\sigma} + \frac{1}{\sigma^3} \sum_{i=1}^{n} [\ln(x_i - \gamma) - \mu]^2$$

$$+ \frac{1}{\sigma} \sum_{j=1}^{k} c_j \xi_j Q_j = 0, \quad (6.4.4)$$

$$\frac{\partial \ln L}{\partial \gamma} = \sum_{i=1}^{n} (x_i - \gamma)^{-1} + \frac{1}{\sigma^2} \sum_{i=1}^{n} \frac{\ln(x_i - \gamma) - \mu}{x_i - \gamma}$$

$$+ \frac{1}{\sigma} \sum_{j=1}^{k} \frac{c_j Q_j}{T_j - \gamma} = 0,$$

where

$$Q_j = Q(\xi_j) = \frac{\phi(\xi_j)}{1 - \Phi(\xi_j)}, \quad \xi_j = \frac{1}{\sigma}[\ln(T_j - \gamma) - \mu], \quad (6.4.5)$$

and where $F(T_j) = \Phi(\xi_j)$.

It is, of course, necessary that we solve the three equations of (6.4.4) simultaneously for the estimates ($\hat{\gamma}$, $\hat{\mu}$, $\hat{\sigma}$). A straightforward trial-and-error iterative procedure for this purpose was outlined by Cohen (1976). An improved procedure, that is applicable to all types of truncated and censored samples as well as to complete samples, was suggested by Giesbrecht and Kempthorne (1976). It was given further consideration by Cohen and Whitten (1988). An adaptation of the Giesbrecht–Kempthorne procedure forms the basis for a subsequent subsection on computational procedures.

6.4.2 Modified Maximum Likelihood Estimators for Censored Samples

Modified maximum likelihood estimating equations for samples from the three-parameter lognormal distribution are $\partial \ln L/\partial \mu = 0$, $\partial \ln L/\partial \sigma = 0$, and $E[\ln (X_{1:N} - \gamma)] = \ln(x_{1:N} - \gamma)$. For a progressively censored sample, the expanded forms of the first two of these equations are identical to the first two equations of (6.4.4). The third of the preceding equations, which in this case replaces the third equation of (6.4.4), can be written in an expanded form as

$$\gamma + \exp[\mu + \sigma E(Z_{1:N})] - x_{1:N} = 0, \quad (6.4.6)$$

where $Z_{1:N} = (Y_{1:N} - \mu)/\sigma$ and $E(Z_{1:N})$ is the expected value of the first-order statistic in the standard normal distribution (0, 1). For some occasions, it is convenient to substitute $\beta = \exp(\mu)$ as the scale parameter, and with this substitution, equation (6.4.6) becomes

$$\gamma + \beta \exp[\sigma E(Z_{1:N})] - x_{1:N} = 0. \quad (6.4.7)$$

The three applicable estimating equations (the first two equations of (6.4.4) plus either (6.4.6) or (6.4.7)) must now be solved simultaneously for the required estimates. This can be accomplished by following essentially the same procedure as that suggested for calculating the LMLE. Values of $E(Z_{1:N})$ required for the calculation of the MMLE can be found in Harter's tables (1961, 1969). Selected values from this source are given here as Table 6.1.

An alternate MMLE for which $E[F(X_{1:N})] = F(x_{1:N}) = 1/(N + 1)$ employs the estimating equation

$$\gamma + \exp(\mu + \sigma z_{1:N}) = x_{1:N} \quad (6.4.8)$$

as a replacement for equation (6.4.6) or the equivalent (6.4.7), where $z_{1:N}$ is the standard normal variate corresponding to the cdf value of $1/(N + 1)$; that is, $z_{1:N} = \Phi^{-1}[1/(N + 1)]$.

Samples from the Lognormal Distribution

Table 6.1 Expected values of the first-order statistic in samples from the standard normal distribution (0, 1)

n	$E(Z_{1,n})$	n	$E(Z_{1,n})$	n	$E(Z_{1,n})$
5	-1.16296	36	-2.11812	100	-2.50759
10	-1.53875	38	-2.14009	125	-2.58634
12	-1.62923	40	-2.16078	150	-2.64925
14	-1.70338	45	-2.20772	175	-2.70148
16	-1.75699	50	-2.24907	200	-2.74604
18	-1.82003	55	-2.28598	225	-2.78485
20	-1.86748	60	-2.31928	250	-2.81918
22	-1.90969	65	-2.34958	280	-2.85572
24	-1.94767	70	-2.37736	300	-2.87777
26	-1.98216	75	-2.40299	315	-2.89327
28	-2.01371	80	-2.42677	350	-2.92651
30	-2.04276	85	-2.44894	375	-2.94810
32	-2.06967	90	-2.46970	400	-2.96818
34	-2.09471	95	-2.48920	1000[a]	-3.09053

[a] Entry approximated as $Z_{1,1000} \doteq \Phi^{-1}(1/1001)$.
Source: Extracted from Harter's (1961) tables.

6.5 COMPUTATIONAL PROCEDURES

An adaptation of the Giesbrecht–Kempthorne procedure provides a simple method for simultaneously solving estimating equations for parameters of the three-parameter lognormal distribution that is applicable for both local maximum likelihood and modified maximum likelihood estimators. Furthermore, this method is applicable for all types of truncated and censored samples.

When the random variable X is distributed lognormally (γ, μ, σ^2), then $Y = \ln(X - \gamma)$ is distributed normally (μ, σ^2). Consequently, the analysis of lognormal data requires only that we make the transformation $\{y_i\} = \{\ln(x_i - \gamma)\}$, $i = 1, 2, \ldots, n$ on the uncensored observations, and the corresponding transformation $\{D_j\} = \{\ln(T_j - \gamma)\}$ on the points of censoring or truncation. The transformed sample then becomes a normal sample to be analyzed by using applicable normal theory. When γ is known, the transformation and any subsequent estimation or other analysis is quite simple. When γ is unknown and must then be estimated from the sample data, we begin with an approximation $\gamma_1 < x_{1:N}$ and use the applicable normal distribution estimators from Chapters 2, 3, and 4 to calculate approximations (i.e., conditional estimates) $\mu_1 = \mu(\gamma_1)$ and $\sigma_1 = \sigma(\gamma_1)$. These approximations must then be "tested" by substitution in a third estimating function. For progressively censored samples, the LMLE test function is the left side of the third equation of (6.4.4) and the MMLE test function is either (6.4.6) or (6.4.7). If the test function equals zero, then $\hat{\gamma} = \gamma_1$, $\hat{\mu} = \mu_1$, $\hat{\sigma} = \sigma_1$, and no further calculations are needed. Otherwise, we select a second approximation γ_2 and repeat the cycle of computations. We continue until we find two approximations γ_i and γ_j in a sufficiently narrow interval such that $TF(\gamma_i) \gtreqless 0 \gtreqless TF(\gamma_j)$, where TF designates the applicable test function. We interpolate between γ_i and γ_j for the final estimates.

Giesbrecht and Kempthorne employed the likelihood function as their test function and required that it should attain its maximum value. However, they sometimes encountered computational problems in ascertaining when a maximum had been reached.

Both LMLE and MMLE test functions for the various sample types of interest here are given below. Definitions of $Q(\xi)$, $\bar{Q}_1(\xi_1, \xi_2)$, $\bar{Q}_2(\xi_1, \xi_2)$, $\Omega_1(\xi_1, \xi_2)$, and $\Omega_2(\xi_1, \xi_2)$, which appear in these functions, are given in Chapter 2, equation (2.2.11), and in Chapter 3, equations (3.2.6) and (3.3.3), respectively.

LMLE Test Functions

For progressively censored samples:

$$\frac{\partial \ln L}{\partial \gamma} = \frac{1}{\sigma^2} \sum_{i=1}^{n} \frac{\ln(x_i - \gamma) - \mu}{x_i - \mu} + \sum_{i=1}^{n} \left(\frac{1}{x_i - \gamma} \right) + \frac{1}{\sigma} \sum_{j=1}^{k} \frac{c_j Q_j}{T_j - \gamma}.$$

For doubly censored samples:

$$\frac{\partial \ln L}{\partial \gamma} = \frac{1}{\sigma^2} \sum_{i=1}^{n} \frac{\ln(x_i - \gamma) - \mu}{x_i - \mu} + \sum_{i=1}^{n} \left(\frac{1}{x_i - \gamma} \right) - \frac{n}{\sigma} \left[\frac{\Omega_1}{T_1 - \gamma} - \frac{\Omega_2}{T_2 - \gamma} \right].$$

For doubly truncated samples:

$$\frac{\partial \ln L}{\partial \gamma} = \frac{1}{\sigma^2} \sum_{i=1}^{n} \frac{\ln(x_i - \gamma) - \mu}{x_i - \gamma} + \sum_{i=1}^{n} \left(\frac{1}{x_i - \gamma} \right) - \frac{n}{\sigma} \left[\frac{\bar{Q}_1}{T_1 - \gamma} - \frac{\bar{Q}_2}{T_2 - \gamma} \right].$$

For singly right censored samples:

$$\frac{\partial \ln L}{\partial \gamma} = \frac{1}{\sigma^2} \sum_{i=1}^{n} \frac{\ln(x_i - \gamma) - \mu}{x_i - \gamma} + \sum_{i=1}^{n} \left(\frac{1}{x_i - \gamma} \right) + \frac{c}{\sigma} \left(\frac{Q(\xi)}{T - \gamma} \right).$$

Samples from the Lognormal Distribution

For singly right truncated samples:

$$\frac{\partial \ln L}{\partial \gamma} = \frac{1}{\sigma_2} \sum_{i=1}^{n} \frac{\ln(x_i - \gamma) - \mu}{x_i - \gamma} + \sum_{i=1}^{n} \left(\frac{1}{x_i - \gamma}\right) + \frac{n}{\sigma}\left(\frac{Q(-\xi)}{T - \gamma}\right).$$

For singly left censored samples:

$$\frac{\partial \ln L}{\partial \gamma} = \frac{1}{\sigma_2} \sum_{i=1}^{n} \frac{\ln(x_i - \gamma) - \mu}{x_i - \gamma} + \sum_{i=1}^{n} \left(\frac{1}{x_i - \gamma}\right) - \frac{c}{\sigma}\left(\frac{Q(-\xi)}{T - \gamma}\right).$$

For singly left truncated samples:

$$\frac{\partial \ln L}{\partial \gamma} = \frac{1}{\sigma_2} \sum_{i=1}^{n} \frac{\ln(x_i - \gamma) - \mu}{x_i - \gamma} + \sum_{i=1}^{n} \left(\frac{1}{x_i - \gamma}\right) - \frac{n}{\sigma}\left(\frac{Q(\xi)}{T - \gamma}\right).$$

For complete samples:

$$\frac{\partial \ln L}{\partial \gamma} = \frac{1}{\sigma_2} \sum_{i=1}^{n} \frac{\ln(x_i - \gamma) - \mu}{x_i - \gamma} + \sum_{x=1}^{n} \left(\frac{1}{x_i - \gamma}\right).$$

MMLE Test Functions

For progressively censored samples:

$$\gamma + \exp[\mu + \sigma E(Z_{1:N})] - x_{1:N},$$

or

$$\gamma + \exp[\mu + \sigma z_{1:N}] - x_{1:N}.$$

For singly right censored samples:

$$\gamma + \exp[\mu + \sigma E(Z_{1:N})] - x_{1:N},$$

or

$$\gamma + \exp[\mu + \sigma z_{1:N}] - x_{1:N}.$$

For complete samples:

$$\gamma + \exp[\mu + \sigma E(Z_{1:N})] - x_{1:N},$$

or

$$\gamma + \exp[\mu + \sigma z_{1:N}] - x_{1:N}.$$

Since the MMLE employ the first-order statistic, they are applicable only for samples that include this value.

6.6 ERRORS OF ESTIMATES

Since the transformed lognormal distribution is normally distributed, the asymptotic variances and covariances of $\hat{\mu}$ and $\hat{\sigma}$ given in Chapters 2, 3, and 4 for the various truncated and censored samples from the normal distribution are also applicable here. The variance of an estimate of the threshold parameter γ based on a right censored sample of total size N should approximately equal the variance of a corresponding estimate from a complete sample. This result was given by Cohen (1951) as

$$V(\hat{\gamma}) = \frac{\sigma^2}{N\omega} \left[\frac{\beta^2}{\omega(1 + \sigma^2) - (1 + 2\sigma^2)} \right]. \qquad (6.6.1)$$

6.7 AN ILLUSTRATIVE EXAMPLE

Example 6.1. In order to illustrate computational procedures involved in calculating parameter estimates from a single right censored sample from a lognormal population, an example originally given by Cohen (1951) and subsequently used by Cohen and Whitten (1988) has been selected. The complete sample consists of $N = 20$ randomly chosen observations from a lognormal distribution in which $\gamma = 100$, $\beta = 50$ ($\mu = \ln \beta = 3.912023$), and $\sigma = 0.4$ ($\omega = 1.3219144$). Individual observations listed in order of magnitude are tabulated as follows:

A Random Sample from a Lognormal Population

127.211	135.880	153.070	166.475
128.709	137.338	155.369	168.554
131.375	144.328	155.680	174.800
132.971	145.788	157.238	184.101
133.143	148.290	164.304	201.415

For the purpose of this illustration, the sample is considered to be Type II censored with $c = 2$ at $T = y_{18:20} = 174.800$. For the censored sample, $N = 20$, $n = 18$, $c = 2$, and $h = c/N = 2/20 = 0.1$. The only information assumed to be known about the two censored observations is that they exceed 174.800. We illustrate only the MMLE as this is usually the preferred estimator when samples are small.

We select a first approximation $\gamma_1 = 110$, make the transformation $\{y_i\} = \ln(x_i - 110)$, $i = 1, 2, \ldots, 18$, $D = \ln(T - 110) = \ln(174.800 -$

Samples from the Lognormal Distribution

110) = 4.171306, and employ the maximum likelihood estimators (2.5.11) for singly censored normal samples. For the transformed data with $\gamma_1 = 110$, we calculate $\bar{y}(110) = 3.5531876$, $s_y(110) = 0.4191069$, and $\alpha(110) = s_y^2/(D - \bar{y})^2 = 0.4191069$. We enter Table 2.3 with $h = 0.1$ and $\alpha(110) = 0.4191069$ and interpolate to obtain $\lambda(110) = 0.1389$. When these values are substituted into (2.5.11), we calculate as conditional estimates $\mu(110) = 3.6390442$ and $\sigma^2(110) = 0.2287201$. Thus, $\sigma(110) = 0.4782469$. With $z_{1:N} = \Phi^{-1}[1/(N + 1)] = \Phi^{-1}(1/21) = -1.67$ from a table of normal curve areas, we substitute in equation (6.4.8) to calculate

$$110 + \exp[3.6390442 + 0.4782469(-1.67)] = 127.122.$$

Since the value thus calculated is less than the smallest sample observation ($x_{1:20} = 127.211$), we select a second approximation $\gamma_2 = 112$ and repeat the preceding calculations to obtain $\mu(112) = 3.5793849$ and $\sigma(112) = 0.5092228$. This time, substitution in (6.4.8) gives

$$112 + \exp[3.5793849 + 0.5092228(-1.67)] = 127.317,$$

a value that exceeds $x_{1:20}$. We therefore conclude that $110 < \hat{\gamma} < 112$, and final estimates are obtained by interpolation as summarized below.

γ	$E(x_{1:20})$	μ	σ
112.00	127.317	3.5793849	0.5092228
110.91	*127.211*	*3.6118151*	*0.4923846*
110.00	127.122	3.6390442	0.4782469

As final estimates we have $\hat{\gamma} = 110.9$, $\hat{\mu} = 3.612$, $\hat{\sigma} = 0.492$, $\hat{\omega} = 1.274$, $\hat{\alpha}_3 = 1.71$, and $\hat{\beta} = 37.03$. These compare favorably with corresponding estimates (104.65, 3.782, 0.407, 1.180, 1.352, and 43.900) which were originally calculated by Cohen (1951) from the complete sample.

Standard errors calculated from equations (2.6.1) with variance factors as given in Table 2.4 are $\sigma_{\hat{\sigma}} = 0.08$ and $\sigma_{\hat{\mu}} = 0.111$. From (6.6.1) we calculate $V(\hat{\gamma}) = 133.15$, and the standard error $\sigma_{\hat{\gamma}} = 11.54$. Although these standard errors are not strictly applicable for the MMLE, previous simulation studies of Cohen and Whitten (1980, 1981) indicate that they are reasonably close to the correct values.

7
Truncated and Censored Samples from the Inverse Gaussian and the Gamma Distributions

7.1 THE INVERSE GAUSSIAN DISTRIBUTION

The inverse Gaussian (IG) is a "long-tailed" positively skewed distribution, similar in shape to the lognormal distribution, that provides a useful alternative to other skewed distributions as a model in life-span and reaction-time studies. Although it was originally derived as the first passage of time distribution of Brownian motion with positive drift, it is perhaps better known today for its role as a model for highly skewed data. It is applicable in the physical, biological, and management sciences.

This distribution has been extensively studied by numerous investigators. Among these are Schroedinger (1915), Smoluchowsky (1915), Wald (1944), Tweedie (1956, 1957a, b), Wasan (1968), Wasan and Roy (1969), Chhikara and Folks (1974, 1988), Folks and Chhikara (1978), Padgett and Wei (1979), Cheng and Amin (1981), Chan et al. (1983, 1984), Cohen and Whitten (1985, 1988), plus many others. An excellent expository account of this distribution is given by Johnson and Kotz (1970).

In this chapter we are primarily concerned with parameter estimation from censored samples. Estimation from complete samples has been adequately considered in the various references cited above.

In the parameterization employed by Tweedie (1956), the two-parameter probability density function (pdf) is

The Inverse Gaussian and Gamma Distributions

$$g(y; \mu, \lambda) = \left(\frac{\lambda}{2\pi y^3}\right)^{1/2} \exp\left[-\frac{\lambda}{2\mu^2}\frac{(y-\mu)^2}{y}\right],$$
$$y > 0, \lambda > 0, \mu > 0, \qquad (7.1.1)$$
$$= 0 \quad \text{elsewhere}.$$

Cheng and Amin (1981) shifted the origin with the transformation $Y = (X - \gamma)$ and wrote the pdf of the three-parameter distribution as

$$f(x; \gamma, \mu, \lambda) = \left(\frac{\lambda}{2\pi(x-\gamma)^3}\right)^{1/2} \exp\left[-\frac{\lambda}{2\mu^2}\frac{(x-\gamma-\mu)^2}{x-\gamma}\right],$$
$$x > \gamma, \lambda > 0, \mu > 0, \qquad (7.1.2)$$
$$= 0 \quad \text{elsewhere}.$$

When we set $\lambda = \mu^3/\sigma^2$, the pdf in the parameterization of Chan et al. (1983) follows from (7.1.2) as

$$f(x; \gamma, \mu, \sigma) = \frac{1}{\sigma\sqrt{2\pi}}\left(\frac{\mu}{x-\gamma}\right)^{3/2}$$
$$\times \exp\left[-\frac{1}{2}\left(\frac{\mu}{x-\gamma}\right)\left(\frac{(x-\gamma)-\mu}{\sigma}\right)^2\right], \quad x > \gamma, \qquad (7.1.3)$$
$$= 0 \quad \text{elsewhere}.$$

The expected value (mean), variance, and third and fourth standard moments are

$$E(X) = \gamma + \mu, \qquad V(X) = \sigma^2,$$
$$\alpha_3(X) = \sqrt{\beta_1} = \frac{3\sigma}{\mu}, \qquad (7.1.4)$$
$$\alpha_4(X) = \beta_2 = 3 + \frac{5}{3}\beta_1.$$

In this notation γ is the threshold parameter, $\gamma + \mu$ is the mean, σ is the standard deviation, and $\alpha_3 = 3\sigma/\mu$ becomes the primary shape parameter. The betas are Pearson's notation for α_3^2 and α_4, respectively. Note that β_2 is a linear function of β_1. The pdf of the standardized IG distribution where $Z = [X - E(X)]/\sigma$, as derived by Chan et al. (1983), is

$$g(z; 0, 1, \alpha_3) = \frac{1}{\sqrt{2\pi}}\left(\frac{3}{3+\alpha_3 z}\right)^{3/2} \exp\left[-\frac{z^2}{2}\left(\frac{3}{3+\alpha_3 z}\right)\right],$$
$$-\frac{3}{\alpha_3} < z < \infty, \qquad (7.1.5)$$
$$= 0 \quad \text{elsewhere}.$$

and the standardized cdf is

$$G(z; 0, 1, \alpha_3) = \Phi\left[\frac{z}{\sqrt{1 + (\alpha_3/3)z}}\right]$$

$$+ \exp\left(\frac{18}{\alpha_3^2}\right) \Phi\left[\frac{-(z + 6/\alpha_3)}{\sqrt{1 + (\alpha_3/3)z}}\right], \quad (7.1.6)$$

where $\Phi(\ )$ is the cdf of the standard normal distribution $(0, 1)$.

It follows from the pdf of (7.1.5) that $g(0) = 1/\sqrt{2\pi}$ for all values of α_3. Furthermore, it is readily noted that the limiting form of (7.1.5) as $\alpha_3 \to 0$ is

$$\lim_{\alpha_3 \to 0} g(z; 0, 1, \alpha_3) = \frac{1}{\sqrt{2\pi}} e^{-z^2/2}, \quad -\infty < z < \infty, \quad (7.1.7)$$

which is recognized as the pdf of the standard normal distribution. The mode of the standardized IG distribution can be obtained by equating the first derivative of (7.1.5) to zero. Accordingly, it follows that

$$\text{Mo}(Z) = \frac{\sqrt{36 + \alpha_3^4} - \alpha_3^2 - 6}{2\alpha_3}. \quad (7.1.8)$$

7.1.1 Maximum Likelihood Estimators for Censored Samples

The loglikelihood function of a progressively censored sample from the IG distribution with pdf (7.1.3) becomes

$$\ln L = -n \ln \sigma + \frac{3n}{2} \ln \mu - \frac{3}{2} \sum_{i=1}^{n} \ln(x_i - \gamma)$$

$$- \frac{\mu}{2\sigma^2} \sum_{i=1}^{n} \frac{(x_i - \gamma - \mu)^2}{x_i - \gamma} + \sum_{j=1}^{k} c_j \ln[1 - F(T_j)] + \text{const}, \quad (7.1.9)$$

where

$$\xi_j = \frac{T_j - \gamma - \mu}{\sigma} \quad \text{and} \quad F(T_j) = G(\xi_j).$$

Maximum likelihood estimating equations are obtained by differentiating $\ln L$ with respect to the parameters being estimated and equating to zero. This requires that we differentiate $F(T_j)$, an operation that involves differentiation under an integral sign. To avoid some of the resulting complications, we set $\sigma = \mu \alpha_3/3$ in equation (7.1.3), which now becomes

The Inverse Gaussian and Gamma Distributions

$$f(x; \gamma, \mu, \alpha_3) = \begin{cases} \dfrac{3\sqrt{\mu}}{\alpha_3\sqrt{2\pi}} (x - \gamma)^{-3/2} \exp\left[-\dfrac{9(x - \gamma - \mu)^2}{2\mu\alpha_3^2(x - \gamma)} \right], & x > \gamma, \\ 0 & \text{elsewhere.} \end{cases} \quad (7.1.10)$$

With this parameterization the loglikelihood function of (7.1.9) becomes

$$\ln L = \frac{n}{2} \ln \mu - n \ln \alpha_3 - \frac{3}{2} \sum_{i=1}^{n} \ln(x_i - \gamma)$$
$$- \frac{9}{2\mu\alpha_3^2} \sum_{i=1}^{n} \frac{(x_i - \gamma - \mu)^2}{x_i - \gamma} + \sum_{j=1}^{k} c_j \ln[1 - G(\xi_j)] + \text{const.} \quad (7.1.11)$$

Maximum likelihood estimating equations $\partial \ln L/\partial\gamma = 0$, $\partial \ln L/\partial\mu = 0$, and $\partial \ln L/\partial\alpha_3 = 0$, are

$$\frac{\partial \ln L}{\partial \gamma} = \frac{3}{2} \sum_{i=1}^{n} (x_i - \gamma)^{-1} + \frac{9n}{2\mu\alpha_3^2} - \frac{9\mu}{2\alpha_3^2} \sum_{i=1}^{n} (x_i - \gamma)^{-2}$$
$$- \sum_{j=1}^{k} \frac{c_j}{1 - G_j} \frac{\partial G_j}{\partial \gamma} = 0,$$

$$\frac{\partial \ln L}{\partial \mu} = \frac{n}{2\mu} + \frac{9}{2\mu^2\alpha_3^2} \sum_{i=1}^{n} (x_i - \gamma) - \frac{9}{2\alpha_3^2} \sum_{i=1}^{n} (x_i - \gamma)^{-1}$$
$$- \sum_{j=1}^{k} \frac{c_j}{1 - G_j} \frac{\partial G_j}{\partial \mu} = 0, \quad (7.1.12)$$

$$\frac{\partial \ln L}{\partial \alpha_3} = -\frac{n}{\alpha_3} + \frac{9}{\mu\alpha_3^2} \sum_{i=1}^{n} \frac{(x_i - \gamma - \mu)^2}{x_i - \gamma}$$
$$- \sum_{j=1}^{k} \frac{c_j}{1 - G_j} \frac{\partial G_j}{\partial \alpha_3} = 0.$$

We recall that

$$G(\xi_j) = \int_{-3/\alpha_3}^{\xi_j} g(z; 0, 1, \alpha_3) \, dz, \quad \text{where } \xi_j = \frac{T_j - \gamma - \mu}{\sigma}$$
$$= \frac{3(T_j - \gamma - \mu)}{\alpha_3\mu} = \frac{3(T_j - \gamma)}{\alpha_3\mu} - \frac{3}{\alpha_3}, \quad (7.1.13)$$

and where $g(z; 0, 1, \alpha_3)$ is given by (7.1.5). Derivatives of $G_j = G(\xi_j)$ which appear in (7.1.12) are obtained as

$$\frac{\partial G_j}{\partial \gamma} = -\frac{3}{\mu\alpha_3}g(\xi_j), \qquad \frac{\partial G_j}{\partial \mu} = -\frac{1}{\mu}\left(\xi_j + \frac{3}{\alpha_3}\right)g(\xi_j),$$

$$\frac{\partial G_j}{\partial \alpha_3} = \int_{-3/\alpha_3}^{\xi_j} \frac{\partial g}{\partial \alpha_3} dz - \frac{\xi_j}{\alpha_3}g(\xi_j).$$
(7.1.14)

We designate the hazard function as $h(\xi_j)$ and the integral function as $A(\xi_j)$. Thus,

$$h_j = h(\xi_j) = \frac{g(\xi_j)}{1 - G(\xi_j)},$$

$$A_j = A(\xi_j) = \int_{-3/\alpha_3}^{\xi_j} \frac{\partial g}{\partial \alpha_3} dz.$$
(7.1.15)

With this notation the three estimating equations of (7.1.12) become

$$\frac{\partial \ln L}{\partial \gamma} = \frac{3}{2}\sum_{i=1}^{n}(x_i - \gamma)^{-1} + \frac{9n}{2\mu\alpha_3^2} - \frac{9\mu}{2\alpha_3^2}\sum_{i=1}^{n}(x_i - \gamma)^{-2}$$

$$+ \frac{3}{\alpha_3\mu}\sum_{j=1}^{k}c_j h_j = 0,$$

$$\frac{\partial \ln L}{\partial \mu} = \frac{n}{2\mu} + \frac{9}{2\mu^2\alpha_3^2}\sum_{i=1}^{n}(x_i - \gamma) - \frac{9}{2\alpha_3^2}\sum_{i=1}^{n}(x_i - \gamma)^{-1}$$

$$+ \frac{1}{\mu}\sum_{j=1}^{n}c_j\left(\xi_j + \frac{3}{\alpha_3}\right)h_j = 0,$$
(7.1.16)

$$\frac{\partial \ln L}{\partial \alpha_3} = -\frac{n}{\alpha_3} + \frac{9}{\mu\alpha_3^2}\sum_{i=1}^{n}\frac{(x_i - \gamma - \mu)^2}{x_i - \gamma}$$

$$+ \frac{1}{\alpha_3}\sum_{j=1}^{k}c_j\xi_jh_j - \sum_{j=1}^{k}c_jA_j = 0.$$

Maximum likelihood estimates ($\hat{\gamma}$, $\hat{\mu}$, $\hat{\alpha}_3$) and subsequently $\hat{\sigma} = \hat{\mu} \hat{\alpha}_3/3$, can be obtained by simultaneously solving the three estimating equations of (7.1.16). This can be accomplished by employing various iterative procedures, but the evaluation of the integral function $A(\xi_j)$ is likely to be troublesome. This problem can be avoided if we bypass the third equation of (7.1.16) and proceed directly to the loglikelihood function (7.1.11). Accordingly, we select a first approximation $\alpha_3^{(1)}$ and solve the first two equations of (7.1.16) simultaneously for conditional estimates $\gamma_1 = \gamma(\alpha_3^{(1)})$ and $\mu_1 = \mu(\alpha_3^{(1)})$. We then substitute the three approximations $\alpha_3^{(1)}$, γ_1, μ_1 into (7.1.11) to calculate $\ln L(\alpha_3^{(1)})$. We repeat these calculations for several values in the vicinity of the maximum and a plot of $\ln L$ as a function of α_3 can enable us to determine the estimate $\hat{\alpha}_3$ that is

The Inverse Gaussian and Gamma Distributions

required. With $\hat{\alpha}_3$ thus calculated, the remaining estimates can be obtained from the first two equations of (7.1.16) or by interpolation.

As noted in earlier chapters of this book, singly censored samples may be regarded as special cases of progressively censored samples. Accordingly, estimating equations (7.1.16) with $k = 1$ are also applicable in singly right censored samples.

7.1.2 Modified Maximum Likelihood Estimators for Censored Samples

Modified estimators, which are often preferred as alternatives to maximum likelihood estimators, employ as estimating equations $\partial \ln L/\partial \gamma = 0$, $\partial \ln L/\partial \mu = 0$, and $E[F(X_{r:N})] = F(x_{r:N}) = r/(N + 1)$. In many applications, it will be convenient to let $r = 1$, but larger values (as large as the sample median when it is available) might sometimes be desirable. The first two equations are the same as the first two equations of (7.1.16).

The computational procedure for calculating these estimates is similar to that employed in calculating maximum likelihood estimates. We begin with a first approximation $\alpha_3^{(1)}$. This value is substituted into the first two equations of (7.1.16), which are then solved simultaneously for corresponding approximations $\gamma(\alpha_3^{(1)})$ and $\mu(\alpha_3^{(1)})$. We enter tables of the cdf of the IG distribution with $\alpha_3 = \alpha_3^{(1)}$ and $F = r/(N + 1)$ and interpolate as necessary to obtain the corresponding approximation to $z_{r:N}$. We calculate $\sigma_1 = \mu_1 \alpha_3^{(1)}/3$ and then calculate an estimate of $X_{r:N}$ based on these approximations as

$$E(X_{r:N}) = \mu_1 + \gamma_1 + \sigma_1 z_{r:N}^{(1)}. \qquad (7.1.17)$$

If the value thus calculated is equal to the corresponding sample value $x_{r:N}$, then $\hat{\alpha}_3 = \alpha_3^{(1)}$, $\hat{\gamma} = \gamma_1$, $\hat{\sigma} = \sigma_1$, and further calculations are unnecessary. Otherwise, we select a second approximation $\alpha_3^{(2)}$ and repeat the cycle of calculations. Calculations are continued until we find two estimates of $X_{r:N}$ in a sufficiently narrow interval and such that

$$E(X_{r:N}^{(i)}) \lesseqgtr x_{r:N} \lesseqgtr E(X_{r:N}^{(j)}).$$

We then interpolate for our final estimates as follows:

α_3	$\gamma(\alpha_3)$	$\mu(\alpha_3)$	$\sigma = \mu\alpha_3/3$	$X_{r:N}$
α_3^i	γ^i	μ^i	σ^i	$E(X_{r:N}^i)$
$\hat{\alpha}_3$	$\hat{\gamma}$	$\hat{\mu}$	$\hat{\sigma}$	$X_{r:N}$
α_3^j	γ^j	μ^j	σ^j	$E(x_{r:N}^j)$

7.1.3 Maximum Likelihood Estimators for Truncated Samples

The loglikelihood function of a singly right truncated sample from a three-parameter inverse Gaussian distribution with pdf (7.1.10) is

$$\ln L = \frac{n}{2} \ln \mu - n \ln \alpha_3 - \frac{3}{2} \sum_{i=1}^{n} \ln (x_i - \gamma)$$
$$- \frac{9}{2\mu\alpha_3^2} \sum_{i=1}^{n} \frac{(x_i - \gamma - \mu)^2}{x_i - \gamma} - n \ln F(T). \quad (7.1.18)$$

Maximum likelihood estimating equations are

$$\frac{\partial \ln L}{\partial \gamma} = \frac{3}{2} \sum_{i=1}^{n} (x_i - \gamma)^{-1} + \frac{9n}{2\mu\alpha_3^2} - \frac{9\mu}{2\alpha_3^2} \sum_{i=1}^{n} (x_i - \gamma)^{-2}$$
$$- \frac{n}{F(T)} \frac{\partial F(T)}{\partial \gamma} = 0,$$

$$\frac{\partial \ln L}{\partial \mu} = \frac{n}{2\mu} + \frac{9}{2\mu^2\alpha_3^2} \sum_{i=1}^{n} (x_i - \gamma) - \frac{9}{2\alpha_3^2} \sum_{i=1}^{n} (x_i - \gamma)^{-1}$$
$$- \frac{n}{F(T)} \frac{\partial F(T)}{\partial \mu} = 0, \quad (7.1.19)$$

$$\frac{\partial \ln L}{\partial \alpha_3} = \frac{n}{\alpha_3} + \frac{9}{\mu\alpha_3^3} \sum_{i=1}^{n} \frac{(x_i - \gamma - \mu)^2}{x_i - \gamma}$$
$$- \frac{n}{F(T)} \frac{\partial F(T)}{\partial \alpha_3} = 0,$$

where the partials of F are

$$\frac{\partial F(T)}{\partial \gamma} = -\frac{3}{\mu\alpha_3} g(\xi), \quad \frac{\partial F(T)}{\partial \mu} = -\frac{1}{\mu}\left(\xi + \frac{3}{\alpha_3}\right) g(\xi). \quad (7.1.20)$$

$$\frac{\partial F(T)}{\partial \alpha_3} = \int_{-3/\alpha_3}^{\xi} \frac{\partial g}{\partial \alpha_3} dz - \frac{\xi}{\alpha_3} g(\xi) \quad \text{and} \quad \xi = \frac{T - E(X)}{\sqrt{V(X)}}.$$

When the partials of (7.1.20) are substituted into (7.1.19), the resulting estimating equations become

$$\frac{\partial \ln L}{\partial \gamma} = \frac{3}{2} \sum_{i=1}^{n} (x_i - \gamma)^{-1} + \frac{9n}{2\mu\alpha_3^2} - \frac{9\mu}{2\alpha_3^2} \sum_{i=1}^{n} (x_i - \gamma)^{-2}$$
$$+ \frac{3n}{\mu\alpha_3} \frac{g(\xi)}{G(\xi)} = 0,$$

The Inverse Gaussian and Gamma Distributions

$$\frac{\partial \ln L}{\partial \mu} = \frac{n}{2\mu} + \frac{9}{2\mu^2 \alpha_3^2} \sum_{i=1}^{n} (x_i - \gamma) - \frac{9}{2\alpha_3^2} \sum_{i=1}^{n} (x_i - \gamma)^{-1}$$
$$+ \frac{n}{\mu}\left(\xi + \frac{3}{\alpha_3}\right) \frac{g(\xi)}{G(\xi)} = 0,$$
(7.1.21)

$$\frac{\partial \ln L}{\partial \alpha_3} = -\frac{n}{\alpha_3} + \frac{9}{\mu \alpha_3^3} \sum_{i=1}^{n} \frac{(x_i - \gamma - \mu)^2}{x_i - \gamma}$$
$$- \frac{n}{G(\xi)} \left[\int_{-3/\alpha_3}^{\xi} \frac{\partial g}{\partial \alpha_3} dz - \frac{\xi}{\alpha_3} g(\xi) \right] = 0.$$

Estimates $\hat{\gamma}$, $\hat{\mu}$, and $\hat{\alpha}_3$ can be calculated as the simultaneous solution of the preceding equations. However, because of computational problems involved in evaluating the third equation of (7.1.21), it is suggested that we employ the procedure previously outlined for the solution of censored sample estimating equations. We select approximations to α_3 which are then substituted into the first two equations of (7.1.21). These equations are then solved for conditional estimates (approximations) $\gamma(\alpha_3)$ and $\mu(\alpha_3)$. Several sets of approximations are calculated in this manner and substituted directly into $\ln L$ in search of a maximum. A plot of $\ln L$ as a function of α_3 will be helpful in determining the required estimate $\hat{\alpha}_3$. The modified maximum likelihood estimates are not considered to be feasible for truncated samples.

7.2 THE GAMMA DISTRIBUTION

Like the Weibull, lognormal, and inverse Gaussian distribution, the gamma distribution is also a positively skewed distribution that is frequently employed as a model in reliability and life-span studies. When $\alpha_3 < 2$, it is bell-shaped with a discernible mode, but when $\alpha_3 \geq 2$, it becomes reverse-J shaped. When $\alpha_3 = 2$, the exponential distribution emerges as a special case of both the Weibull and the gamma distribution. Karl Pearson included the three-parameter gamma distribution as Type III of his generalized system of frequency distributions. Although numerous writers have previously considered this distribution, we will mention only a few at this time. Johnson and Kotz (1970) present an excellent expository account of the gamma distribution and its properties along with 130 references. In a recent monograph, Bowman and Shenton (1988) also provide an extensive list of references.

Our primary concern in this section is with estimation based on censored and truncated samples from the three-parameter gamma distribution. Some of the same regularity problems that arise in connection with maximum likelihood estimation in the Weibull, lognormal, and inverse Gaussian distributions must also be faced in estimating gamma parameters. As a consequence, modified estimators become attractive as alternatives to the MLE. We consider both maximum likelihood and modified maximum likelihood estimators.

7.2.1 The Gamma Density Function and Its Characteristics

The density function of the three-parameter gamma distribution is

$$f(x; \gamma, \beta, \rho) = \frac{\beta^{-\rho}}{\Gamma(\rho)} (x - \gamma)^{\rho - 1} \exp\left(-\frac{x - \gamma}{\beta}\right),$$
$$\gamma \leq x < \infty, \rho > 0, \qquad (7.2.1)$$
$$= 0 \quad \text{elsewhere}.$$

The expected value, variance, third and fourth standard moments, and mode (when $\rho > 1$) are

$$E(X) = \gamma + \rho\beta, \qquad V(X) = \rho\beta^2,$$
$$\alpha_3(X) = 2/\sqrt{\rho}, \qquad \alpha_4(X) = 3 + \frac{3}{2}\alpha_3^2, \qquad (7.2.2)$$
$$\text{Mo}(X) = \gamma + \beta(\rho - 1).$$

In this notation, γ is the threshold parameter, β is the scale parameter, and ρ is the shape parameter. When we make the standardizing transformation $Z = [X - E(X)]/\sqrt{V(X)}$, the resulting standard distribution $(0, 1, \rho)$ with ρ as the shape parameter becomes

$$g(z; 0, 1, \rho) = \frac{(\sqrt{\rho})^\rho}{\Gamma(\rho)} (z + \sqrt{\rho})^{\rho - 1} \exp[-\sqrt{\rho}(z + \sqrt{\rho})],$$
$$-\sqrt{\rho} \leq z < \infty, \qquad (7.2.3)$$
$$= 0 \quad \text{elsewhere}.$$

With α_3 as the shape parameter, the preceding pdf may be written as

$$g(z; 0, 1, \alpha_3) = \frac{(2/\alpha_3)^{4/\alpha_3^2}}{\Gamma(4/\alpha_3^2)} \left(z + \frac{2}{\alpha_3}\right)^{(4/\alpha_3^2) - 1}$$
$$\times \exp\left[-\frac{2}{\alpha_3}\left(z + \frac{2}{\alpha_3}\right)\right], \quad -\frac{2}{\alpha_3} \leq z < \infty, \qquad (7.2.4)$$
$$= 0 \quad \text{elsewhere}.$$

The Inverse Gaussian and Gamma Distributions

7.2.2 Maximum Likelihood Estimators for Censored Samples

The loglikelihood function of a progressively censored sample from a gamma distribution with pdf (7.2.1) becomes

$$\ln L = n \ln/\Gamma(\rho) - n\rho \ln \beta + (\rho - 1) \sum_{i=1}^{n} \ln(x_i - \gamma)$$

$$- \frac{1}{\beta} \sum_{i=1}^{n} (x_i - \gamma) + \sum_{j=1}^{k} c_j \ln[1 - F_j] + \text{const.} \quad (7.2.5)$$

When $\rho > 1$, maximum likelihood estimating equations may be obtained by differentiating $\ln L$ with respect to the parameters and equating to zero. Thus we obtain

$$\frac{\partial \ln L}{\partial \gamma} = \frac{n}{\beta} - (\rho - 1) \sum_{i=1}^{n} (x_i - \gamma)^{-1} - \sum_{j=1}^{k} \frac{c_j}{1 - F_j} \frac{\partial F_j}{\partial \gamma} = 0,$$

$$\frac{\partial \ln L}{\partial \beta} = -\frac{n\rho}{\beta} + \frac{1}{\beta^2} \sum_{i=1}^{n} (x_i - \gamma) - \sum_{j=1}^{k} \frac{c_j}{1 - F_j} \frac{\partial F_j}{\partial \beta} = 0, \quad (7.2.6)$$

$$\frac{\partial \ln L}{\partial \rho} = -n\psi(\rho) - n \ln \beta + \sum_{i=1}^{n} \ln(x_i - \gamma) - \sum_{j=1}^{k} \frac{c_j}{1 - F_j} \frac{\partial F_j}{\partial \rho} = 0,$$

where $\psi(\rho)$ is the digamma function, $\psi(\rho) = \partial \ln \Gamma(\rho)/\partial \rho = \Gamma'(\rho)/\Gamma(\rho)$.

The standardized cdf is

$$G(z; 0, 1, \rho) = \int_{-\sqrt{\rho}}^{z} g(t; 0, 1, \rho) \, dt, \quad (7.2.7)$$

where $g(z; 0, 1, \rho)$ is given by (7.2.3). When $z = \xi_j$, then

$$\xi_j = \frac{T_j - \gamma - \rho\beta}{\beta\sqrt{\rho}} = \frac{T - \gamma}{\beta\sqrt{\rho}} - \sqrt{\rho} \quad \text{and} \quad F(T_j) = G(\xi_j). \quad (7.2.8)$$

The partials of F may now be expressed as

$$\frac{\partial F_j}{\partial \gamma} = \frac{\partial G_j}{\partial \xi_j} \frac{\partial \xi_j}{\partial \gamma} = -\frac{1}{\beta\sqrt{\rho}} g(\xi_j),$$

$$\frac{\partial F_j}{\partial \beta} = \frac{\partial G_j}{\partial \xi_j} \frac{\partial \xi_j}{\partial \beta} = -\frac{1}{\beta}(\xi_j + \sqrt{\rho})g(\xi_j), \quad (7.2.9)$$

$$\frac{\partial F_j}{\partial \rho} = \frac{\partial G_j}{\partial \xi_j} \frac{\partial \xi_j}{\partial \rho} = \int_{-\sqrt{\rho}}^{\xi_j} \frac{\partial g}{\partial \rho} dz - \left(\frac{\xi_j}{2\rho} + \frac{1}{\sqrt{\rho}}\right)g(\xi_j).$$

When the results from (7.2.9) are substituted into (7.2.6), the estimating equations become

$$\frac{\partial \ln L}{\partial \gamma} = \frac{n}{\beta} - (\rho - 1) \sum_{i=1}^{n} (x_i - \gamma)^{-1}$$

$$+ \frac{1}{\beta \sqrt{\rho}} \sum_{j=1}^{n} c_j h_j = 0,$$

$$\frac{\partial \ln L}{\partial \beta} = -\frac{n\rho}{\beta} + \frac{1}{\beta^2} \sum_{i=1}^{n} (x_i - \gamma) \quad (7.2.10)$$

$$+ \frac{1}{\beta} \sum_{j=1}^{k} c_j(\xi_j + \sqrt{\rho}) h_j = 0,$$

$$\frac{\partial \ln L}{\partial \rho} = n\psi(\rho) - n \ln \beta + \sum_{i=1}^{n} \ln(x_i - \gamma)$$

$$- \sum_{j=1}^{k} \frac{c_j}{1 - G_j} \frac{\partial G_j}{\partial \xi_j} \frac{\partial \xi_j}{\partial \rho} = 0,$$

where

$$h_j = h(\xi_j) = \frac{g(\xi_j)}{1 - G(\xi_j)}.$$

Solution of the three equations of (7.2.10) for the three maximum likelihood estimates ($\hat{\gamma}$, $\hat{\beta}$, $\hat{\rho}$) can be carried out by following the same procedure as that outlined for the solution of the inverse Gaussian estimating equations (7.1.16). Evaluation of the third equation of (7.2.10) involves complications that are similar to those involved in evaluation of the third equation of (7.1.16). Here, for the gamma distribution, with ρ and thus α_3, fixed, we solve the first two equations of (7.2.10) for conditional estimates of γ and β and substitute directly into the loglikelihood equation (7.2.5) until we find the value of ρ or ($\alpha_3 = 2/\sqrt{\rho}$) that maximizes this function, just as we did in calculating maximum likelihood estimates of IG parameters.

As previously noted, singly right censored samples are a special case of progressively censored samples in which $k = 1$. Accordingly, with $k = 1$, estimators for progressively censored samples are also applicable for singly right censored samples.

7.2.3 Modified Maximum Likelihood Estimators for Censored Samples

Modified maximum likelihood estimators for parameters of the gamma distribution are quite similar to those for the IG parameters. Estimating equations are

The Inverse Gaussian and Gamma Distributions 117

$\partial \ln L/\partial \gamma = 0$, $\partial \ln L/\partial \beta = 0$, and $E[F(X_{r:n})] = F(x_{r:n}) = r/(N+1)$. As previously noted, in many applications, we set $r = 1$, but it may sometimes be desirable to select a larger value (even as large as $r = N/2$ when the median is included in the sample observations). The first two of these estimating equations for a progressively censored sample are identical to the first two equations of (7.2.10).

Calculation of these estimates differs very little from the calculation of MMLE for IG samples. We select a first approximation ρ_1 or $(\alpha_3^1 = 2/\sqrt{\rho_1})$, which is then substituted into the first two equations of (7.2.10). These equations are then solved simultaneously for approximations $\gamma(\rho_1)$ and $\beta(\rho_1)$. We enter a table of the gamma distribution cdf with the approximation $\alpha_3^{(1)} = 2/\sqrt{\rho_1}$ and $F = r/(N+1)$ and interpolate to read an approximation $z_{r:N}^{(1)}$. A corresponding estimate of the rth ordered variate $X_{r:N}$ is then calculated as

$$E(X_{r:N}^{(1)}) = \gamma_1 + \rho_1\beta_1 + \beta_1\sqrt{\rho_1}\,z_{r:N}^{(1)}. \quad (7.2.11)$$

If $E(X_{n:N}) = X_{n:N}$, then no further calculations are needed. Otherwise we continue with additional iterations until we are able to interpolate as shown below:

ρ	$\beta(\rho)$	$\gamma(\rho)$	$X_{r:N}$
ρ_{i_k}	β_{i_k}	γ_{i_k}	$E(X_{r:N}^{(i)})$
$\hat{\rho}$	$\hat{\beta}$	$\hat{\gamma}$	$x_{r:N}$
ρ_{j_k}	β_{j_k}	γ_{j_k}	$E(X_{r:N}^{(j)})$

7.2.4 Maximum Likelihood Estimators for Truncated Samples

The likelihood function of a singly right truncated sample from a three-parameter gamma distribution with pdf (7.2.1) is

$$\ln L = -n \ln \Gamma(\rho) - n\rho \ln \beta - \frac{1}{\beta}\sum_{i=1}^{n}(x_i - \gamma) \quad (7.2.12)$$
$$+ (\rho - 1)\sum_{i=1}^{n}\ln(x_i - \gamma) - n \ln F(T).$$

Maximum likelihood estimating equations applicable when $\rho > 1$ are

$$\frac{\partial \ln L}{\partial \gamma} = \frac{n}{\beta} - (\rho - 1)\sum_{i=1}^{n}(x_i - \gamma)^{-1} - \frac{n}{F(T)}\frac{\partial F(T)}{\partial \gamma} = 0,$$
$$\frac{\partial \ln L}{\partial \beta} = -\frac{n\rho}{\beta} + \frac{1}{\beta^2}\sum_{i=1}^{n}(x_i - \gamma) - \frac{n}{F(T)}\frac{\partial F(T)}{\partial \beta} = 0, \quad (7.2.13)$$

$$\frac{\partial \ln L}{\partial \rho} = -n\psi(\rho) - n \ln \beta + \sum_{i=1}^{n} \ln (x_i - \gamma) - \frac{n}{F(T)} \frac{\partial F(T)}{\partial \rho} = 0,$$

where

$$\frac{\partial F}{\partial \gamma} = \frac{-1}{\beta\sqrt{\rho}} g(\xi), \quad \frac{\partial F}{\partial \beta} = -\frac{1}{\beta}(\xi + \sqrt{\rho})g(\xi),$$

$$\frac{\partial F}{\partial \rho} = \int_{-\sqrt{\rho}}^{\xi} \frac{\partial g(z; 0, 1, \rho)}{\partial \rho} dz - \left(\frac{\xi}{2\rho} + \frac{1}{\sqrt{\rho}}\right) g(\xi; 0, 1, \rho),$$

(7.2.14)

and where

$$\xi = \frac{T - \gamma - \rho\beta}{\beta\sqrt{\rho}} = \frac{T - \gamma}{\beta\sqrt{\rho}} - \sqrt{\rho}.$$

When the partials given in (7.2.14) are substituted into (7.2.13), the resulting estimating equations become

$$\frac{\partial \ln L}{\partial \gamma} = \frac{n}{\beta} - (\rho - 1)\sum_{i=1}^{n}(x_i - \gamma)^{-1} + \frac{n}{\beta\sqrt{\rho}} \frac{g(\xi)}{G(\xi)} = 0,$$

$$\frac{\partial \ln L}{\partial \beta} = -\frac{n\rho}{\beta} + \frac{1}{\beta^2}\sum_{i=1}^{n}(x_i - \gamma) + \frac{n}{\beta}(\xi + \sqrt{\rho})\frac{g(\xi)}{G(\xi)} = 0, \quad (7.2.15)$$

$$\frac{\partial \ln L}{\partial \rho} = -n\psi(\rho) - n \ln \beta + \sum_{i=1}^{n} \ln(x_i - \gamma) - \frac{n}{G(\xi)}\frac{\partial F(T)}{\partial \rho} = 0.$$

Estimates $\hat{\gamma}$, $\hat{\beta}$, and $\hat{\rho}$ can be calculated as the simultaneous solution of the three equations of (7.2.15). However, to avoid the complications involved in evaluating the third of these equations, we follow the procedure that was employed in calculating estimates from censored samples. We begin with a first approximation ρ_1 or $(\alpha_3^{(1)} = 2/\sqrt{\rho_1})$. We substitute this value into the first two equations of (7.2.15) and solve for the conditional estimates (approximations) $\gamma(\rho_1)$ and $\beta(\rho_1)$. These approximations are then substituted directly into the loglikelihood equation (7.2.12) in search of a maximum. It will be necessary to repeat these calculations for several different values of ρ and perhaps plot a graph of $\ln L$ as a function of ρ, in order to determine the value of ρ for which $\ln L$ attains its maximum. Since the number of missing observations is unknown in truncated samples, the modified estimators that were presented for censored samples are not deemed feasible here.

7.2.5 Estimation in the Two-Parameter Cases

When the origin is known and can therefore be set at zero, the two-parameter distributions become special cases of the more general three-parameter distri-

The Inverse Gaussian and Gamma Distributions

butions. With one less parameter to be estimated, computations are somewhat simpler.

Progressively Censored Samples from the IG Distribution

The loglikelihood function of a progressively censored sample from a two-parameter IG distribution follows from (7.1.9) with $\gamma = 0$, as

$$\ln L = -n \ln \sigma + \frac{3n}{2} \ln \mu - \frac{3}{2} \sum_{i=1}^{n} \ln x_i - \frac{\mu}{2\sigma^2} \sum_{i=1}^{n} \frac{(x_i - \mu)^2}{x_i} \quad (7.2.16)$$
$$+ \sum_{j=1}^{k} c_j \ln[1 - F(T_j)] + \text{const.}$$

With α_3 assumed to be known, the remaining maximum likelihood estimating equation becomes

$$\frac{\partial \ln L}{\partial \mu} = \frac{n}{2\mu} + \frac{9}{2\mu^2 \alpha_3^2} \sum_{i=1}^{n} x_i - \frac{9}{2\alpha_3^2} \sum_{i=1}^{n} x_i^{-1} \quad (7.2.17)$$
$$+ \sum_{j=1}^{k} \frac{c_j}{\mu} \left(\xi_j + \frac{3}{\alpha_3} \right) h_j = 0.$$

It is relatively easy to solve this equation for the conditional estimate $\mu(\alpha_3)$ by trial and error plus interpolation. Both MLE and MMLE can then be calculated as described for the three-parameter IG distribution.

Truncated Samples from the IG Distribution

The loglikelihood function of a singly right truncated sample from the two-parameter IG distribution follows from (7.2.12) with $\gamma = 0$ as

$$\ln L = \frac{n}{2} \ln \mu - n \ln \alpha_3 - \frac{3}{2} \sum_{i=1}^{n} \ln x_i \quad (7.2.18)$$
$$- \frac{9}{2\mu \alpha_3^2} \sum_{i=1}^{n} \frac{(x_i - \mu)^2}{x_i} - n \ln F(T).$$

With α_3 known, the only remaining maximum likelihood estimating equation becomes

$$\frac{\partial \ln L}{\partial \mu} = \frac{n}{2\mu} + \frac{9}{2\mu^2 \alpha_3^2} \sum_{i=1}^{n} x_i - \frac{9}{2\alpha_3^2} \sum_{i=1}^{n} x_i^{-1} \quad (7.2.19)$$
$$+ \frac{1}{\mu} \left(\xi + \frac{3}{\alpha_3} \right) \frac{g(\xi)}{G(\xi)} = 0.$$

For given or assumed values of α_3, we substitute into (7.2.19) and calculate conditional estimates $\mu(\ )(\alpha_3)$. As we did in the three-parameter case, the conditional values (approximations) are substituted directly into the loglikelihood function, which in this case is (7.2.18), in search of a maximum. We make as many trials as necessary to find the value of α_3 that produces a maximum for $\ln L$.

Progressively Censored Samples from the Gamma Distribution

The loglikelihood function of a progressively censored sample from a two-parameter gamma distribution follows from (7.2.5) with $\gamma = 0$ as

$$\ln L = n \ln \Gamma(\rho) - n\rho \ln \beta + (\rho - 1) \sum_{i=1}^{n} \ln x_i \quad (7.2.20)$$
$$- \frac{1}{\beta} \sum_{i=1}^{n} x_i + \sum_{j=1}^{k} c_j \ln[1 - F_j] + \text{const.}$$

With α_3 known (or assumed to be known), the remaining maximum likelihood estimating equation becomes

$$\frac{\partial \ln L}{\partial \beta} = -\frac{n\rho}{\beta} + \frac{1}{\beta^2} \sum_{i=1}^{n} x_i + \frac{1}{\beta} \sum_{j=1}^{k} c_j(\xi_j + \sqrt{\rho})h_j = 0. \quad (7.2.21)$$

For given values of ρ, we substitute into (7.2.21) and calculate conditional estimates $\beta(\rho)$. Both MLE and MMLE can then be calculated as described for the three-parameter gamma distribution.

Truncated Samples from the Gamma Distribution

The loglikelihood function of a singly right truncated sample from the two-parameter gamma distribution follows from (7.2.11) with $\gamma = 0$ as

$$\ln L = -n \ln \Gamma(\rho) - n\rho \ln \beta - \frac{1}{\beta} \sum_{i=1}^{n} x_i$$
$$+ (\rho - 1) \sum_{i=1}^{n} \ln x_i - n \ln F(T). \quad (7.2.22)$$

With ρ known, the only remaining estimating equation becomes

$$\frac{\partial \ln L}{\partial \beta} = -\frac{n\rho}{\beta} + \frac{1}{\beta^2} \sum_{i=1}^{n} x_i + \frac{n}{\beta}(\xi + \sqrt{\rho})\frac{g(\xi)}{G(\xi)} = 0. \quad (7.2.23)$$

For given approximations ρ_i, we substitute into (7.2.23) and calculate corresponding conditional estimates β_i which are subsequently substituted into

The Inverse Gaussian and Gamma Distributions

(7.2.22) in search of a maximum. We make as many trials as necessary to determine the required maximum.

7.3 A PSEUDO-COMPLETE SAMPLE TECHNIQUE FOR PARAMETER ESTIMATION FROM CENSORED SAMPLES

A pseudo-complete sample technique of Whitten et al. (1988), whereby censored samples are "completed," offers an attractive alternative to maximum likelihood and modified maximum likelihood estimators for singly censored samples. This procedure is applicable in all types of distributions, and it is particularly useful in calculating estimates of IG and gamma distribution parameters from singly right censored samples. In these cases the computational complications involved with differentiation under integral signs is avoided, and we are able to employ our choice of complete sample estimators.

This technique is quite simple in concept. An iterative procedure is employed to estimate values of the c censored observations, x_{n+i}, $i = 1, 2, \ldots, c$. These estimated observations are then added to the n complete observations to form a pseudo-complete sample of size $N = n + c$, and parameter estimates are then calculated by employing moment, maximum likelihood, modified, or other complete sample estimators.

Estimating equations for completed values of the censored observations are

$$E[F(X_{n+i})] = F(\hat{x}_{n+i}) = \frac{n+i}{N+1}. \qquad (7.3.1)$$

This equation is equivalent to

$$\hat{x}_{n+i} = \mu + \hat{\sigma} z_{n+i} = T + \hat{\sigma}(z_{n+i} - z_n), \qquad (7.3.2)$$

where the z_{n+i} are order statistics of the standard distribution $(0, 1, \alpha_3)$, and where T is the point of censoring; that is, $z_{n+i} = [x_{n+i} - E(X)]/\sigma$, $i = 0, 1, \ldots, c$. It follows that

$$G(z_{n+i}; 0, 1, \alpha_3) = \frac{n+i}{N+1}, \qquad (7.3.3)$$

where $G(\ )$ is the cdf of the standard distribution involved. At this time, we are concerned with the IG and the gamma distributions. Equation (7.3.3) is employed to obtain values of z_{n+i}, and it might be more convenient to express this equation as

$$z_{n+i} = G^{-1}\left[\frac{n+i}{N+1}, \alpha_3\right]. \qquad (7.3.4)$$

We let $x_{n+i}^{(j)}$, $i = 1, 2, \ldots, c$, designate the jth iteration to $\hat{x}_{n+i}$. For the first iteration (i.e., the first approximation), it is convenient to let $x_{n+1}^{(1)} = x_{n+2}^{(1)} = \cdots = x_{n+c}^{(1)} = T$. Of course, T is a lower bound on all x_{n+i}, and we could begin with first approximations that exceed this value. However, in most applications the choice of T for first approximations is convenient, and it usually results in reasonably rapid convergence to the final estimates.

First approximations to the mean and standard deviation of the pseudo-complete sample of size N are

$$\bar{x}_N^{(1)} = \frac{1}{N}\left[\sum_{i=1}^{n} x_i + cT\right], \qquad (7.3.5)$$

$$s_N^{(1)} = \sqrt{\frac{1}{N-1}\left[\sum_{i=1}^{n}(x_i - \bar{x}_N^{(1)})^2 + c(T - \bar{x}_N^{(1)})^2\right]}.$$

For subsequent approximations, that is, for $j = 2, 3, \ldots, k$, we have

$$\bar{x}_N^{(j)} = \frac{1}{N}\left[\sum_{i=1}^{n} x_i + \sum_{i=1}^{c} x_{n+i}^{(j)}\right], \qquad (7.3.6)$$

$$s_N^{(j)} = \sqrt{\frac{1}{N-1}\left[\sum_{i=1}^{n}(x_i - \bar{x}_N^{(j)})^2 + \sum_{i=1}^{c}(x_{n+i}^{(j)} - \bar{x}_N^{(j)})^2\right]}.$$

We must calculate $\{x_{n+i}^{(j)}\}$, $i = 1, 2, \ldots, c$, before $\bar{x}_N^{(j)}$ and $s_N^{(j)}$ can be evaluated, and these calculations depend on the type of distribution. For the IG, gamma, and other skewed distributions, we must also estimate α_3 at each stage of iteration. This might be done in various ways. Perhaps the simplest procedure would be to calculate the "complete" sample moment estimate, but as an alternative we might employ MLE, or modified estimators.

After calculating a first approximation to α_3, we calculate corresponding first approximations to ξ_{n+i} as

$$\xi_{n+i}^{(1)} = G^{-1}\left[\frac{n+i}{N+1}, \alpha_3^{(1)}\right], \qquad (7.3.7)$$

where $G(z: 0, 1, \alpha_3)$ designates the cdf of the standardized parent population. For samples from the Weibull, lognormal, IG, and gamma distributions, inverse interpolation in the cdf tables included in the appendixes will facilitate the calculation of $\xi_{n+i}^{(j)}$.

Second approximations $x_{n+i}^{(2)}$, $i = 1, 2, \ldots, c$ then follow as

$$x_{n+i}^{(2)} = T + s_N^{(1)}(\xi_{n+i}^{(1)} - \xi_n^{(1)}). \qquad (7.3.8)$$

We repeat the cycle of calculations until for all i, the absolute differences between successive approximations,

The Inverse Gaussian and Gamma Distributions

$$\left|\hat{x}_{n+i}^{(j+1)} - \hat{x}_{n+i}^{(j)}\right| \quad \text{and} \quad \left|\hat{s}_N^{(j+1)} - \hat{s}_N^{(j)}\right|$$

are less than some prescribed maximum error. Parameter estimates are then calculated from the "completed" sample by employing appropriate complete sample estimators.

As computational aids for the calculation of complete sample estimates, graphs of α_3 as a function of z_1 and n for the inverse Gaussian distribution have been included as Figure 7.1. A similar chart for the gamma distribution has been included as Figure 7.2. These charts can be used to simplify calculation of modified moment estimates from complete samples. With $\bar{x}$, s, x_1, and n available from sample data, we calculate $z_1 = (x_1 - \bar{x})/s$ and enter the applicable chart with z_1 and n to read the estimate of $\hat{\alpha}_3$. More complete accounts of these calculations have been given by Cohen and Whitten (1985, 1986, 1988).

7.4 AN ILLUSTRATIVE EXAMPLE

To illustrate the practical application of the pseudo-complete sample technique in the calculation of estimates of inverse Gaussian parameters, we give an example from Whitten et al. (1988).

Example 7.1. A random sample of size $N = 100$ was generated from an inverse Gaussian population with $\gamma = 10$, $\mu = 6$, and $\sigma = 5$, and thus with $\alpha_3 = 2.5$. The complete sample data are listed in order of magnitude in Table 7.1. For the complete sample, $\bar{x} = 15.9236$, $s = 5.1460$, $a_3 = 2.3142$, and $x_1 = 10.8884$. For illustrative purposes, we create singly right censored samples by censoring at $x_{70} = 16.1922$ with $c = 30$ ($h = 0.30$); at $x_{75} = 17.2458$ with $c = 25$ ($h = 0.25$); at $x_{80} = 18.3316$ with $c = 20$ ($h = 0.20$); at $x_{85} = 19.0801$ with $c = 15$ ($h = 0.15$); at $x_{90} = 23.0403$ with $c = 10$ ($h = 0.10$); and at $x_{95} = 25.8634$ with $c = 5$ ($h = 0.05$).

PC estimates were calculated from each of these censored samples as described for the IG distribution, and they are displayed in Table 7.2 along with corresponding complete sample modified moment estimates.

As an illustration of the iterative process, results of the nine cycles of iteration required to reach final estimates for $h = 0.05$ are entered in Table 7.3 along with a summary of corresponding complete sample results.

7.4.1 Closing Comments

The author's experience to date indicates that PC estimates are satisfactory in both skewed and normal distributions provided censoring is not too severe. For

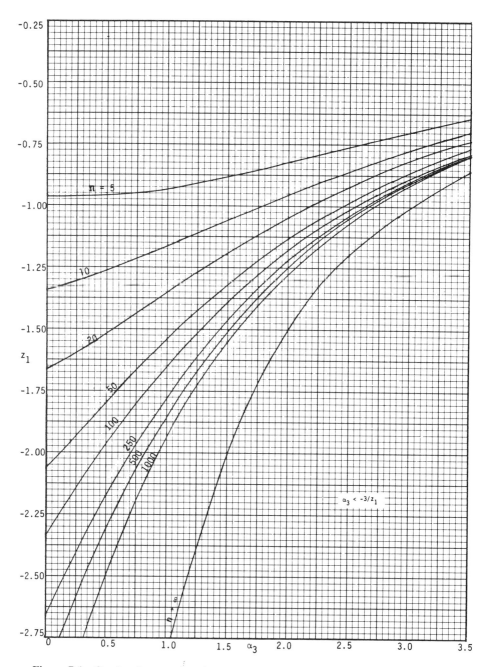

Figure 7.1 Graphs of α_3 as a function of z_1 and n in complete samples from the inverse Gaussian distribution. Reprinted from Cohen and Whitten (1985), Fig. 1, p. 151, with permission of the American Society for Quality Control.

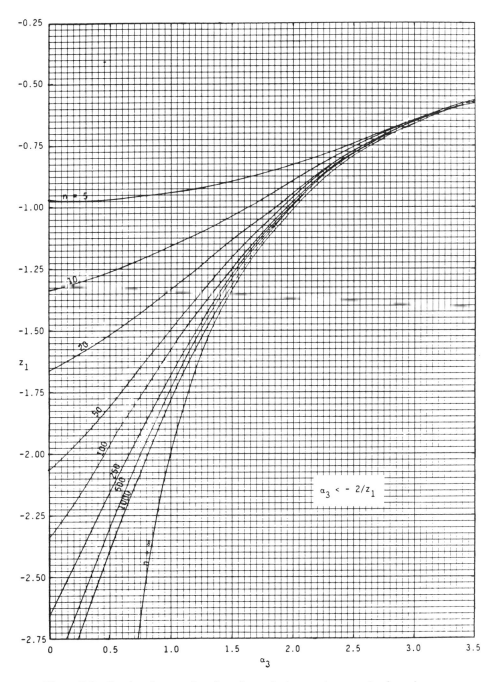

Figure 7.2 Graphs of α_3 as a function of z_1 and n in complete samples from the gamma distribution. Reprinted from Cohen and Whitten (1986), Fig. 1, p. 55, with permission of the American Society for Quality Control.

Table 7.1 A Random Sample from an Inverse Gaussian Distribution: $\gamma = 10$, $\mu = 6$, $\sigma = 5$ ($\alpha_3 = 2.5$)

10.8884	11.1417	11.1562	11.3311	11.3493
11.4726	11.5578	11.5839	11.6037	11.6113
11.6640	11.7501	11.8600	12.0766	12.1246
12.1868	12.2967	12.3286	12.3296	12.3327
12.3504	12.4423	12.5053	12.6288	12.6423
12.6492	12.6268	12.7462	12.8156	12.8532
13.1523	13.1642	13.1807	13.2550	13.3709
13.4310	13.5195	13.5336	13.5786	13.5987
13.6522	13.6539	13.6596	13.6825	13.7392
13.7657	13.7764	13.8136	13.8846	14.0068
14.0490	14.0568	14.3218	14.5941	14.6496
14.6811	14.7629	15.0891	15.2358	15.3520
15.4604	15.5547	15.6059	15.7032	15.7977
15.8906	15.8931	15.9440	16.1099	16.1922
16.2001	16.4352	16.4605	17.2224	17.2458
17.3007	17.4365	17.8224	18.1382	18.3316
18.4063	18.6581	18.8229	19.0166	19.0801
19.2521	19.3555	21.0255	22.1120	23.0403
24.3057	24.9244	25.2042	25.2134	25.8634
26.3125	27.5814	29.3053	32.5742	42.4409

Source: From Sundaraiyer (1986), p. 64, with permission of the author.

Table 7.2 Summary of Estimates of Inverse Gaussian Parameters

Estimator		PARAMETER				
		γ	μ	$E(X)$	σ	α_3
Population		10	6	16	5	2.5
MME (Complete Sample)		10.009	5.914	15.924	5.146	2.610
P.C.	(h=.05)	9.905	5.954	15.859	4.828	2.433
P.C.	(h=.10)	9.911	5.940	15.851	4.829	2.439
P.C.	(h=.15)	9.517	5.872	15.389	3.702	1.892
P.C.	(h=.20)	9.576	5.878	15.454	3.837	1.958
P.C.	(h=.25)	9.461	5.851	15.312	4.565	1.828
P.C.	(h=.30)	9.207	5.835	15.042	3.094	1.591

Source: From Whitten et al. (1988), Table VI, p. 2253, by courtesy of Marcel Dekker, Inc.

The Inverse Gaussian and Gamma Distributions

Table 7.3 Successive Iterations of Censored Observations for Pseudo-complete Samples from the Inverse Gaussian Distribution: $\gamma = 10$, $\mu = 6$, $\alpha = 5$, $E(X) = 16$, $\alpha_3 = 2.5$, $N = 100$, $h = .05$

Item	Iteration Number, (j)									Complete Sample Summary
	1	2	3	4	5	6	7	8	9	
$\bar{x}_N^{(j)}$	15.6350	15.8134	15.8484	15.8567	15.8587	15.8592	15.8593	15.8593	15.8593	15.924
$\mu^{(j)}$	5.9760	5.9590	5.9553	5.9545	5.9544	5.9543	5.9543	5.9543	5.9543	5.915
$s_N^{(j)}$	5.1485	4.7899	4.7899	4.8186	4.8257	4.6274	4.8278	4.8279	4.8280	5.146
$\gamma^{(j)}$	9.6590	9.8931	9.8931	9.9021	9.9043	9.9049	9.9050	9.9050	9.9050	10.009
$\alpha_3^{(j)}$	2.0826	2.3515	2.4129	2.4277	2.4314	2.4322	2.4324	2.4325	2.4325	2.610
$z_1^{(j)}$	-1.1442	-1.0544	-1.0355	-1.0311	-1.0300	-1.0297	-1.0296	-1.0296	-1.0296	-0.979
$\xi_{95}^{(j)}$	1.7737	1.7648	1.7619	1.7612	1.7610	1.7610	1.7609	1.7609	1.7609	
$\xi_{96}^{(j)}$	1.9576	1.9580	1.9572	1.9570	1.9569	1.9569	1.9569	1.9569	1.9569	
$\xi_{97}^{(j)}$	2.1947	2.1979	2.1998	2.2003	2.2004	2.2004	2.2004	2.2004	2.2004	
$\xi_{98}^{(j)}$	2.4809	2.5123	2.5183	2.5197	2.5200	2.5201	2.5201	2.5201	2.5201	
$\xi_{99}^{(j)}$	2.9043	2.9648	2.9771	2.9801	2.9808	2.9809	2.9810	2.9810	2.9810	
$\xi_{100}^{(j)}$	3.6431	3.7609	3.7860	3.7920	3.7935	3.7938	3.7939	3.7940	3.7940	
$x_{96}^{(j)}$	25.8634	26.6263	26.7661	26.7988	26.8067	26.8087	26.8092	26.8093	26.8093	26.3125
$x_{97}^{(j)}$	25.8634	27.5687	27.8865	27.9610	27.9791	27.9836	27.9847	27.9849	27.9850	27.5814
$x_{98}^{(j)}$	25.8634	28.7973	29.3552	29.4863	29.5182	29.5260	29.5280	29.5284	29.5285	29.3053
$x_{99}^{(j)}$	25.8634	30.5537	31.4684	31.6841	31.7367	31.7495	31.7527	31.7535	31.7537	32.5742
$x_{100}^{(j)}$	25.8634	33.6188	31.1869	35.5587	35.6493	35.6715	35.6770	35.6783	35.6787	42.4409

Source: Adapted from Whitten et al. (1988), Table VII, p. 2253, by courtesy of Marcel Dekker, Inc.

$h < 0.30$, convergence has been reasonably rapid in most applications. The only convergence problem of any consequence that has been encountered thus far, has involved estimation in the IG and gamma distributions when α_3 is near zero. This is unlikely to prove troublesome in practical applications, however, since a value of α_3 that is close to zero suggests that the normal rather than a skewed distribution is a more appropriate model. A more complete account of the pseudo-complete sample technique is given by Whitten et al. (1988).

8

Truncated and Censored Samples from the Exponential and the Extreme Value Distributions

8.1 THE EXPONENTIAL DISTRIBUTION
8.1.1 Introduction

The exponential distribution, sometimes referred to as the negative exponential distribution, is a special case of the Weibull and also of the gamma distribution for which the Weibull shape parameter $\delta = 1$ or the gamma shape parameter $\rho = 1$. It is a positively skewed reverse J-shaped distribution. The two-parameter pdf is

$$f(x; \gamma, \beta) = \frac{1}{\beta} \exp - \left(\frac{x - \gamma}{\beta}\right), \quad \gamma < x < \infty, \beta > 0, \quad (8.1.1)$$
$$= 0 \quad \text{elsewhere.}$$

and the cdf is

$$F(x; \gamma, \beta) = 1 - \exp\left[-\left(\frac{x - \gamma}{\beta}\right)\right]. \quad (8.1.2)$$

In this notation, γ is the threshold parameter and β is the scale parameter. In many applications the origin is at zero, that is, $\gamma = 0$, and in these cases the distribution is completely characterized by the single parameter β. Basic characteristics of the exponential distribution are

The Exponential and Extreme Value Distributions

$$E(X) = \gamma + \beta, \quad V(X) = \beta^2,$$
$$\text{Me}(X) = \gamma + \beta \ln 2, \quad (8.1.3)$$
$$\alpha_3(X) = 2, \quad \alpha_4(X) = 9,$$

and the hazard function is

$$h(x) = \frac{1}{\beta}. \quad (8.1.4)$$

The cumulative hazard function is

$$H(x) = \int_\gamma^x h(t)\, dt = \left(\frac{x - \gamma}{\beta}\right), \quad (8.1.5)$$

which is a linear function of x.

This distribution has been widely employed as a model in life-span distributions and in various reliability distributions. Since the hazard function is constant for all values of the random variable, it is a suitable model for lifetime data where used items are considered to be as good as new ones. In many applications where the Weibull, gamma, or other distribution might be a more appropriate model, the exponential is used as an approximation because of its simplicity and the ease with which calculations can be made.

Graphs of the pdf and the cdf of this distribution are given in Figures 8.1 and 8.2.

8.1.2 Censored Sample Estimators

The loglikelihood function of a k-stage progressively censored sample from a two-parameter exponential distribution with pdf (8.1.1) where $N = n + \sum_{j=1}^k c_j$ is

$$\ln L = -n \ln \beta - \sum_{i=1}^n \left(\frac{x_i - \gamma}{\beta}\right) - \sum_{j=1}^k c_j \left(\frac{T_j - \gamma}{\beta}\right) + \text{const}. \quad (8.1.6)$$

This is an increasing function of γ. The maximum likelihood estimate of γ is therefore the largest permissible value. Since $\gamma \leq x_{1:N}$, it follows that the MLE for γ is

$$\hat{\gamma} = x_{1:N}. \quad (8.1.7)$$

The estimating equation for β is $\partial \ln L / \partial \beta = 0$, which follows from (8.1.6) as

$$\frac{\partial \ln L}{\partial \beta} = -\frac{n}{\beta} + \frac{1}{\beta^2}\left[\sum_{i=1}^n (x_i - \gamma) + \sum_{j=1}^k c_j(T_j - \gamma)\right] = 0. \quad (8.1.8)$$

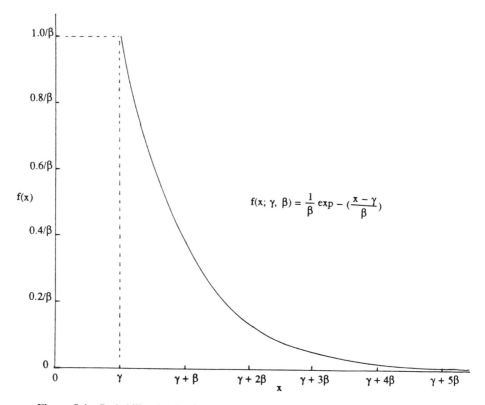

Figure 8.1 Probability density function of the exponential distribution.

Accordingly, the maximum likelihood estimators are

$$\hat{\gamma} = x_{1:N}, \qquad (8.1.9)$$

$$\hat{\beta} = \frac{1}{n}\left[\sum_{i=1}^{n}(x_i - \hat{\gamma}) + \sum_{j=1}^{k} c_j(T_j - \hat{\gamma})\right].$$

To eliminate bias in the estimation of γ, we are led to consider modified maximum likelihood estimators with estimating equations $\partial \ln L/\partial \beta = 0$ and $E(X_{1:N}) = x_{1:N}$. The pdf of the first-order statistic in a sample of size N from a two-parameter exponential distribution with pdf (8.1.1) is

$$f(x_{1:N}; \gamma, \beta) = \frac{N}{\beta}\exp\left[-\frac{N}{\beta}(x_{1:N} - \gamma)\right], \qquad \gamma < x_{1:N}, \qquad (8.1.10)$$

and the expected value is

$$E(X_{1:N}) = \gamma + \beta/N. \qquad (8.1.11)$$

The Exponential and Extreme Value Distributions

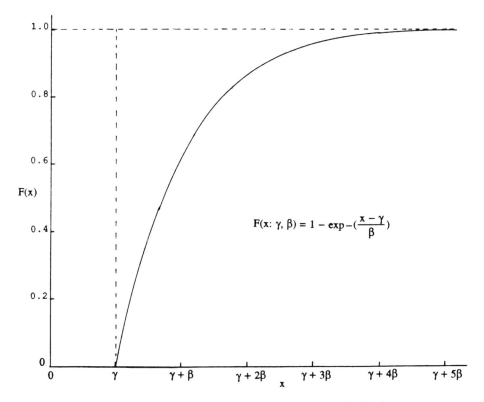

Figure 8.2 Cumulative probability function of the exponential distribution.

It thus follows that estimating equations for the modified estimators are

$$\gamma + \beta/N = x_{1:N}, \tag{8.1.12}$$

$$\beta = \frac{1}{n}\left[\sum_{i=1}^{n}(x_i - \gamma) + \sum_{j=1}^{k}c_j(T_j - \gamma)\right].$$

We let

$$\text{ST} = \sum_{i=1}^{n} x_i + \sum_{j=1}^{k} c_j T_j, \tag{8.1.13}$$

and the MMLE can be expressed as

$$\hat{\beta} = \frac{\text{ST} - Nx_{1:N}}{n-1},$$

$$\hat{\gamma} = \frac{nx_{1:N} - \text{ST}/N}{n-1}. \tag{8.1.14}$$

For a complete (uncensored) sample, $c = 0$ for all j, $N = n$ and $ST = n\bar{x}$. It then follows that

$$\hat{\beta} = \frac{n(\bar{x} - x_{1:n})}{n - 1}$$

$$\hat{\gamma} = \frac{nx_{1:n} - \bar{x}}{n - 1}$$
(8.1.15)

These estimators are recognized as the best linear unbiased (BLUE) and the minimum variance unbiased (MVUE) for complete samples from the two-parameter exponential distribution.

8.1.3 Truncated Sample Estimators

The pdf of a two-parameter exponential distribution that is truncated on the right at $x = T$ is

$$f_T(x; \gamma, \beta) = \frac{1}{\beta} \exp\left[-\frac{1}{\beta}(x - \gamma)\right] [F(T)]^{-1}. \tag{8.1.16}$$

The loglikelihood function of a random sample of size n from a truncated distribution with pdf (8.1.16) is

$$\ln L = -n \ln \beta - \frac{1}{\beta} \sum_{i=1}^{n} (x_i - \gamma)$$
$$- n \ln \left\{ 1 - \exp\left[-\frac{1}{\beta}(T - \gamma)\right] \right\}. \tag{8.1.17}$$

Maximum Likelihood Estimators

Estimating equations, in this case, are $\gamma = x_1$ and $\partial \ln L / \partial \beta = 0$. Thus

$$\hat{\gamma} = x_1,$$

$$\frac{\partial \ln L}{\partial \beta} = -\frac{n}{\beta} + \frac{1}{\beta^2} \sum_{i=1}^{n} (x_i - \gamma) \tag{8.1.18}$$

$$+ \frac{n(T - \gamma)}{\beta^2} \left[\frac{1}{\exp(1/\beta)(T - \gamma) - 1}\right] = 0.$$

In order to simplify the notation, we let

$$J(T; \gamma, \beta) = \left[\exp\frac{1}{\beta}(T - \gamma) - 1\right]^{-1}, \tag{8.1.19}$$

The Exponential and Extreme Value Distributions

and the estimating equations of (8.1.18) become

$$\hat{\gamma} = x_1, \qquad (8.1.20)$$
$$\hat{\beta} = (\bar{x} - \hat{\gamma}) + (T - \hat{\gamma}) J(T; \hat{\beta}, \hat{\gamma}).$$

With $\hat{\gamma} = x_1$, the second of these equations can be solved for $\hat{\beta}$ by employing a "trial and error" iterative procedure.

The first-order statistic is a biased estimator of γ, and in small samples, the bias can be substantial. In an effort to eliminate or at least to reduce bias, we are led to consider modified estimators, just as we did for censored samples.

Modified Maximum Likelihood Estimators

Estimating equations for the MMLE are $E(X_{1:N}) = x_{1:N}$ and $\partial \ln L/\partial \beta = 0$. However, in truncated samples, the number of missing observations and thus N are unknown. Only n is known. We must, therefore, be content with an estimate of N which we express as

$$E(N) = \frac{n}{F(T)} = \frac{n}{1 - e^{-(T-\gamma)/\beta}}. \qquad (8.1.21)$$

Accordingly, the estimating equations become

$$\gamma + \frac{\beta}{n}(1 - e^{-(T-\gamma)/\beta}) = x_1$$
$$\beta = (\bar{x} - \gamma) + (T - \gamma)J(\beta, \gamma), \qquad (8.1.22)$$

where the second of the preceding equations is identical with the second equation of (8.1.20) for the MLE. The simultaneous solution of the two equations of (8.1.22) yields the required estimates $\hat{\beta}$ and $\hat{\gamma}$.

Estimation in the One-Parameter Truncated Exponential Distribution

When γ is known and may therefore be set equal to zero, the second equation of (8.1.22) and also the second equation of (8.1.20) become

$$\beta - T \cdot J(\beta) = \bar{x}, \qquad (8.1.23)$$

where $J(\beta) = 1/[e^{T/\beta} - 1]$. In this case, the solution of (8.1.23) for β is relatively simple.

Asymptotic Variance

In the one-parameter case the asymptotic variance of $\hat{\beta}$ is

$$V(\hat{\beta}) = \left[E\left(\frac{\partial^2 \ln L}{\partial \beta^2}\right) \right]^{-1}, \qquad (8.1.24)$$

where

$$\frac{\partial^2 \ln L}{\partial \beta^2} = \frac{n}{\beta^2} - \frac{2n\bar{x}}{\beta^3} + \frac{nT}{\beta^2} J'(\beta) - \frac{2nT}{\beta^3} J(\beta),$$

and

$$J'(\beta) = \frac{\partial J}{\partial \beta} = \frac{T}{\beta^2} e^{T/\beta} J^2(\beta). \tag{8.1.25}$$

For moderate to large-sized samples, satisfactory approximations to the variance can be obtained when we replace the expected value of the second partial with actual sample values; that is,

$$E\left(\frac{\partial^2 \ln L}{\partial \beta^2}\right) \doteq \left.\frac{\partial^2 \ln L}{\partial \beta^2}\right|_{\beta = \hat{\beta}}. \tag{8.1.26}$$

8.1.4 Hazard Plot Estimates

Nelson (1969) proposed a simple graphical procedure based on a plot of the cumulative hazard function for both determining the distribution type and for estimating its parameters from progressively censored samples. This procedure can be used for a number of different distributions. It is particularly well suited to estimation in the Weibull and the exponential distributions.

Since the cumulative hazard function, $H(x)$, of the exponential distribution as given by (8.1.5) is linear, a plot of this function on ordinary rectangular coordinate graph paper provides a simple graphical technique for estimating β and γ.

From (8.1.5), we note that

$$H(\gamma + k\beta) = k, \quad H(\gamma) = 0, \quad H(\gamma + \beta) = 1. \tag{8.1.27}$$

We therefore need only select two points with coordinates $[(\gamma + k_1\beta), k_1]$ and $[(\gamma + k_2\beta), k_2]$, respectively, and solve for the required estimates. For example, the points $(\gamma, 0)$ and $((\gamma + \beta), 1)$ will often be convenient selections. The estimates then follow as

$$\gamma^* = H^{-1}(0), \quad \gamma^* + \beta^* = H^{-1}(1), \quad \beta^* = H^{-1}(1) - H^{-1}(0). \tag{8.1.28}$$

Of course, it is first necessary that the sample data be used to plot the cumulative hazard function. Nelson (1969) outlined a simple practical procedure for accomplishing this task for a progressively censored sample. For a given sample, full-term (complete) and partial-term (censored) observations are ordered with respect to magnitude and assigned reverse rank numbers. The first (smallest) observation is assigned the reverse rank N. The next is assigned the reverse rank $N - 1$. The process is continued until the last (largest) observation is assigned reverse rank 1. The reverse ranks thus represent the number of survivors immediately prior to censoring of a corresponding sample item. Estimates of the hazard or instantaneous failure rate (in the context of a life test) as percentages are then calculated at times x_i, $i = 1, \ldots, n$, as

$$h(x_i) = 100 \left(\frac{1}{m_i}\right), \tag{8.1.29}$$

where m_i is the reverse rank corresponding to x_i. The cumulative hazard is calculated as

$$H(x_i) = \sum_{j=1}^{i} h(x_j). \tag{8.1.30}$$

A plot of the cumulative hazard function is then made on appropriate hazard cross section graph paper. There will be a plotted point for each full-term observation. If the correct paper is chosen, the plotted points will lie on a straight line. As previously noted, ordinary rectangular cross-section paper is proper for the exponential distribution.

The Weibull Distribution

At this point, we digress to consider some similarities between the exponential and the Weibull distribution. As previously noted, the exponential is a special case of the Weibull distribution in which the shape parameter is $\delta = 1$. The cumulative hazard function of the Weibull distribution is

$$H(x) = \left(\frac{x - \gamma}{\beta}\right)^{\delta} \quad \text{and} \quad H(\gamma + k\beta) = k^{\delta}. \tag{8.1.31}$$

On taking logarithms of both sides of the first equation of (8.1.31), we obtain

$$\ln H(x) = \delta \ln (x - \gamma) - \delta \ln \beta. \tag{8.1.32}$$

Accordingly, $\ln H(x)$ is a linear function of $\ln (x - \gamma)$. Consequently, a plot of $H(x)$ on log-log coordinate paper will be a straight line with the shape parameter δ as the slope. To calculate estimates γ^* and β^*, it may be convenient to make a plot on rectangular coordinate paper in order to read $\gamma^* = H^{-1}(0)$. Of course, $(\gamma^* + B^*) = H^{-1}(1)$ can be read from either a rectangular or a log-log plot. We then calculate $\beta^* = H^{-1}(1) - H^{-1}(0)$, and it follows from (8.1.31) that

$$\delta^* = \frac{\ln H(\gamma^* + k\beta^*)}{\ln k}. \tag{8.1.33}$$

This procedure can be carried out regardless of the type of coordinate paper used for the plot, but a straight-line plot might be more accurate than one that is curved. There is less bunching of plotted points on log-log paper in the vicinity of the origin, and this is a factor in improving accuracy. Nelson (1969) explored the subject of hazard plotting in considerable detail for various distributions in addition to the exponential and the Weibull. For the extreme value distribution, he pointed out that the appropriate hazard plotting paper is semilog with lifespans plotted on the linear scale and $H(x)$ on the logarithmic scale. Special plotting papers which Nelson describes in detail were developed for other distributions.

The underlying theory of hazard plots is based on the assumption that items selected for censoring are chosen at random from the survivors at the time of censoring. It is therefore important in the application of these procedures that items not be selected for censoring because of indications that they are about to fail. Otherwise, estimates obtained might be invalid.

Example 8.1.1. To illustrate the hazard plotting technique and parameter estimation from progressively censored samples, we have selected a sample that was originally used as an illustration by Nelson (1969). Sample observations consist of times to failure and/or times to censoring for 70 generator fans. Sample data together with calculated values of the hazard function $h(x)$ and the cumulative hazard function $H(x)$ are entered in Table 8.1. In summary, $N = 70$, $n = 12$, $x_{1:70} = 4500$, $ST = 3,444,400$, $ST/n = 287,033.33$, $\Sigma_{i=1}^{12} x_i = 365,700$, and $\Sigma_{j=1}^{k} c_j T_j = 3,078,700$. A rectangular coordinate plot of the cumulative hazard function for this example is given as Figure 8.3 with a straight line fitted by eye. The straight-line fit indicates that the exponential distribution is the appropriate model for these data. From the fitted line, we read $H(268,000) = 100\%$ and $H(450) = 0$. Accordingly $\gamma^* = 450$, $\gamma^* + \beta^* = 268,000$, and $\beta^* = 275,550$. We let $k = 1/2$ in equation (8.1.33) and use this equation to calculate

$$\delta^* = -\ln 2 \ln H\left[\left(\frac{\beta^*}{2}\right) + \gamma^*\right] = -1.442695 \ln(0.5) = 1.00.$$

This result confirms our earlier conclusion that this sample is from an exponential distribution. For comparison with the hazard plot estimates, we employ equations (8.1.14) to calculate modified maximum likelihood estimates as

$$\hat{\gamma} = \frac{12(4500) - 3,444,400/70}{11} = 436,$$

$$\hat{\beta} = \frac{3,444,400 - 70(4500)}{11} = 284,491.$$

Estimates of the mean and standard deviation are $\hat{E}(X) = \hat{\gamma} + \hat{\beta} = 284,927$ and $\hat{\sigma}_x = \hat{\beta} = 284,491$. Differences between these estimates and the hazard

Table 8.1 A Progressively Censored Sample Consisting of Life-Span Observations of 70 Generator Fans

Rank	Reverse rank (m)	Time 1000 hr (x)	h(x) (%)	H(x) (%)	Rank	Reverse rank (m)	Time 1000 hr (x)	h(x) (%)	H(x) (%)
1	70	4.5	1.43	1.43	36	35	43.0+		
2	69	4.6+			37	34	46.0	2.94	18.78
3	68	11.5	1.47	2.90	38	33	48.5+		
4	67	11.5	1.49	4.39	39	32	48.5+		
5	66	15.6+			40	31	48.5+		
6	65	16.0	1.54	5.93	41	30	48.5+		
7	64	16.6+			42	29	50.0+		
8	63	18.5+			43	28	50.0+		
9	62	18.5+			44	27	50.0+		
10	61	18.5+			45	26	61.0+		
11	60	18.5+			46	25	61.0	4.00	22.78
12	59	18.5+			47	24	61.0+		
13	58	20.3+			48	23	61.0+		
14	57	20.3+			49	22	63.0+		
15	56	20.3+			50	21	64.5+		
16	55	20.7	1.82	7.75	51	20	64.5+		
17	54	20.7	1.85	9.60	52	19	67.0+		
18	53	20.8	1.89	11.49	53	18	74.5+		
19	52	22.0+			54	17	78.0+		
20	51	30.0+			55	16	78.0+		
21	50	30.0+			56	15	81.0+		
22	50	30.0+			57	14	81.0+		
23	48	30.0+			58	13	82.0+		
24	47	31.0	2.13	13.62	59	12	85.0+		
25	46	32.0+			60	11	85.0+		
26	45	34.5	2.22	15.84	61	10	85.0+		
27	44	37.5+			62	9	87.5+		
28	43	37.5+			63	8	87.5	12.50	35.28
29	42	41.5+			64	7	87.5+		
30	41	41.5+			65	6	94.0+		
31	40	41.5+			66	5	99.0+		
32	39	41.5+			67	4	101.0+		
33	38	43.0+			68	3	101.0+		
34	37	43.0+			69	2	101.0+		
35	36	43.0+			70	1	115.0+		

Note: $h(x) = (100/m)$; $H(x) = \Sigma h(x)$.

Source: Adapted from Nelson (1969), Table 1, p 29, with permission of the American Society for Quality Control.

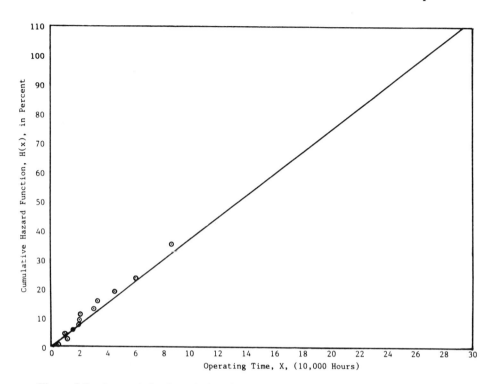

Figure 8.3 A cumulative hazard plot of generator fan data. Reprinted from Cohen and Whitten (1988), Fig. 7.3, p. 125, by courtesy of Marcel Dekker, Inc.

plot estimates are small and simply reflect the limitations of graphical methods.

The data for this example were collected to answer questions raised by the manufacturer concerning the feasibility of offering an 80,000 hour warranty on his fans. He wished to know the magnitude of the resulting repair or replacement obligation if this warranty should become effective. To make this determination, we substitute the MMLE $\hat{\gamma} = 436$ and $\hat{\beta} = 284{,}491$ into the cdf (8.1.2) and calculate $P[X < 80{,}000] = 1 - \exp[-(80{,}000 - 436)/284{,}491] = 0.244$. We therefore conclude that an 80,000 hour warranty without an improvement in quality to increase the life-span might require repair or replacement of as many as 24 percent of all fans sold.

This example originally appeared in a General Electric technical report by Nelson prior to publication in *J. Qual. Tech.* (Nelson, 1969). Additional details have been presented by Nelson (1972, 1982). Further consideration is given by Nelson in "How to Analyze Reliability Data," *ASQC Basic References in Quality Control Statistical Techniques*.

8.2 THE EXTREME VALUE DISTRIBUTION

Extreme value distribution play an important role in the analysis of lifetime data. They are also useful in the analysis of rainfall, flood flow, earthquake, and other meteorological data. They are applicable as models in strength of material, corrosion, and other "weakest link" studies. The designation of extreme value is commonly applied to three types of limiting distributions, which approximate distributions of extremes (least or greatest) in large random samples. The cdfs of the three types of greatest extreme values are given below. Limiting distributions for the three types of least extreme values can be obtained by replacing X with $-X$.

Type I

$$F(x; \mu, \alpha) = \exp\left[-\exp\left(-\frac{x-\mu}{\alpha}\right)\right], \quad -\infty < x < \infty. \quad (8.2.1)$$

Type II

$$F(x; \mu, \alpha, \eta) = \exp\left(-\frac{x-\mu}{\alpha}\right)^{-\eta}, \quad x \geq \mu,$$
$$= 0 \quad \text{elsewhere.} \quad (8.2.2)$$

Type III

$$F(x; \mu, \alpha, \eta) = \exp\left(\frac{x-\mu}{\alpha}\right)^{\eta}, \quad x \leq \mu,$$
$$= 1 \quad \text{otherwise,} \quad (8.2.3)$$

where μ, α (>0), and η (>0) are parameters.

Of these three distributions, Type I is the one most often referred to in discussions of "extreme value" distributions. Many authors consider it to be "the" extreme value distribution, and primary consideration here is devoted to this type. Detailed discussions of these distributions can be found in books by Johnson and Kotz (1970), Lawless (1982), Mann et al. (1974), and Elandt-Johnson and Johnson (1980). There are, of course, many other sources of information about these distributions. Bortkiewicz (1922) was perhaps the first to present a general theory in connection with a study of the range in samples from the normal distribution. For a thorough account of the theory of extreme values in all of its facets, readers are referred to Gumbel (1958) and to a classic expository paper on this subject by Gnedenko (1943). It was Gumbel who pioneered application of the Type I distribution as a model for the analysis of lifetime data. In recognition of his contributions, this distribution is often referred to as the Gumbel distribution. Because of the functional form, the Type I distribution is sometimes called the double exponential distribution. On some occasions it is referred to as the log-Weibull distribution as a consequence of its relationship to the Weibull distribution. If Y has a Weibull distribution (β, δ) then $X = \ln Y$ has an extreme value distribution (μ, α). Conversely, if X has an extreme value distribution (μ, α) then $Y = \exp(X)$ has a Weibull distribution (β, δ). Accordingly, results derived in terms of one distribution are easily transferred to the other. In some applications, it is easier to work with the Weibull distribution and in others the extreme value distribution is easier to work with.

8.2.1 Some Fundamentals

As given in Chapter 5, the pdf of the two-parameter Weibull distribution is

$$f(y; \beta, \delta) = \frac{\delta}{\beta^\delta} y^{\delta-1} \exp\left(-\frac{y}{\beta}\right)^\delta, \quad 0 < y < \infty, \delta > 0, \beta > 0,$$
$$= 0 \quad \text{elsewhere,} \tag{8.2.4}$$

and the pdf of the Type I distribution of greatest extreme values is

$$f(x; \mu, \alpha) = \frac{1}{\alpha} \exp\left(-\frac{x-\mu}{\alpha}\right) \exp\left[-\exp\left(-\frac{x-\mu}{\alpha}\right)\right],$$
$$-\infty < x < \infty, \quad \alpha > 0, \tag{8.2.5}$$

where μ and α are location and scale parameters, respectively, and

$$\mu = \ln \beta, \quad \alpha = \frac{1}{\delta}. \tag{8.2.6}$$

Note that although δ is the shape parameter of the Weibull distribution, its reciprocal becomes the scale parameter of the extreme value distribution. To obtain the pdf of the Type I distribution of least extreme values, it is necessary only that x be replaced with $-x$.

The extreme value distribution is unimodal with inflection points at $\mu \pm 0.96242\alpha$. As given by Johnson and Kotz (1970), the moment-generating function of the Type I distribution of greatest extreme values is

$$M_x(t) = E(e^{tX}) = e^{\mu t}\Gamma(1 - \alpha t), \quad \alpha|t| < 1, \tag{8.2.7}$$

and the cumulant-generating function is

$$\text{CGF}_x(t) = \mu t + \ln \Gamma(1 - \alpha t). \tag{8.2.8}$$

The Exponential and Extreme Value Distributions

The cumulants follow as

$$\kappa_1(X) = \mu - \alpha\psi(1) = \mu + 0.57722\alpha, \quad (8.2.9)$$
$$\kappa_r(X) = (-\alpha)^r \psi^{(r-1)}(1), \quad r \geq 2.$$

where $\psi(\ )$ is the digamma function. The expected value is

$$E(X) = \mu + 0.57722\alpha, \quad (8.2.10)$$

where 0.57722 is recognized as Euler's constant.

Other characteristics of interest are

$$V(X) = \frac{\pi^2}{6}\alpha^2 = 1.64493\alpha^2,$$

$$\alpha_3^2(X) = \beta_1(X) = 1.29857$$

$$\alpha_4(X) = \beta_2(X) = 5.4, \quad (8.2.11)$$

$$\text{Mo}(X) = \mu,$$

$$\text{Me}(X) = \mu - \alpha \ln(\ln 2) = \mu + 0.36651\alpha.$$

The distribution of greatest extremes is positively skewed with $\alpha_3 = 1.139548$, whereas the distribution of least extremes is negatively skewed with $\alpha_3 = -1.139548$.

If we let $\mu = 0$ and $\alpha = 1$ in (8.2.5), we obtain the following standard pdfs:

Standard Type I PDF of Greatest Extreme Values

$$f(y) = e^{-y} \exp(-e^{-y}). \quad (8.2.12)$$

Standard Type I PDF of Least Extreme Values

$$f(y) = e^{y} \exp(-e^{y}). \quad (8.2.13)$$

Graphs of these pdfs are shown in Figure 8.4.

8.2.2 Parameter Estimation

Singly Right Censored Samples

Let $y_1 \leq y_2 \leq \ldots y_n \leq T$ be the n smallest observations in a random sample of N from a two-parameter Weibull distribution and let $x_1 \leq x_2 \leq \ldots \leq x_n \leq \ln T$ where $x_i = \ln y_i$ designate equivalent observations in a sample from the Type I distribution of smallest extremes. For a Type II censored sample, $T = x_{n:N}$, the nth-order statistic in a sample of size N. For a Type I censored sample

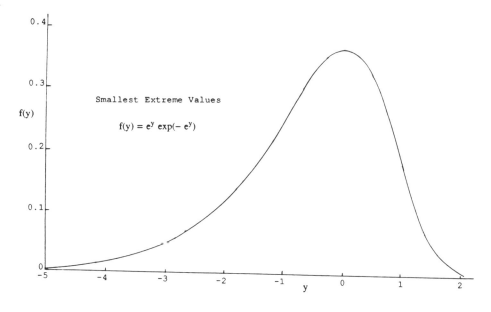
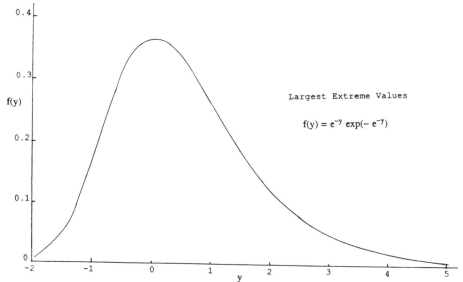

Figure 8.4 Probability density functions of extreme value distributions.

The Exponential and Extreme Value Distributions

$T > x_{n:N}$. Let $c = (N - n)$ designate the number of censored observations, and the likelihood function of a sample as thus described from a distribution that is Type I of least extreme values is

$$L = \frac{N!}{c!} \left[\prod_{i=1}^{n} \frac{1}{\alpha} e^{(x_i - \mu)/\alpha} \exp\left(-e^{(x_i - \mu)/\alpha}\right) \right] [\exp(-e^{(\ln T - \mu)/\alpha})]^c, \quad (8.2.14)$$

where for any sequence

$$\Sigma^* z_i = \sum_{i=1}^{n} z_i + c z_{n:N}. \quad (8.2.15)$$

The loglikelihood function then becomes

$$\operatorname{Ln} L = -n \ln \alpha + \frac{1}{\alpha} \sum_{i=1}^{n} (x_i - \mu) - \Sigma^* \exp\left(\frac{x_i - \mu}{\alpha}\right). \quad (8.2.16)$$

Maximum likelihood estimating equations, $\partial \ln L/\partial \mu = 0$, and $\partial \ln L/\partial \alpha = 0$ follow as

$$\frac{\partial \ln L}{\partial \mu} = -\frac{n}{\alpha} + \frac{1}{\alpha} \Sigma^* \exp\left(\frac{x_i - \mu}{\alpha}\right) = 0,$$

$$\frac{\partial \ln L}{\partial \alpha} = -\frac{n}{\alpha} - \frac{1}{\alpha} \sum_{i=1}^{n} \left(\frac{x_i - \mu}{\alpha}\right) \quad (8.2.17)$$

$$+ \frac{1}{\alpha} \Sigma^* \left(\frac{x_i - \mu}{\alpha}\right) \exp\left(\frac{x_i - \mu}{\alpha}\right) = 0.$$

Maximum likelihood estimates $\hat{\mu}$ and $\hat{\alpha}$ are subsequently obtained as the simultaneous solution of these two equations. Since explicit solutions cannot be found, we must resort to iterative techniques such as Newton's method or perhaps a "trial and error" procedure. Otherwise, we might have obtained estimates $\hat{\beta}$ and $\hat{\delta}$ of Weibull parameters as described in Section 5.3. Extreme value parameter estimates then follow from (8.2.6) as $\hat{\mu} = \ln \hat{\beta}$ and $\hat{\alpha} = 1/\hat{\delta}$.

The solution of estimating equations (8.2.17) can be made easier if we first eliminate μ between these two equations and thereby obtain

$$\frac{\Sigma^* x_i \exp(x_i/\alpha)}{\Sigma^* \exp(x_i/\alpha)} - \alpha = \frac{1}{n} \sum_{i=1}^{n} x_i. \quad (8.2.18)$$

It is relatively easy to solve this equation for $\hat{\alpha}$ by "trial and error" or by employing other standard iterative procedures such as Newton's method. In the process of eliminating μ between the two equations of (8.2.17), we obtained

$$e^\mu = \left[\frac{1}{n} \Sigma^* \exp \frac{x_i}{\alpha}\right]^\alpha, \quad (8.2.19)$$

from the first of these equations. With $\hat{\alpha}$ determined from equation (8.2.18), we can now employ (8.2.19) to calculate $\hat{\mu}$. Note that equation (8.2.18) corresponds to equation (5.3.6) for calculating the Weibull estimate $\hat{\delta}$.

Progressively Censored Samples

We consider a progressively censored sample of size N from a Type I distribution of least extreme values that is censored at points $\ln T_j, j = 1, \ldots, k$, where at each point of censoring c_j items are removed (censored) from further observation. The number of complete (full-term) observations then is $n = N - \sum_{j=1}^{k} c_j$. The likelihood function for a random sample as described is

$$L = K \left[\prod_{i=1}^{n} \frac{1}{\alpha} e^{(x_i - \mu)/\alpha} \exp(-e^{(x_i - \mu)/\alpha}) \right] \left[\prod_{j=1}^{k} \{\exp(-e^{(\ln T_j - \mu)/\alpha}\}^{c_j} \right], \quad (8.2.20)$$

where K is an ordering constant. On taking logarithms, we have

$$\operatorname{Ln} L = -n \ln \alpha + \sum_{i=1}^{n} \left(\frac{x_i - \mu}{\alpha} \right) - \sum_{i=1}^{n} \exp\left(\frac{x_i - \mu}{\alpha} \right)$$
$$- \sum_{j=1}^{k} c_j \exp\left(\frac{\ln T_j - \mu}{\alpha} \right) + \text{const.} \quad (8.2.21)$$

Maximum likelihood estimating equations, $\partial \ln L/\partial \mu = 0$ and $\partial \ln L/\partial \alpha = 0$, in this case follow as

$$\frac{\partial \ln L}{\partial \mu} = -\frac{n}{\alpha} + \frac{1}{\alpha} \sum_{i=1}^{n} \exp\left(\frac{x_i - \mu}{\alpha} \right)$$
$$+ \frac{1}{\alpha} \sum_{j=1}^{k} c_j \exp\left(\frac{\ln T_j - \mu}{\alpha} \right) = 0,$$

$$\frac{\partial \ln L}{\partial \alpha} = -\frac{n}{\alpha} - \frac{1}{\alpha} \sum_{i=1}^{n} \left(\frac{x_i - \mu}{\alpha} \right) \quad (8.2.22)$$
$$+ \frac{1}{\alpha} \sum_{i=1}^{n} \left(\frac{x_i - \mu}{\alpha} \right) \exp\left(\frac{x_i - \mu}{\alpha} \right)$$
$$+ \frac{1}{\alpha} \sum_{j=1}^{k} c_j \left(\frac{\ln T_j - \mu}{\alpha} \right) \exp\left(\frac{\ln T_j - \mu}{\alpha} \right) = 0.$$

Note that estimating equations (8.2.17) for singly censored samples can be obtained as a special case of these equations with $j = 1$.

We can eliminate μ between the two equations of (8.2.22) to obtain an equation in α that corresponds to equation (8.2.18) for singly censored samples. For the

The Exponential and Extreme Value Distributions

calculation of $\hat{\mu}$, the first of the preceding equations can be simplified to a form that corresponds to (8.2.19).

8.2.3 An Illustrative Example

Example 8.2.1 Mann and Fertig (1973) gave failure times of certain airplane components subjected to a life test. From a total of 13 randomly selected test items, 10 were observed until failure and the test was terminated when the tenth failure occurred. The sample was thus Type II singly right censored. Logarithms of life-spans of the 10 failed items are -1.514, -0.693, -0.128, 0, 0.278, 0.285, 0.432, 0.565, 0.916, and 1.099. In summary, $N = 13$, $n = 10$, $c = 3$, and $\ln T = x_{10:13} = 1.099$. This sample is assumed to be from a Type I distribution of least extremes, and the maximum likelihood estimates $\hat{\mu}$ and $\hat{\alpha}$ can be calculated from equations (8.2.18 and (8.2.19). Lawless (1982) also used this example as an illustration, and he employed these estimators to calculate $\hat{\mu} = 0.821$ and $\hat{\alpha} = 0.706$. Mann and Fertig used tables of coefficients included in their 1973 paper to calculate linear estimates from logarithms of the ordered sample observations as $\bar{\mu} = 0.873$ and $\bar{\alpha} = 0.715$.

For alternate calculations of the MLE, we make the transformation $y_i = \exp x_i$, to restore the original equivalent Weibull sample. Accordingly, failure times in original time units of the 10 failed sample units are 0.22, 0.50, 0.88, 1.00, 1.32, 1.33, 1.54, 1.76, 2.50, and 3.00. Maximum likelihood estimates of Weibull parameters, calculated as described in Chapter 5 from these data by using equations (5.3.6), (5.3.7), and (5.3.8) are $\hat{\delta} = 1.417$, $\hat{\theta} = 3.201$, and $\hat{\beta} = 2.273$. Estimates of the extreme value parameters follow from (8.2.6) as

$$\hat{\mu} = \ln \hat{\beta} = 0.821 \quad \text{and} \quad \hat{\alpha} = \frac{1}{\hat{\delta}} = 0.706,$$

in agreement with maximum likelihood estimates calculated by Lawless directly from the corresponding extreme value sample.

9
Truncated and Censored Samples from the Rayleigh Distribution

9.1 INTRODUCTION

The Rayleigh distribution is a positively skewed distribution, originally derived by Lord J. W. S. Rayleigh (1919) in connection with a study of acoustical problems. In its most general form, this distribution may be considered as the distribution of the distance X from the origin to a point $(Y_1, Y_2, \ldots, Y_p)$ in a p-dimensional Euclidean space, where the components Y_i are independent random variables, each of which is normally distributed $(0, \sigma^2)$. The random variable X thus is

$$X = \left[\sum_{i=1}^{p} Y_i\right]^{1/2}. \qquad (9.1.1)$$

The pdf of the Rayleigh distribution (i.e., the pdf of X) follows as

$$f(x; p, \sigma) = \frac{2x^{p-1} \exp[-x^2/2\sigma^2]}{(2\sigma^2)^{p/2} \, \Gamma(p/2)}, \quad 0 < x < \infty, \qquad (9.1.2)$$
$$= 0 \quad \text{elsewhere.}$$

With $\sigma = 1$, (9.1.2) is the pdf of the chi distribution with p degrees of freedom.

Some important characteristics of the Rayleigh distribution are as follows:

Samples from the Rayleigh Distribution

The cdf is $F[p/2, x^2/2\sigma^2]$ where

$$F(a, x) = \frac{1}{\Gamma(a)} \int_0^x t^{a-1} e^{-t} \, dt. \tag{9.1.3}$$

The kth moment about the origin is

$$\mu_k' = \frac{2^{k/2} \sigma^k \Gamma\left(\frac{k+p}{2}\right)}{\Gamma(p/2)}. \tag{9.1.4}$$

The expected value, variance, and mode are

$$E(X) = \frac{\sigma\sqrt{2}\,\Gamma[(1+p)/2]}{\Gamma(p/2)},$$

$$V(X) = 2\sigma^2 \left[\frac{\Gamma(2+p)/2}{\Gamma(p/2)} - \left(\frac{\Gamma(1-p)/2}{\Gamma(p/2)}\right)^2\right], \tag{9.1.5}$$

$$\mathrm{Mo}(X) = \sigma\sqrt{p-1}.$$

Readers are reminded that $\Gamma(1/2) = \sqrt{\pi}$, $\Gamma(1+z) = z\Gamma(z)$, and thus $\Gamma(3/2) = \sqrt{\pi}/2$, $\Gamma(5/2) = 3\sqrt{\pi}/4$, etc. These results are needed in evaluations of (9.1.5) for specific values of p.

Estimation in this chapter will be limited to truncated and censored samples. The calculation of estimates from complete samples has been adequately covered elsewhere. First, however, we examine special cases of interest to engineers, physicists, and other scientists as models for the analysis of data resulting from studies of wave propagation, radiation, target errors, velocities of gas particles, and related fields of inquiry.

9.2 SOME SPECIAL CASES

Applications of the Rayleigh distribution for $p > 3$ are somewhat limited, but the special cases in which $p = 1, 2$, and 3 are important in various scientific applications.

9.2.1 The Folded Normal Distribution, $p = 1$

The one-dimensional Rayleigh distribution is sometimes known as the folded normal, the folded Gaussian, or the half normal distribution. With $p = 1$, the pdf (9.1.2), in this case becomes

$$f(x; 1, \sigma) = \frac{1}{\sigma}\sqrt{\frac{2}{\pi}} \exp\left(-\frac{x^2}{2\sigma^2}\right), \quad 0 < x < \infty, \quad (9.2.1)$$
$$= 0 \quad \text{elsewhere.}$$

The expected value, variance, coefficient of variation, α_3, α_4, and mode of this distribution are

$$E(X) = \sigma\sqrt{\frac{2}{\pi}} = 0.797885\sigma,$$

$$V(X) = \frac{\sigma^2(\pi - 2)}{\pi} = 0.363380\sigma^2,$$

$$CV(X) = \sqrt{\frac{\pi - 2}{2}} = 0.755511, \quad (9.2.2)$$

$$\alpha_3(X) = \frac{\sqrt{2}(4 - \pi)}{(\pi - 2)^{3/2}} = 0.995272,$$

$$\alpha_4(X) = \frac{3\pi^2 - 4\pi - 12}{(\pi - 2)^2} = 3.869177,$$

$$Mo(X) = 0.$$

The cdf can be expressed in terms of the cdf of the standard normal distribution as

$$F(x; 1, \sigma) = 2\Phi\left(\frac{x}{\sigma}\right) - 1, \quad (9.2.3)$$

where $\Phi()$ is the cdf of the standard normal distribution $(0, 1)$.

9.2.2 The Two-Dimensional Rayleigh Distribution, $p = 2$

This special case is perhaps the most important of all the Rayleigh distributions. It is the one that received the most attention from its discoverer, and may users are inclined to consider it as the only Rayleigh distribution. It is not only a special case of the general Rayleigh distribution, it is also a special case of the Weibull distribution. In connection with applications concerning the analysis of target errors, it is sometimes referred to as the circular normal distribution

With $p = 2$, the pdf of (9.1.2) becomes

$$f(x; 2, \sigma) = \frac{x}{\sigma^2} \exp\left(-\frac{x^2}{2\sigma^2}\right), \quad 0 < x < \infty, \quad (9.2.4)$$
$$= 0 \quad \text{elsewhere.}$$

Samples from the Rayleigh Distribution

This same pdf follows from the two-parameter Weibull distribution (β, δ) with pdf (5.2.1) when we set $\delta = 2$ and $\beta^2 = 2\sigma^2$. Moments and related properties of this distribution can be obtained from either the Weibull properties or the p-dimensional Rayleigh properties. The Weibull parameterization (β, 2) enables the pdf to be written as

$$f(x; 2, \beta) = \frac{2x}{\beta^2} \exp\left(-\frac{x^2}{\beta^2}\right), \quad 0 < x < \infty \quad (9.2.5)$$
$$= 0 \quad \text{elsewhere},$$

or with $\theta = \beta^2$, as

$$f(x; 2, \theta) = \frac{2x}{\theta} \exp\left(-\frac{x^2}{\theta}\right), \quad 0 < x < \infty, \quad (9.2.6)$$
$$= 0 \quad \text{elsewhere}.$$

Thus, we have three equivalent parameterizations for this distribution. The cdf with σ as the parameter is

$$F(x; 2, \sigma) = 1 - \exp\left(-\frac{x^2}{2\sigma^2}\right). \quad (9.2.7)$$

The kth moment about the origin is

$$\mu'_k = 2^{k/2}\sigma^k \Gamma_k = \left(\frac{\sigma}{\sqrt{2}}\right)^k \left(\frac{k}{2}\right) \Gamma\left(\frac{k}{2}\right). \quad (9.2.8)$$

where, as in Chapter 3, $\Gamma_k = \Gamma(1 + k/2)$. Other characteristics of interest are

$$E(X) = \sigma \sqrt{\frac{\pi}{2}} = 1.25331\sigma,$$

$$V(X) = \frac{\sigma^2(4 - \pi)}{2} = 0.429204\sigma^2,$$

$$CV(X) = \frac{\sqrt{V(X)}}{E(X)} = \sqrt{\frac{4 - \pi}{\pi}} = 0.522723,$$

$$\alpha_3(X) = \frac{2(\pi - 3)\sqrt{\pi}}{(4 - \pi)^{3/2}} = 0.631110, \quad (9.2.9)$$

$$\alpha_4(X) = \frac{32 - 3\pi^2}{(4 - \pi)^2} = 3.245089,$$

$$Mo(X) = \sigma,$$

$$Me(X) = \sigma\sqrt{\ln 4} = 1.17741\sigma.$$

The rth percentile is

$$x_r = [-2\sigma^2 \ln(1 - r)]^{1/2}, \quad (9.2.10)$$

and the hazard function $h(x) = f(x)/[1 - F(x)]$, which is of interest in reliability analysis, is the increasing function

$$h(x) = \frac{x}{\sigma^2}. \quad (9.2.11)$$

9.2.3 The Maxwell–Boltzmann Distribution, $p = 3$

With $p = 3$ in the pdf (9.1.2) becomes

$$\begin{aligned} f(x; 3, \sigma) &= \frac{1}{\sigma}\sqrt{\frac{2}{\pi}}\left(\frac{x}{\sigma}\right)^2 \exp\left(-\frac{x^2}{2\sigma^2}\right), \quad 0 < x < \infty, \\ &= 0 \quad \text{elsewhere.} \end{aligned} \quad (9.2.12)$$

Moments and other characteristics of interest in this case are

$$\begin{aligned} E(X) &= 2\sigma\sqrt{\frac{2}{\pi}} = 1.595769\sigma, \\ V(X) &= \frac{\sigma^2(3\pi - 8)}{\pi} = 0.453521\sigma^2, \\ \mathrm{CV}(X) &= \sqrt{\frac{3\pi - 8}{8}} = 0.422016, \\ \alpha_3(X) &= \frac{\sqrt{2}(32 - 10\pi)}{(3\pi - 8)^{3/2}} = 0.485693, \\ \alpha_4(X) &= \frac{15\pi^2 + 16\pi - 192}{(3\pi - 8)^2} = 3.108164, \\ \mathrm{Mo}(X) &= \sigma\sqrt{2}, \end{aligned} \quad (9.2.13)$$

This distribution is of special interest to engineers and physicists. It was originally derived by Maxwell (1860) as the distribution of the velocity of gas particles in three-dimensional space. Boltzmann (1877) used Maxwell's results in explaining the thermodynamics of gases on the basis of kinetic theory in which gases were viewed as particles undergoing movement at different velocities and

Samples from the Rayleigh Distribution

collisions according to the principles of mechanics. In recognition of his contributions, Boltzmann's name has also become associated with this distribution.

9.2.4 The Two-Parameter Rayleigh Distribution

With the addition of a threshold parameter, γ, to the pdf (9.2.4) of the two-dimensional Rayleigh distribution, we have

$$f(x; 2, \sigma, \gamma) = \left(\frac{x - \gamma}{\sigma^2}\right) \exp\left[-\frac{(x - \gamma)^2}{2\sigma^2}\right],$$

$$= 0 \quad \text{elsewhere.}$$
(9.2.14)

This distribution might compete with other positively skewed distributions as a model for use in life-span and related applications where the origin is different from zero. In a similar manner, a threshold parameter might be added to other members of the Rayleigh family of distributions, and this is left as an exercise for readers.

With the addition of the threshold parameter, our interest is aroused in the distribution of the first-order statistic. As noted in our study of the Weibull distribution in Chapter 5, when $X \sim W(\gamma, \beta, \delta)$, the first-order statistic in a random sample of size n is $X_{1:n} \sim W(\gamma, \beta', \delta)$ where $\beta' = \beta/n^{1/\delta}$. Since, in the special case under consideration here, $\beta^2 = 2\sigma^2$, and $\delta = 2$, it follows that $\beta' = \sigma\sqrt{2/n}$. Accordingly $X_{1:n} \sim W(\gamma, \beta', 2)$, and the pdf is

$$f(x_{1:n}; \gamma, \sigma) = \frac{n}{\sigma^2}(x_{1:n} - \gamma) \exp\left[-\frac{n}{2\sigma^2}(x_{1:n} - \gamma)^2\right],$$

$$\gamma < x_{1:n} < \infty,$$
(9.2.15)

$$= 0 \quad \text{elsewhere,}$$

and the expected value becomes

$$E(X_{1:n}) = \gamma + \sigma\sqrt{\frac{\sigma}{2n}}.$$
(9.2.16)

9.3 PARAMETER ESTIMATION

9.3.1 Parameter Estimation in Singly Right Censored Samples

We consider a random sample that consists of a total of N observations, of which n are less than or at most equal to T plus c censored observations that exceed

T. The likelihood function of a sample as described from the *p*-dimensional Rayleigh distribution with pdf (9.1.2) is

$$L = 2^{-n(p-2)/2} \sigma^{-np} \left[\Gamma\left(\frac{p}{2}\right)\right]^{-n} \left[\prod_{i=1}^{n} x_i^{p-1}\right]$$
$$\times \left[\exp\left(-\frac{1}{2\sigma^2}\sum_{i=1}^{n} x_i^2\right)\right] [1 - F_p(T)]^c K, \quad (9.3.1)$$

where *K* is an ordering constant that does not involve parameters to be estimated and $F_p(X)$ is the cdf of *X*.

On taking logarithms of (9.3.1), the loglikelihood function becomes

$$\ln L = -np \ln \sigma + (p-1)\sum_{i=1}^{n} \ln x_i - \frac{1}{2\sigma^2}\sum_{i=1}^{n} x_i^2$$
$$+ c \ln[1 - F_p(T)] + \text{const.} \quad (9.3.2)$$

If we let $Z = X/\sigma$, the pdf of (9.1.2) becomes

$$g_p(z) = \frac{2^{-(p-2)/2}}{\Gamma(p/2)} z^{p-1} \exp\left(-\frac{z^2}{2}\right), \quad 0 < z < \infty$$
$$= 0 \quad \text{elsewhere.} \quad (9.3.3)$$

When $X = T$, then $Z = T/\sigma = \xi$, and

$$F(T) = G(\xi) = \int_0^\xi g(z)\, dz. \quad (9.3.4)$$

In making the transformation from *X* to *Z*, ξ has become the parameter to be estimated. After $\hat{\xi}$ has been calculated, it then follows that $\hat{\sigma} = T/\hat{\xi}$.

The maximum likelihood estimating equation $\partial \ln L/\partial \sigma = 0$ follows from (9.3.2) as

$$\frac{\partial \ln L}{\partial \sigma} = -\frac{np}{\sigma} + \frac{1}{\sigma^3}\sum_{i=1}^{n} x_i^2 - \frac{c}{1 - F(T)}\frac{\partial F}{\partial \sigma} = 0, \quad (9.3.5)$$

where

$$\frac{\partial F}{\partial \sigma} = \frac{\partial G}{\partial \sigma} = \frac{\partial G}{\partial \xi}\frac{\partial \xi}{\partial \sigma} = -g(\xi)\left(\frac{T}{\sigma^2}\right). \quad (9.3.6)$$

We substitute (9.3.6) into (9.3.5) and at the same time replace σ with T/ξ and simplify to obtain

$$\frac{1}{nT^2}\sum_{i=1}^{n} x_i^2 = \frac{1}{\xi}\left[\frac{p}{\xi} - \left(\frac{c}{n}\right)\frac{g_p(\xi)}{1 - G_p(\xi)}\right] = H_p\left(\xi, \frac{c}{n}\right), \quad (9.3.7)$$

Samples from the Rayleigh Distribution

where as a result of the transformation from X to Z, ξ has now become the parameter to be estimated. We calculate $\hat{\xi}$ as the solution of (9.3.7), and $\hat{\sigma}$ follows as

$$\hat{\sigma} = \frac{T}{\hat{\xi}}. \tag{9.3.8}$$

In order to evaluate $H_p(\xi, c/n)$ as defined in (9.3.7), we need to evaluate the integral

$$G_p(z) = \int_0^z g_p(t)\,dt = \int_0^z \frac{2^{-(p-2)/2}}{\Gamma(p/2)} t^{p-1} e^{-t^2/2}\,dt. \tag{9.3.9}$$

This can be done by repeated integration by parts. For $p = 2$, and 3, we find

$$G_2(z) = \sqrt{2\pi}\,[\phi(0) - \phi(z)], \qquad G_3(z) = 2[\Phi(z) - z\phi(z)] - 1, \tag{9.3.10}$$

where $\phi(z)$ and $\Phi(z)$ are the pdf and the cdf, respectively, of the standard normal distribution $(0,1)$. With $p = 2$ and 3, respectively, in (9.3.2), the standardized pdfs become

$$g_2(z) = \sqrt{2\pi}\,z\,\phi(z) \qquad \text{and} \qquad g_3(z) = 2z^2\,\phi(z). \tag{9.3.11}$$

Censored Sample Estimator; p = 2

For $p = 2$, we substitute $g()$ and $G()$ from (9.3.11) and (9.3.10) into (9.3.7), and simplify to obtain

$$H_2\left(\xi, \frac{c}{n}\right) = \frac{2}{\xi^2} - \frac{c}{n} = \frac{2\sigma^2}{T^2} - \frac{c}{n}. \tag{9.3.12}$$

It then follows from (9.3.7) that, in this case,

$$\hat{\sigma} = \sqrt{\frac{1}{2n}\left[\sum_{i=1}^n x_i^2 + cT^2\right]}. \tag{9.3.13}$$

This estimator might be derived in a more direct manner with $F_2(x; \sigma) = 1 - \exp[-x^2/2\sigma^2]$ by returning to (9.3.2) and writing

$$\ln L = -2n \ln \sigma + \sum_{i=1}^n \ln x_i - \frac{1}{2\sigma^2}\sum_{i=1}^n x_i^2 - \frac{cT^2}{2\sigma^2} + \text{const.} \tag{9.3.14}$$

Accordingly,

$$\frac{\partial \ln L}{\partial \sigma} = -\frac{2n}{\sigma} + \frac{1}{\sigma^3}\sum_{i=1}^n x_i^2 + \frac{1}{\sigma^3}cT^2 = 0, \tag{9.3.15}$$

and the estimator of (9.3.13) follows immediately without introducing ξ.

Censored Sample Estimator, p = 3

For $p = 3$, we substitute $g_3()$ and $G_3()$ from (9.3.11) and (9.3.10) into (9.3.7) and simplify to obtain

$$H_3\left(\xi, \frac{c}{n}\right) = \frac{3}{\xi^2} - \frac{c}{n}\left[\frac{\xi\phi(\xi)}{1 - \Phi(\xi) + \xi\phi(\xi)}\right], \qquad (9.3.16)$$

and the estimate for ξ follows from (9.3.7) as

$$H_3\left(\hat{\xi}, \frac{c}{n}\right) = \frac{1}{nT^2}\sum_{i=1}^{n} x_i^2. \qquad (9.3.17)$$

Censored Sample from the Two-Parameter Rayleigh Distribution, p = 2

The loglikelihood function in this case is

$$\ln L = -2n \ln \sigma + \sum_{i=1}^{n} \ln(x_i - \gamma) - \frac{1}{2\sigma^2}\sum_{i=1}^{n}(x_i - \gamma)^2$$
$$- \frac{c}{2\sigma^2}(T - \gamma)^2 + \text{const}, \qquad (9.3.18)$$

and the maximum likelihood estimating equations are

$$\frac{\partial \ln L}{\partial \gamma} = -\sum_{i=1}^{n}(x_i - \gamma)^{-1} + \frac{1}{\sigma^2}\sum_{i=1}^{n}(x_i - \gamma) + \frac{c}{\sigma^2}(T - \gamma) = 0,$$
$$\frac{\partial \ln L}{\partial \sigma} = -\frac{2n}{\sigma} + \frac{1}{\sigma^3}\sum_{i=1}^{n}(x_i - \gamma)^2 + \frac{c}{\sigma^3}(T - \gamma)^2 = 0. \qquad (9.3.19)$$

We eliminate σ between these two equations and simplify to obtain the following equation in γ only:

$$\frac{\sum_{i=1}^{n} x_i + cT - N\gamma}{\sum_{i=1}^{n}(x_i - \gamma)^{-1}} - \frac{1}{2n}\left[\sum_{i=1}^{n}(x_i - \gamma)^2 + c(T - \gamma)^2\right] = 0. \qquad (9.3.20)$$

This equation can be solved iteratively for the MLE, $\hat{\gamma}$. Then $\hat{\sigma}^2$ follows from the second equation of (9.3.19) as

$$\hat{\sigma}^2 = \frac{1}{2n}\left[\sum_{i=1}^{n}(x_i - \hat{\gamma})^2 + c(T - \hat{\gamma})^2\right]. \qquad (9.3.21)$$

Modified maximum likelihood estimators that employ the second equation of (9.3.19) plus the equation $E(X_{1:N}) = x_{1:N}$ exhibit a smaller bias than the MLE

Samples from the Rayleigh Distribution

and may therefore be preferred over the MLE in most applications. For these estimators, in place of the first equation of (9.3.19), we have

$$E(X_{1:n}) = \gamma + \frac{\sigma\sqrt{\pi}}{2N} = x_{1:N}. \qquad (9.3.22)$$

We estimate σ between the two applicable estimating equations to obtain

$$\frac{2N}{\pi}(x_{1:N} - \gamma)^2 - \frac{1}{2n}\left[\sum_{i=1}^{n}(x_i - \gamma)^2 + c(T - \gamma)^2\right] = 0, \qquad (9.3.23)$$

in which γ is the only unknown. We solve this equation for $\hat{\gamma}$, and $\hat{\sigma}^2$ then follows from (9.3.21).

9.3.2 Parameter Estimation in Singly Right Truncated Samples

The loglikelihood function of a truncated sample that consists of n randomly selected observations from a population with pdf (9.1.2) when each observation is subject to the restriction $0 < x \leq T$ is

$$\ln L = -np \ln \sigma + (p - 1)\sum_{i=1}^{n} \ln x_i - \frac{1}{2\sigma^2}\sum_{i=1}^{n} x_i^2$$
$$- n \ln F_p(T) + \text{const}, \qquad (9.3.24)$$

where $F()$, $G()$, $g()$ are as previously defined in connection with censored sampling.

The maximum likelihood estimating equation in this case is

$$\frac{\partial \ln L}{\partial \sigma} = -\frac{np}{\sigma} + \frac{1}{\sigma^3}\sum_{i=1}^{n} x_i^2 + \frac{n\xi}{\sigma}\frac{g_p(\xi)}{G_p(\xi)} = 0. \qquad (9.3.25)$$

We substitute $\xi = T/\sigma$ into this equation and simplify to obtain

$$J_p(\hat{\xi}) = \frac{1}{nT^2}\sum_{i=1}^{n} x_i^2, \qquad (9.3.26)$$

where

$$J_p(\xi) = \frac{1}{\xi}\left[\frac{p}{\xi} - \frac{g_p(\xi)}{G_p(\xi)}\right]. \qquad (9.3.27)$$

Estimating equation (9.3.26) can be solved for $\hat{\xi}$ by standard iterative methods or by interpolation in applicable tables of the estimating function $J_p(\xi)$.

For the special cases of $p = 2$ and 3, respectively, the estimating function of (9.3.27) becomes

$$J_2(\xi) = \frac{2}{\xi^2} - \frac{\phi(\xi)}{\phi(0) - \phi(\xi)}, \qquad (9.3.28)$$

$$J_3(\xi) = \frac{3}{\xi^2} - \frac{\xi\phi(\xi)}{\Phi(\xi) - \xi\phi(\xi) - 0.5}.$$

To facilitate the calculation of estimates, tables of the two functions of (9.3.28) are included as Tables 9.1 and 9.2. In addition, graphs are included as Figure 9.1. A table of the censored sample estimating function $H_3[\xi, (c/n)]$ is included as Table 9.3.

9.4 RELIABILITY OF ESTIMATES

The asymptotic variance of the maximum likelihood estimate $\hat{\sigma}$ is given by

$$\text{Asy Var}(\hat{\sigma}) = -\left[E\left(\frac{\partial^2 \ln L}{\partial \sigma^2}\right)\right]^{-1}. \qquad (9.4.1)$$

For truncated and censored samples, this leads to specific results, which follow.

For Truncated Samples

$$E\left(\frac{\partial^2 \ln L}{\partial \sigma^2}\right) = -\frac{2np}{\sigma^2}\left[1 - \frac{\xi g_p(\xi)}{2pG_p(\xi)}\left\{\xi^2 - (p-2) + \frac{\xi g_p(\xi)}{G_p(\xi)}\right\}\right]. \qquad (9.4.2)$$

In the two-dimensional case, it then follows that

$$\text{Asy Var}(\hat{\sigma}) = \frac{\sigma^2}{4n}\left[1 - \left(\frac{\xi^4}{4}\right)\frac{\phi(0)\phi(\xi)}{[\phi(0) - \phi(\xi)]}\right]^{-1}. \qquad (9.4.3)$$

For Censored Samples

In this case the expected value of the second partial of the loglikelihood function is

$$E\left(\frac{\partial^2 \ln L}{\partial \sigma^2}\right) = -\frac{2pE(n)}{\sigma^2}\left[1 - \frac{\xi g_p(\xi)}{2pG_p(\xi)}\left\{\xi^2 - (p-2) - \frac{\xi g_p(\xi)}{1 - G_p(\xi)}\right\}\right], \qquad (9.4.4)$$

where $E(n) = NG_p(\xi)$ in Type I censored samples and $E(n) = n$ in Type II samples.

Samples from the Rayleigh Distribution

Table 9.1 Truncated Sample Estimating Function $J_2(z)$, for Rayleigh Distribution, $J_2(z) = 2/z^2 - \phi(z)/[\phi(0) - \phi(z)]$

z	.00	.01	.02	.03	.04	.05	.06	.07	.08	.09
.0	.50000	.50000	.49998	.49996	.49993	.49990	.49985	.49980	.49973	.49966
.1	.49958	.49950	.49940	.49930	.49918	.49906	.49893	.49880	.49865	.49850
.2	.49833	.49816	.49798	.49780	.49760	.49740	.49718	.49696	.49673	.49650
.3	.49625	.49600	.49573	.49546	.49518	.49490	.49460	.49430	.49398	.49366
.4	.49333	.49300	.49265	.49230	.49193	.49156	.49118	.49080	.49040	.49000
.5	.48959	.48917	.48874	.48830	.48785	.48740	.48694	.48647	.48599	.48550
.6	.48501	.48450	.48399	.48347	.48295	.48241	.48186	.48131	.48075	.48018
.7	.47960	.47902	.47842	.47782	.47721	.47659	.47597	.47533	.47469	.47404
.8	.47338	.47271	.47204	.47135	.47066	.46996	.46925	.46854	.46781	.46708
.9	.46634	.46559	.46484	.46407	.46330	.46252	.46174	.46094	.46014	.45933
1.0	.45851	.45768	.45684	.45600	.45515	.45429	.45343	.45255	.45167	.45078
1.1	.44989	.44898	.44807	.44715	.44623	.44529	.44435	.44340	.44245	.44148
1.2	.44051	.43953	.43855	.43756	.43656	.43555	.43453	.43351	.43248	.43145
1.3	.43041	.42936	.42830	.42724	.42617	.42509	.42401	.42292	.42182	.42072
1.4	.41961	.41850	.41737	.41624	.41511	.41397	.41282	.41167	.41051	.40934
1.5	.40817	.40699	.40581	.40462	.40342	.40222	.40102	.39981	.39859	.39737
1.6	.39614	.39490	.39366	.39242	.39117	.38992	.38866	.38740	.38613	.38485
1.7	.38358	.38229	.38101	.37972	.37842	.37712	.37582	.37451	.37319	.37188
1.8	.37056	.36923	.36791	.36658	.36524	.36390	.36256	.36122	.35987	.35852
1.9	.35717	.35581	.35445	.35309	.35172	.35035	.34898	.34761	.34624	.34486
2.0	.34348	.34210	.34072	.33933	.33795	.33656	.33517	.33378	.33239	.33100
2.1	.32960	.32821	.32681	.32541	.32402	.32262	.32122	.31982	.31842	.31702
2.2	.31562	.31422	.31282	.31142	.31002	.30863	.30723	.30583	.30443	.30304
2.3	.30164	.30024	.29885	.29746	.29607	.29468	.29329	.29190	.29052	.28913
2.4	.28775	.28637	.28499	.28361	.28224	.28087	.27950	.27813	.27677	.27540
2.5	.27404	.27269	.27133	.26998	.26863	.26729	.26595	.26461	.26327	.26194
2.6	.26061	.25929	.25796	.25665	.25533	.25402	.25271	.25141	.25011	.24882
2.7	.24753	.24624	.24496	.24368	.24241	.24114	.23987	.23861	.23736	.23611
2.8	.23486	.23362	.23238	.23115	.22992	.22870	.22748	.22627	.22506	.22386
2.9	.22267	.22147	.22029	.21911	.21793	.21676	.21559	.21444	.21328	.21213
3.0	.21099	.20985	.20872	.20759	.20647	.20535	.20424	.20314	.20204	.20095
3.1	.19986	.19878	.19770	.19663	.19557	.19451	.19346	.19241	.19137	.19033
3.2	.18930	.18828	.18726	.18625	.18524	.18424	.18324	.18225	.18127	.18029
3.3	.17932	.17835	.17739	.17644	.17549	.17454	.17361	.17267	.17175	.17083
3.4	.16991	.16900	.16810	.16720	.16631	.16542	.16454	.16367	.16280	.16193
3.5	.16107	.16022	.15937	.15853	.15769	.15686	.15604	.15521	.15440	.15359
3.6	.15278	.15199	.15119	.15040	.14962	.14884	.14807	.14730	.14654	.14578
3.7	.14503	.14428	.14354	.14280	.14207	.14134	.14061	.13990	.13918	.13848
3.8	.13777	.13707	.13638	.13569	.13501	.13433	.13365	.13298	.13231	.13165
3.9	.13099	.13034	.12969	.12905	.12841	.12778	.12714	.12652	.12590	.12528
4.0	.12466	.12405	.12345	.12285	.12225	.12166	.12107	.12048	.11990	.11933
4.1	.11875	.11818	.11762	.11706	.11650	.11595	.11539	.11485	.11431	.11377
4.2	.11323	.11270	.11217	.11165	.11112	.11061	.11009	.10958	.10907	.10857
4.3	.10807	.10757	.10708	.10659	.10610	.10562	.10514	.10466	.10418	.10371
5.0	.08000	.07968	.07936	.07905	.07873	.07842	.07811	.07780	.07750	.07719

Source: From Cohen and Whitten (1988), Table 10.1, p. 193, by courtesy of Marcel Dekker, Inc.

Table 9.2 Truncated Sample Estimating Function $J_3(z)$, for Rayleigh Distribution, $J_3(z) = 3/z^2 - z\phi(z)/[\Phi(z) - z\phi(z) - 0.5]$

z	.00	.01	.02	.03	.04	.05	.06	.07	.08	.09
.0	.60000	.60043	.60001	.59997	.59995	.59991	.59988	.59983	.59978	.59972
.1	.59966	.59959	.59951	.59942	.59933	.59923	.59912	.59901	.59889	.59876
.2	.59863	.59849	.59834	.59818	.59802	.59785	.59768	.59750	.59731	.59711
.3	.59691	.59670	.59648	.59626	.59603	.59579	.59554	.59529	.59503	.59477
.4	.59450	.59422	.59393	.59364	.59333	.59303	.59271	.59239	.59206	.59173
.5	.59138	.59103	.59068	.59031	.58994	.58956	.58918	.58878	.58838	.58798
.6	.58756	.58714	.58671	.58628	.58584	.58539	.58493	.58446	.58399	.58351
.7	.58303	.58254	.58203	.58153	.58101	.58049	.57996	.57942	.57888	.57833
.8	.57777	.57720	.57663	.57605	.57546	.57487	.57426	.57365	.57304	.57241
.9	.57178	.57114	.57049	.56984	.56918	.56851	.56783	.56715	.56646	.56576
1.0	.56505	.56434	.56362	.56289	.56215	.56141	.56065	.55990	.55913	.55836
1.1	.55757	.55679	.55599	.55519	.55437	.55355	.55273	.55189	.55105	.55020
1.2	.54935	.54848	.54761	.54673	.54585	.54495	.54405	.54314	.54223	.54130
1.3	.54037	.53943	.53848	.53753	.53657	.53560	.53462	.53364	.53265	.53165
1.4	.53064	.52963	.52861	.52758	.52654	.52550	.52445	.52339	.52233	.52126
1.5	.52018	.51909	.51800	.51690	.51579	.51467	.51355	.51242	.51128	.51014
1.6	.50899	.50783	.50666	.50549	.50431	.50313	.50193	.50073	.49953	.49832
1.7	.49710	.49587	.49464	.49340	.49215	.49090	.48964	.48837	.48710	.48582
1.8	.48453	.48324	.48194	.48064	.47933	.47801	.47669	.47536	.47403	.47269
1.9	.47134	.46999	.46864	.46727	.46590	.46453	.46315	.46177	.46037	.45898
2.0	.45758	.45617	.45476	.45335	.45192	.45050	.44907	.44763	.44619	.44475
2.1	.44330	.44184	.44039	.43892	.43746	.43599	.43451	.43303	.43155	.43006
2.2	.42857	.42708	.42558	.42408	.42258	.42107	.41956	.41805	.41653	.41501
2.3	.41349	.41197	.41044	.40891	.40738	.40584	.40431	.40277	.40123	.39968
2.4	.39814	.39659	.39504	.39350	.39194	.39039	.38884	.38728	.38573	.38417
2.5	.38261	.38106	.37950	.37794	.37638	.37482	.37326	.37170	.37014	.36858
2.6	.36702	.36546	.36390	.36234	.36078	.35922	.35767	.35611	.35456	.35300
2.7	.35145	.34990	.34835	.34680	.34526	.34371	.34217	.34063	.33909	.33756
2.8	.33602	.33449	.33296	.33143	.32991	.32839	.32687	.32535	.32384	.32233
2.9	.32082	.31932	.31782	.31632	.31482	.31333	.31185	.31037	.30889	.30741
3.0	.30594	.30447	.30301	.30155	.30010	.29865	.29720	.29576	.29432	.29289
3.1	.29146	.29004	.28862	.28721	.28580	.28439	.28300	.28160	.28021	.27883
3.2	.27745	.27608	.27471	.27335	.27199	.27064	.26930	.26796	.26662	.26529
3.3	.26397	.26265	.26134	.26004	.25874	.25744	.25615	.25487	.25359	.25232
3.4	.25106	.24980	.24855	.24730	.24606	.24483	.24360	.24238	.24116	.23995
3.5	.23875	.23755	.23636	.23518	.23400	.23282	.23166	.23050	.22934	.22820
3.6	.22705	.22592	.22479	.22367	.22255	.22144	.22034	.21924	.21815	.21706
3.7	.21598	.21491	.21384	.21278	.21173	.21068	.20964	.20860	.20757	.20655
3.8	.20553	.20452	.20352	.20252	.20152	.20053	.19955	.19858	.19761	.19664
3.9	.19569	.19473	.19379	.19285	.19191	.19099	.19006	.18915	.18823	.18733
4.0	.18643	.18553	.18464	.18376	.18288	.18201	.18115	.18028	.17943	.17858
4.1	.17773	.17689	.17606	.17523	.17441	.17359	.17277	.17197	.17116	.17037
4.2	.16957	.16878	.16800	.16722	.16645	.16568	.16492	.16416	.16341	.16266
4.3	.16192	.16118	.16045	.15972	.15899	.15827	.15756	.15684	.15614	.15544
5.0	.11999	.11951	.11903	.11856	.11809	.11762	.11716	.11670	.11624	.11578

Source: From Cohen and Whitten (1988), Table 10.2, p. 195, by courtesy of Marcel Dekker, Inc.

Samples from the Rayleigh Distribution

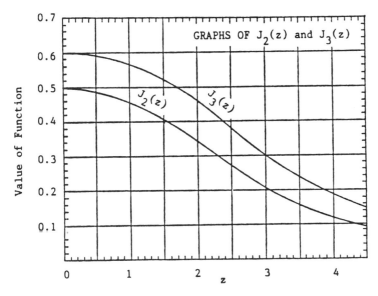

Figure 9.1 Graphs of estimating functions for truncated samples from two- and three-dimensional Rayleigh distributions. Adapted from Cohen (1955b), Fig. 1, p. 1127, with permission of the American Statistical Association.

For the two-dimensional distribution,

$$\text{Asy Var}(\hat{\sigma}) = \frac{\sigma^2}{4E(n)} \qquad (9.4.5)$$

9.5 ILLUSTRATIVE EXAMPLES

To illustrate computational procedures in practical applications of estimators presented in this chapter, two examples have been selected from Cohen (1955). These examples are of no particular significance other than as vehicles for the illustration of estimate calculations.

Example 9.1. With the aid of a table of random numbers, a sample of size $N = 27$ was generated from a two-dimensional Rayleigh distribution for which $\sigma = 10$. By appropriately truncating and subsequently censoring, the same basic sample data is employed to illustrate parameter estimation in truncated, censored, and complete samples. To create a truncated sample, the two observations exceeding the terminus, $T = 26$, were eliminated.

Table 9.3 Censored Sample Estimating Function $H_3(h, z)$, for Rayleigh Distribution, $H_3(h, z) = 3/z^2 - hz\phi(z)/[z\phi(z) + 1 - \Phi(z)]; h = n_0/n$

z \ h	.01	.02	.03	.04	.05	.06	.07	.08	.09
.6	8.32912	8.32490	8.32068	8.31647	8.31225	8.30804	8.30382	8.29960	8.29539
.7	6.11770	6.11296	6.10821	6.10346	6.09872	6.09397	6.08923	6.08448	6.07973
.8	4.68228	4.67705	4.67183	4.66660	4.66138	4.65615	4.65093	4.64571	4.64048
.9	3.69805	3.69240	3.68674	3.68109	3.67543	3.66978	3.66412	3.65847	3.65282
1.0	2.99396	2.98792	2.98188	2.97584	2.96980	2.96376	2.95772	2.95168	2.94564
1.1	2.47295	2.46657	2.46018	2.45380	2.44741	2.44103	2.43464	2.42826	2.42187
1.2	2.07664	2.06994	2.06325	2.05656	2.04986	2.04317	2.03647	2.02978	2.02308
1.3	1.76818	1.76121	1.75423	1.74726	1.74029	1.73332	1.72635	1.71938	1.71241
1.4	1.52339	1.51617	1.50896	1.50174	1.49452	1.48730	1.48008	1.47286	1.46564
1.5	1.32589	1.31845	1.31101	1.30357	1.29613	1.28869	1.28125	1.27380	1.26636
1.6	1.16423	1.15659	1.14895	1.14131	1.13367	1.12603	1.11839	1.11075	1.10311
1.7	1.03024	1.02242	1.01460	1.00678	.99896	.99114	.98332	.97550	.96768
1.8	.91794	.90996	.90198	.89400	.88602	.87803	.87005	.86207	.85409
1.9	.82300	.81477	.80664	.79851	.79039	.78226	.77413	.76600	.75787
2.0	.74174	.73348	.72522	.71696	.70870	.70044	.69218	.68392	.67566
2.1	.67189	.66351	.65513	.64675	.63838	.63000	.62162	.61324	.60486
2.2	.61135	.60286	.59437	.58588	.57740	.56891	.56042	.55193	.54344
2.3	.55852	.54993	.54135	.53276	.52417	.51559	.50700	.49841	.48983
2.4	.51216	.50348	.49480	.48613	.47745	.46877	.46010	.45142	.44274
2.5	.47124	.46248	.45372	.44496	.43621	.42745	.41869	.40993	.40117
2.6	.43495	.42612	.41728	.40845	.39962	.39078	.38195	.37311	.36428
2.7	.40262	.39372	.38481	.37591	.36701	.35810	.34920	.34030	.33140
2.8	.37369	.36472	.35575	.34679	.33782	.32886	.31989	.31092	.30196
2.9	.34769	.33867	.32964	.32062	.31160	.30257	.29355	.28452	.27550
3.0	.32426	.31518	.30610	.29702	.28794	.27886	.26979	.26071	.25163
3.1	.30305	.29392	.28479	.27566	.26654	.25741	.24828	.23915	.23002
3.2	.28380	.27462	.26545	.25627	.24710	.23793	.22875	.21958	.21041
3.3	.26627	.25705	.24783	.23862	.22940	.22018	.21097	.20175	.19254
3.4	.25026	.24100	.23175	.22249	.21324	.20398	.19473	.18547	.17621
3.5	.23561	.22631	.21702	.20773	.19844	.18914	.17985	.17056	.16127
3.6	.22216	.21283	.20350	.19418	.18485	.17552	.16620	.15687	.14754
3.7	.20978	.20042	.19106	.18171	.17235	.16299	.15363	.14427	.13491
3.8	.19837	.18898	.17959	.17021	.16082	.15143	.14204	.13265	.12327
3.9	.18782	.17841	.16899	.15958	.15016	.14075	.13133	.12191	.11250
4.0	.17806	.16862	.15918	.14973	.14029	.13085	.12141	.11197	.10253
4.1	.16900	.15953	.15007	.14060	.13114	.12167	.11221	.10274	.09327
4.2	.16058	.15109	.14160	.13211	.12263	.11314	.10365	.09416	.08467
4.3	.15274	.14323	.13372	.12421	.11470	.10519	.09568	.08617	.07666
4.4	.14543	.13590	.12637	.11684	.10731	.09778	.08825	.07872	.06919
4.5	.13860	.12905	.11950	.10995	.10040	.09085	.08131	.07176	.06221
4.6	.13221	.12264	.11308	.10351	.09394	.08438	.07481	.06524	.05568
4.7	.12622	.11664	.10706	.09747	.08789	.07831	.06872	.05914	.04956
4.8	.12061	.11101	.10141	.09181	.08221	.07261	.06301	.05341	.04381
4.9	.11533	.10572	.09610	.08649	.07688	.06726	.05765	.04803	.03842
5.0	.11037	.10074	.09111	.08149	.07186	.06223	.05260	.04297	.03334

Source: From Cohen and Whitten (1988), Table 10.3, pp. 198–200, by courtesy of Marcel Dekker, Inc.

Samples from the Rayleigh Distribution

Table 9.3 *Continued*

z \ h	.10	.15	.20	.25	.30	.35	.40	.45	.50
.6	8.29117	8.27009	8.24901	8.22792	8.20684	8.18576	8.16468	8.14360	8.12252
.7	6.07499	6.05126	6.02753	6.00380	5.98007	5.95634	5.93260	5.90887	5.88514
.8	4.63526	4.60914	4.58301	4.55689	4.53077	4.50465	4.47853	4.45241	4.42629
.9	3.64716	3.61889	3.59062	3.56235	3.53408	3.50581	3.47754	3.44926	3.42099
1.0	2.93960	2.90940	2.87920	2.84900	2.81881	2.78861	2.75841	2.72821	2.69801
1.1	2.41549	2.38356	2.35164	2.31971	2.28778	2.25586	2.22393	2.19201	2.16008
1.2	2.01639	1.98292	1.94945	1.91598	1.88250	1.84903	1.81556	1.78209	1.74862
1.3	1.70544	1.67058	1.63573	1.60087	1.56602	1.53116	1.49631	1.46145	1.42660
1.4	1.45842	1.42233	1.38623	1.35014	1.31405	1.27795	1.24186	1.20576	1.16967
1.5	1.25892	1.22172	1.18451	1.14730	1.11010	1.07289	1.03569	.99848	.96128
1.6	1.09547	1.05726	1.01906	.98086	.94265	.90445	.86625	.82804	.78984
1.7	.95986	.92076	.88166	.84256	.80346	.76435	.72525	.68615	.64705
1.8	.84611	.80620	.76629	.72638	.68647	.64656	.60665	.56674	.52683
1.9	.74875	.70911	.66847	.62783	.58719	.54655	.50591	.46527	.42463
2.0	.66740	.62610	.58480	.54351	.50221	.46091	.41961	.37831	.33701
2.1	.59648	.55458	.51269	.47079	.42889	.38699	.34510	.30320	.26130
2.2	.53496	.49252	.45008	.40764	.36520	.32276	.28032	.23788	.19544
2.3	.48124	.43831	.39538	.35244	.30951	.26658	.22364	.18071	.13778
2.4	.43407	.39068	.34730	.30392	.26053	.21715	.17377	.13038	.08700
2.5	.39241	.34862	.30482	.26103	.21724	.17344	.12965	.08585	.04206
2.6	.35545	.31128	.26711	.22294	.17877	.13460	.09043	.04626	.00209
2.7	.32249	.27798	.23346	.18895	.14443	.09992	.05540	.01089	-.03363
2.8	.29299	.24816	.20333	.15850	.11366	.06883	.02400	-.02083	-.06566
2.9	.26647	.22135	.17623	.13110	.08598	.04086	-.00426	-.04939	-.09451
3.0	.24255	.19716	.15177	.10638	.06098	.01559	-.02980	-.07519	-.12058
3.1	.22090	.17526	.12962	.08398	.03834	-.00730	-.05294	-.09858	-.14422
3.2	.20123	.15536	.10949	.06363	.01776	-.02811	-.07398	-.11985	-.16572
3.3	.18332	.13724	.09116	.04508	-.00100	-.04709	-.09317	-.13925	-.18533
3.4	.16696	.12068	.07440	.02812	-.01815	-.06443	-.11071	-.15699	-.20327
3.5	.15198	.10551	.05905	.01259	-.03387	-.08033	-.12679	-.17325	-.21972
3.6	.13822	.09159	.04495	-.00168	-.04831	-.09494	-.14157	-.18821	-.23484
3.7	.12556	.07877	.03198	-.01482	-.06161	-.10840	-.15519	-.20198	-.24877
3.8	.11388	.06694	.02000	-.02694	-.07388	-.12082	-.16776	-.21469	-.26163
3.9	.10308	.05601	.00893	-.03815	-.08523	-.13230	-.17938	-.22646	-.27354
4.0	.09309	.04588	-.00133	-.04854	-.09574	-.14295	-.19016	-.23736	-.28457
4.1	.08381	.03648	-.01085	-.05818	-.10551	-.15283	-.20016	-.24749	-.29482
4.2	.07518	.02774	-.01970	-.06714	-.11459	-.16203	-.20947	-.25691	-.30436
4.3	.06715	.01960	-.02795	-.07550	-.12305	-.17060	-.21815	-.26569	-.31324
4.4	.05966	.01201	-.03564	-.08329	-.13094	-.17859	-.22624	-.27389	-.32154
4.5	.05266	.00491	-.04283	-.09058	-.13832	-.18606	-.23381	-.28155	-.32930
4.6	.04611	-.00172	-.04956	-.09739	-.14523	-.19306	-.24089	-.28873	-.33656
4.7	.03997	-.00795	-.05586	-.10378	-.15170	-.19962	-.24754	-.29545	-.34337
4.8	.03421	-.01378	-.06178	-.10978	-.15778	-.20577	-.25377	-.30177	-.34977
4.9	.02880	-.01927	-.06734	-.11541	-.16349	-.21156	-.25963	-.30770	-.35578
5.0	.02371	-.02443	-.07257	-.12072	-.16886	-.21700	-.26515	-.31329	-.36144

Table 9.3 *Continued*

z \ h	.55	.60	.65	.70	.75	.80	.85	.90	.95
.6	8.10143	8.08035	8.05927	8.03819	8.01711	7.99602	7.97494	7.95386	7.93278
.7	5.86141	5.83768	5.81395	5.79022	5.76649	5.74276	5.71903	5.69530	5.67157
.8	4.40017	4.37404	4.34792	4.32180	4.29568	4.26956	4.24344	4.21732	4.19119
.9	3.39272	3.36445	3.33618	3.30791	3.27964	3.25137	3.22310	3.19482	3.16655
1.0	2.66781	2.63761	2.60741	2.57721	2.54701	2.51681	2.48662	2.45642	2.42622
1.1	2.12815	2.09623	2.06430	2.03238	2.00045	1.96853	1.93660	1.90467	1.87275
1.2	1.71515	1.68168	1.64820	1.61473	1.58126	1.54779	1.51432	1.48085	1.44738
1.3	1.39174	1.35689	1.32203	1.28718	1.25232	1.21747	1.18261	1.14776	1.11290
1.4	1.13357	1.09748	1.06138	1.02529	.98920	.95310	.91701	.88091	.84482
1.5	.92407	.88686	.84966	.81245	.77525	.73804	.70084	.66363	.62642
1.6	.75163	.71343	.67523	.63702	.59882	.56062	.52241	.48421	.44601
1.7	.60795	.56885	.52975	.49065	.45155	.41245	.37334	.33424	.29514
1.8	.48692	.44701	.40710	.36719	.32728	.28737	.24746	.20755	.16765
1.9	.38399	.34336	.30272	.26208	.22144	.18080	.14016	.09952	.05888
2.0	.29571	.25441	.21311	.17181	.13052	.08922	.04792	.00662	-.03468
2.1	.21941	.17751	.13561	.09372	.05182	.00992	-.03197	-.07387	-.11577
2.2	.15300	.11056	.06812	.02568	-.01676	-.05920	-.10164	-.14408	-.18652
2.3	.09484	.05191	.00898	-.03396	-.07689	-.11982	-.16276	-.20569	-.24862
2.4	.04362	.00024	-.04315	-.08653	-.12991	-.17330	-.21668	-.26006	-.30345
2.5	-.00174	-.04553	-.08932	-.13312	-.17691	-.22071	-.26450	-.30829	-.35209
2.6	-.04208	-.08625	-.13042	-.17460	-.21877	-.26294	-.30711	-.35128	-.39545
2.7	-.07814	-.12266	-.16717	-.21169	-.25620	-.30072	-.34523	-.38975	-.43426
2.8	-.11049	-.15533	-.20016	-.24499	-.28982	-.33465	-.37948	-.42431	-.46915
2.9	-.13963	-.18476	-.22988	-.27500	-.32012	-.36525	-.41037	-.45549	-.50062
3.0	-.16597	-.21136	-.25675	-.30215	-.34754	-.39293	-.43832	-.48371	-.52910
3.1	-.18986	-.23550	-.28114	-.32678	-.37242	-.41805	-.46369	-.50933	-.55497
3.2	-.21159	-.25746	-.30332	-.34919	-.39506	-.44093	-.48680	-.53267	-.57854
3.3	-.23141	-.27749	-.32357	-.36965	-.41573	-.46182	-.50790	-.55398	-.60006
3.4	-.24955	-.29582	-.34210	-.38838	-.43466	-.48094	-.52721	-.57349	-.61977
3.5	-.26618	-.31264	-.35910	-.40556	-.45202	-.49848	-.54495	-.59141	-.63787
3.6	-.28147	-.32810	-.37473	-.42136	-.46800	-.51463	-.56126	-.60789	-.65452
3.7	-.29556	-.34235	-.38914	-.43593	-.48272	-.52951	-.57630	-.62309	-.66989
3.8	-.30857	-.35551	-.40245	-.44939	-.49633	-.54327	-.59021	-.63714	-.68408
3.9	-.32061	-.36769	-.41477	-.46185	-.50892	-.55600	-.60308	-.65016	-.69723
4.0	-.33178	-.37898	-.42619	-.47340	-.52061	-.56781	-.61502	-.66223	-.70943
4.1	-.34215	-.38948	-.43681	-.48413	-.53146	-.57879	-.62612	-.67345	-.72078
4.2	-.35180	-.39924	-.44668	-.49413	-.54157	-.58901	-.63645	-.68390	-.73134
4.3	-.36079	-.40834	-.45589	-.50344	-.55099	-.59854	-.64609	-.69364	-.74119
4.4	-.36919	-.41684	-.46449	-.51214	-.55979	-.60744	-.65509	-.70274	-.75039
4.5	-.37704	-.42479	-.47253	-.52028	-.56802	-.61577	-.66351	-.71126	-.75900
4.6	-.38440	-.43223	-.48006	-.52790	-.57573	-.62356	-.67140	-.71923	-.76707
4.7	-.39129	-.43921	-.48713	-.53504	-.58296	-.63088	-.67880	-.72672	-.77463
4.8	-.39776	-.44576	-.49376	-.54176	-.58975	-.63775	-.68575	-.73375	-.78174
4.9	-.40385	-.45192	-.49999	-.54807	-.59614	-.64421	-.69228	-.74036	-.78843
5.0	-.40958	-.45772	-.50587	-.55401	-.60215	-.65030	-.69844	-.74658	-.79473

Samples from the Rayleigh Distribution

Truncated sample. For this sample, $n = 25$, $T = 26$, $\sum_{i=1}^{25} x_i^2 = 3561$, and $\sum_{i=1}^{25} x_i^2/nT^2 = 3561/25(26)^2 = 0.21071$. We solve the estimating equation $J_2(\hat{\xi}) = 0.21071$ by interpolation in Table 9.1, and thereby obtain $\hat{\xi} = 3.003$, a value that could be read with only slightly less accuracy from the graph of Figure 9.1. It subsequently follows that $\hat{\sigma} = T/\hat{\xi} = 26/3.003 = 8.66$. The asymptotic variance is calculated from (9.4.3) with 8.66 substituted for σ, as $V(\hat{\sigma}) = 1.275$.

Censored sample. The additional information that $c = 2$ observations were censored at $T = 26$ and are known to exceed this value is added to the truncated sample data. We thus have a Type I censored sample with $N = 27$, $n = 25$, $c = 2$, $T = 26$, and $\sum_{i=1}^{25} x_i^2 = 3561$. Substitution of these values into (9.3.13) yields the required estimator as

$$\hat{\sigma} = \sqrt{\frac{3561 + 2(26)^2}{2(25)}} = 9.91.$$

From (9.4.5), we calculate $V(\hat{\sigma}) = 0.983$.

Complete sample. Actual measurements of the two previously censored observations are added to the sample data to produce a complete sample for which $N = 27$ and $\sum_{i=1}^{27} x_i^2 = 5019$. In this case, the estimate $\hat{\sigma}$ is calculated as

$$\hat{\sigma} = \sqrt{\frac{\sum_{i=1}^{27} x_i^2}{2N}} = \sqrt{\frac{5019}{2(27)}} = 9.64.$$

Estimation from complete samples has been examined in more detail by Cohen (1955) and by Cohen and Whitten (1988), who give the variance from this sample as $V(\hat{\sigma}) = 0.852$. It, however, seems appropriate to note here that the complete sample estimator for σ is a special case of the censored sample estimator of (9.3.13) in which $c = 0$ and $N = n$.

Example 9.2. Data for this sample consist of radial errors of 20 throws of a hand-held dart against a circular target of radius 3.5. As originally proposed, errors were measured from the center of impact. For the purpose of this illustration, they are assumed to be measured from the center of the target. Out of $N = 20$ throws, $n = 19$ resulted in measured errors, $c = 1$ missed the target and were thereby declared to be censored. This resulted in a Type I censored sample. For the 19 measured errors, $\sum_{i=1}^{19} x_i^2 = 80.91$, and from (9.3.13), we calculate

$$\hat{\sigma} = \sqrt{\frac{80.91 + 1(3.5)^2}{2(19)}} = 1.57.$$

In standard units the point of censoring is $\hat{\xi} = T/\hat{\sigma} = 3.5/1.57 = 2.23$.

9.6 SOME CONCLUDING REMARKS

Since the two-dimensional Rayleigh distribution is also a special case of the Weibull distribution, many of the estimators presented in Chapter 5 for Weibull parameters can be used to estimate Rayleigh parameters. Furthermore, estimate variances and covariances obtained for estimates of Weibull parameters might be found useful in calculating confidence intervals for estimates of Rayleigh parameters.

10
Truncated and Censored Samples from the Pareto Distribution

10.1 INTRODUCTION

The Pareto distribution is a highly skewed reverse J-shaped distribution that holds a special attraction for economists and for social scientists. It is named for Vilfredo Pareto, an Italian-born Swiss professor of economics who formulated it (1897) as a model for the distribution of incomes. In recent years it has been the subject of investigations by numerous writers. Among these are Zipf (1949), Muniruzzaman (1950, 1957), Hagstroem (1960), Harris (1968), Mandelbrodt (1960), Srivastava (1965), Malik (1966, 1970), Cirillo (1979), and Arnold (1983). Johnson and Kotz (1970, Vol. 1) present an excellent expository account of this distribution together with discussions of some of its applications and a lengthy list of references.

Maximum likelihood and modified maximum likelihood estimators for parameters of this distribution based on complete (unrestricted) samples were examined by Cohen and Whitten (1988), who also briefly discussed estimation from truncated and censored samples. In this chapter, primary attention is focused on samples that are singly right truncated and singly right censored.

10.2 SOME FUNDAMENTALS

In the parameterization of interest here, the pdf of the Pareto distribution is

$$f(x; \gamma, \alpha) = \alpha \gamma^\alpha x^{-(\alpha+1)}, \quad 0 \leq \gamma \leq x, \alpha > 0, \quad (10.2.1)$$
$$= 0 \quad \text{elsewhere,}$$

and the cdf is

$$F(x; \gamma, \alpha) = 1 - \left(\frac{\gamma}{x}\right)^\alpha, \qquad (10.2.2)$$

where γ is a threshold parameter and α is a shape parameter (sometimes referred to as the Paretian index). Johnson and Kotz (1970) point out that the Pareto distribution is a special case of Pearson's Type VI distribution. The rth moment of this distribution about the origin is

$$\mu'_r = \frac{\alpha \gamma^r}{\alpha - r}, \qquad r < \alpha. \qquad (10.2.3)$$

The moments thus fail to exist unless $\alpha > r$. The expected value, variance, α_3, α_4, mode, median, and mean deviation are

$$E(X) = \frac{\alpha \gamma}{\alpha - 1}, \qquad \alpha > 1,$$

$$V(X) = \frac{\alpha \gamma^2}{(\alpha - 1)^2 (\alpha - 2)}, \qquad \alpha > 2,$$

$$\alpha_3(X) = 2\left(\frac{\alpha + 1}{\alpha - 3}\right)\sqrt{\frac{\alpha - 2}{\alpha}}, \qquad \alpha > 3,$$

$$\alpha_4(X) = \frac{3(\alpha - 2)(3\alpha^2 + \alpha + 2)}{\alpha(\alpha - 3)(\alpha - 4)}, \qquad \alpha > 4, \qquad (10.2.4)$$

$$\text{Mo}(X) = \gamma, \qquad \text{Me}(X) = 2^{1/\alpha}\gamma,$$

$$\text{MD}(X) = 2\gamma(\alpha - 1)^{-1}(1 - \alpha^{-1})^{\alpha - 1}, \qquad \alpha > 1.$$

The hazard or failure rate function for this distribution is the decreasing function

$$H(x) = \frac{\alpha}{x}. \qquad (10.2.5)$$

Consequently, the Pareto distribution might be a suitable life-span model in situations where product or systems development results in improved performance as development proceeds. Note that the mode of this distribution occurs at the origin with $f(\gamma) = \alpha/\gamma$.

Limiting values of $E(x)$, $V(x)$, $\alpha_3(X)$, $\alpha_4(X)$, and $\text{Me}(X)$ as $\alpha \to \infty$ are

$$\lim_{\alpha \to \infty} E(X) = \gamma, \qquad \lim_{\alpha \to \infty} \text{Me}(X) = \gamma,$$

$$\lim_{\alpha \to \infty} V(X) = 0, \qquad (10.2.6)$$

$$\lim_{\alpha \to \infty} \alpha_3(X) = 2, \qquad \lim_{\alpha \to \infty} \alpha_4(X) = 9.$$

Samples from the Pareto Distribution

Thus, the limiting form of the Pareto distribution as $\alpha \to \infty$ is a degenerate form of the two-parameter exponential distribution (8.1.1) in which $\beta = 0$. Graphs of the pdf and the cdf with $\alpha = 4.1$ are given as illustrations in Figures 10.1 and 10.2, respectively.

As an additional display of important distinguishing characteristics of this distribution, tabulations of $E(X)$, $Me(X)$, $V(X)$, $\alpha_3(X)$ and $\alpha_4(X)$ as functions of α are included in Table 10.1 for selected values of this argument. More complete tables of these characteristics are given by Cohen and Whitten (1988).

10.2.1 The First-Order Statistic

The first-order statistic appears in the modified estimators. The cdf of this statistic in a sample of size N is

$$F(x_{1:N}) = 1 - \left(\frac{\gamma}{x_{1:N}}\right)^{N\alpha}, \qquad x_{1:N} \geq \gamma, \tag{10.2.6}$$

and the corresponding pdf is

$$f(x_{1:N}) = N\alpha \gamma^{N\alpha} (x_{1:N})^{-(N\alpha+1)}. \tag{10.2.7}$$

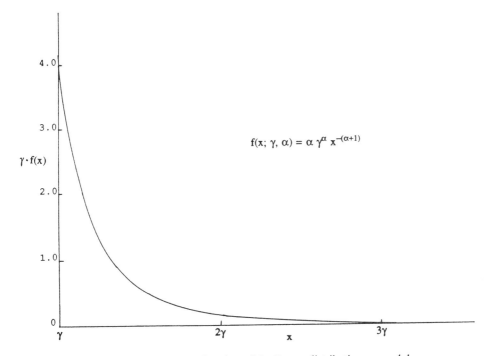

Figure 10.1 Probability density function of the Pareto distribution: $\alpha = 4.1$.

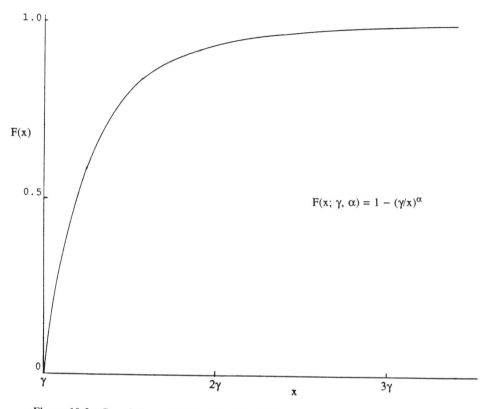

Figure 10.2 Cumulative probability function of the Pareto distribution: $\alpha = 4.1$.

It thus follows that if X is distributed in accordance with a Pareto distribution (γ, α), then $X_{1:N}$ has a Pareto distribution $(\gamma, N\alpha)$ with expected value and variance

$$E(X_{1:N}) = \frac{N\alpha\gamma}{N\alpha - 1} \quad \text{and} \quad V(X_{1:N}) = \frac{N\alpha\gamma^2}{(N\alpha - 1)^2(N\alpha - 2)}. \quad (10.2.8)$$

Other characteristics of interest follow from (10.2.4) when α is replaced with $N\alpha$.

10.3 PARAMETER ESTIMATION
10.3.1 Singly Right Truncated Samples

The loglikelihood function of a random sample of size n from the Pareto distribution that is truncated at a point T such that $x \leq T$ is

Samples from the Pareto Distribution

Table 10.1 Some Characteristics of the Pareto Distribution

α	$E(X)/\gamma$	$M_e(X)/\gamma$	$V(X)/\gamma^2$	$\alpha_3(X)$	$\alpha_4(X)$
1.1	11.00000	1.87786			
2.1	1.90909	1.39107	17.35537		
3.1	1.47619	1.25057	0.63904	48.84604	
4.1	1.32258	1.18419	0.20316	6.63629	789.66519
5.0	1.25000	1.14870	0.10417	4.64758	73.80000
6.0	1.20000	1.12246	0.60000	3.81032	38.66667
7.0	1.16667	1.10409	0.03889	3.38062	27.85714
8.0	1.14286	1.09051	0.02721	3.11769	22.72500
9.0	1.12500	1.08006	0.02009	2.93972	19.75556
10.0	1.11111	1.07177	0.01543	2.81106	17.82857
15.0	1.07143	1.04729	0.00589	2.48253	13.63030
20.0	1.05263	1.03526	0.00308	2.34381	12.13015
25.0	1.04167	1.02811	0.00189	2.17992	11.36260
30.0	1.03448	1.02337	0.00127	2.21843	12.13015
40.0	1.02564	1.01748	0.00069	2.16010	10.36014
50.0	1.02041	1.01395	0.00043	2.12637	10.06002
100.0	1.01010	1.00696	0.00010	2.06154	9.50385
250.0	1.00402	1.00278	0.000016	2.00798	9.19571
500.0	1.00200	1.00138	0.000004	2.01206	9.09692
1000.0	1.00100	1.00069	0.000001	2.00602	9.04823
∞	1.00000	1.00000	0	2.00000	9.00000

$$\ln L = n \ln \alpha + n\alpha \ln \gamma - (\alpha + 1) \sum_{i=1}^{n} \ln x_i - n \ln \left[1 - \left(\frac{\alpha}{T}\right)^{\alpha} \right]. \quad (10.3.1)$$

This is an increasing function of γ, and therefore the maximum likelihood estimate of this parameter must be its largest permissible value. However, since $\gamma \leq x_1$, we choose

$$\hat{\gamma} = x_1. \quad (10.3.2)$$

For a second estimating equation, we differentiate $\ln L$ with respect to α and equate to zero. We then have

$$\frac{\partial \ln L}{\partial \alpha} = \frac{n}{\alpha} + n \ln \gamma - \sum_{i=1}^{n} \ln x_i + \frac{n(\gamma/T)^{\alpha} \ln(\gamma/T)}{1 - (\gamma/T)^{\alpha}} = 0. \quad (10.3.3)$$

This is subsequently simplified to

$$\frac{1}{\hat{\alpha}} + \frac{(\hat{\gamma}/T)^{\hat{\alpha}} \ln(\hat{\gamma}/T)}{1 - (\hat{\gamma}/T)^{\hat{\alpha}}} = \frac{1}{n} \sum_{i=1}^{n} \ln x_i - \ln \hat{\gamma}. \quad (10.3.4)$$

When $\hat{\gamma}$ is known or set equal to x_1 for the MLE, it is a relatively simple matter to solve (10.3.4) for $\hat{\alpha}$ by employing standard iterative methods or even by using the "trial-and-error" procedure.

10.3.2 Singly Right Censored Samples

We consider samples for which N is the total sample size, n is the number of complete observations less than or at most equal to T, and c is the number of censored observations greater than T. The loglikelihood function of a sample of this type from a Pareto distribution with pdf (10.2.1) is

$$\ln L = n \ln \alpha + n\alpha \ln \gamma - (\alpha + 1) \sum_{i=1}^{n} \ln x_i$$
$$+ c\alpha \ln \left(\frac{\gamma}{T}\right) + \text{const.} \quad (10.3.5)$$

Maximum Likelihood Estimators

Applicable estimating equations are

$$\hat{\gamma} = x_{1:N} \quad (10.3.6)$$
$$\frac{\partial \ln L}{\partial \alpha} = \frac{n}{\alpha} + n \ln \gamma - \sum_{i=1}^{n} \ln x_i + c \ln \left(\frac{\gamma}{T}\right) = 0$$

When γ is either known or equated to $x_{1:N}$, the estimator of α becomes

$$\hat{\alpha} = n \left[\sum_{i=1}^{n} \ln x_i - N \ln \hat{\gamma} + c \ln T \right]^{-1}. \quad (10.3.7)$$

Unfortunately, the MLE for γ, given in the first equation of (10.3.6), is biased. Although bias diminishes as the sample size increases, it can be quite troublesome in small and even in moderate-sized samples. In an effort to eliminate the effect of bias, we are led to consider modified maximum likelihood estimators that employ the expected value of the first-order statistic.

Modified Maximum Likelihood Estimators

In this case, the applicable estimating equations are $\partial \ln L/\partial \alpha = 0$ and $E(X_{1:N}) = x_{1:N}$. The first of these equations was given in (10.3.6), and the second equation now becomes

$$\frac{N\alpha\gamma}{N\alpha - 1} = x_{1:N}. \quad (10.3.8)$$

Samples from the Pareto Distribution

We eliminate α between the two estimating equations to obtain the following equation in which $\hat{\gamma}$ is the only unknown.

$$\frac{x_{1:N}}{N(x_{1:N} - \hat{\gamma})} = n\left[\sum_{i=1}^{n} \ln x_i - N \ln \hat{\gamma} + c \ln T\right]^{-1}. \qquad (10.3.9)$$

Equation (10.3.9) can be solved by "trial and error" or by employing any of the various standard iterative methods, and the estimate of α follows from (10.3.8) as

$$\hat{\alpha} = \frac{x_{1:N}}{N(x_{1:N} - \hat{\gamma})}. \qquad (10.3.10)$$

Complete sample estimators can be obtained as special cases of the censored sample estimators by setting $c = 0$ and $N = n$. The complete sample MLE can be obtained in this manner from (10.3.7) and the corresponding MMLE can be obtained from (10.3.9) and (10.3.10).

10.4 ESTIMATE RELIABILITY

As previously noted, maximum likelihood estimators are subject to certain regularity conditions that sometimes limit their usefulness. The loglikelihood function (10.3.1) is unbounded with respect to γ, and since γ is a lower bound on X, it follows that the value of γ that maximizes $\ln L$ subject to this constraint is $\hat{\gamma} = x_{1:N}$. Quant (1966) has shown that the MLE $\hat{\gamma}$ and $\hat{\alpha}$ are consistent in complete samples. They are also consistent in censored samples, although bias is quite large in small samples. When γ is known, the asymptotic variance of $\hat{\alpha}$ can be obtained by inverting the expected value of the second partial derivative. We accordingly differentiate (10.3.6) a second time and subsequently obtain the following approximate variance.

$$\text{Asy Var}(\hat{\alpha}) = \left[E\left(\frac{\partial^2 \ln L}{\partial \alpha^2}\right)\right]^{-1} = E\left(\frac{\hat{\alpha}^2}{n}\right) \doteq \frac{\hat{\alpha}^2}{n}. \qquad (10.4.1)$$

Johnson and Kotz (1970) show that for a complete unrestricted sample, $2n\alpha/\hat{\alpha}$ is distributed as chi-square with $2n$ degrees of freedom, and thus the expected value and variance of $\hat{\alpha}$ are

$$E(\hat{\alpha}) = \frac{n\alpha}{n-2} \quad \text{and} \quad V(\hat{\alpha}) = \frac{n^2\alpha^2}{(n-2)^2(n-3)}. \qquad (10.4.2)$$

These results should also hold approximately for censored samples if the degree of censoring is not excessive. Furthermore, it is to be noted that the asymptotic variance of (10.4.1) differs very little from that given in (10.4.2) when $n > 100$.

In the majority of practical applications of the Pareto distribution, sample sizes are likely to equal or exceed this magnitude. A $100(1 - \eta)\%$ confidence interval on $\hat{\alpha}$ (at least in the case of complete samples) can be calculated from the chi-square distribution as

$$\frac{\hat{\alpha}\chi^2_{2n,\eta/2}}{2n} < \alpha < \frac{\hat{\alpha}\chi^2_{2n,(1-\eta/2)}}{2n}. \qquad (10.4.3)$$

The variance of the MLE $\hat{\gamma} = x_{1:N}$ can be calculated from (10.2.8) as

$$V(\hat{\gamma}) = V(x_{1:N}) = \frac{N\alpha\gamma^2}{(N\alpha - 1)^2(N\alpha - 2)}. \qquad (10.4.4)$$

For large values of N, we might take advantage of the approach to normality to approximate a 95% CI as

$$\hat{\gamma} - 1.96 \frac{\hat{\gamma}}{N\hat{\alpha} - 1} \sqrt{\frac{N\hat{\alpha}}{N\hat{\alpha} - 2}} \leq \gamma \leq \hat{\gamma}$$

$$+ 1.96 \frac{\hat{\gamma}}{N\hat{\alpha} - 1} \sqrt{\frac{N\hat{\alpha}}{N\hat{\alpha} - 2}}. \qquad (10.4.5)$$

This same confidence interval should also provide a satisfactory approximation when it is based on modified maximum likelihood estimates.

10.5 AN ILLUSTRATIVE EXAMPLE

To illustrate the practical application of estimators presented in this chapter, we consider a singly right censored sample consisting of a total of $N = 25$ observations of which $c = 5$ are censored at $X = T = 2700$. The $n = 20$ complete observations are tabulated as follows:

1010	1098	1270	1505
1035	1109	1315	1600
1049	1130	1332	1652
1062	1180	1450	1813
1076	1210	1480	1951

In summary, for this example $N = 25$, $c = 5$, $n = 20$, $x_{1:25} = 1010$, $T = 2700$, $\Sigma_{i=1}^{20} \ln x_i = 143.2632$, $\ln x_{1:25} = 6.9177$, and $\ln T = 7.9010$. Although this sample could be a sample of payroll checks from some specified population of office workers, it actually is of no particular significance except as a vehicle for the illustration of computing procedures.

Samples from the Pareto Distribution

Maximum Likelihood Estimates

From the first equation of (10.3.6), we have $\hat{\gamma} = x_{1:25} = 1010$, and from (10.3.7), we calculate

$$\hat{\alpha} = \frac{20}{143.2632 - 25(6.9177) + 5(7.9010)} = 2.035.$$

The asymptotic variance of $\hat{\alpha}$ as calculated from (10.4.1) is $V(\hat{\alpha}) = \hat{\alpha}^2/n = (2.035)^2/20 = 0.2071$, and $\sigma_{\hat{\alpha}} = \sqrt{0.2071} = 0.455$. An approximate 95% confidence interval based on a limiting normal distribution is

$$1.143 < \alpha < 3.783.$$

To obtain a similar result for $\hat{\gamma}$, we employ (10.4.4) and calculate the variance of $\hat{\gamma}$ as

$$V(\hat{\gamma}) = V(x_{1:25}) = \frac{25(2.035)(1010)^2}{[25(2.035) - 1]^2[25(2.035) - 2]} = 426.87.$$

It follows that $\sigma_{\hat{\gamma}} = 20.66$, and the approximate 95% CI is

$$969.5 < \gamma < 1050.5.$$

A 95% confidence interval on $\hat{\alpha}$ based on the chi-square distribution is calculated from (10.4.3) to be

$$1.243 < \alpha < 3.019.$$

Although this interval is slightly narrower than the interval based on the limiting normal distribution, it tends to confirm the validity of our calculations. However, our sample is not sufficiently large for any firm conclusions.

Modified Maximum Likelihood Estimates

For the MMLE, we substitute in (10.3.9) and solve iteratively to obtain $\hat{\gamma} = 988.9$. We then calculate $\hat{\alpha}$ from (10.3.10) as

$$\hat{\alpha} = \frac{1010}{25(1010 - 988.9)} = 1.915.$$

It is noted that differences between the MLE and the MMLE calculated from this sample are quite small. In most applications, the MMLE might be preferred because of its smaller bias. However, readers are again reminded that the only purpose in selecting this particular example has been to illustrate computing procedures. Caution should be exercised in formulating any firm conclusions concerning the relative merits of estimators on the basis of these calculations.

11
Higher-Moment Estimators of Pearson Distribution Parameters from Truncated Samples

11.1 INTRODUCTION

The presentation in this chapter is based on previous results of Cohen (1941, 1951). A method of higher moments is introduced and employed in the estimation of Pearson distribution parameters from truncated samples with known points of truncation. In the four-parameter distributions, the first six moments of a doubly truncated sample are required, but only the first five moments are needed from a singly truncated sample. The order of the highest-order moment involved in these estimators is equal to the number of distribution parameters plus the number of tails truncated. As a consequence of employing higher-order moments, sampling errors of estimates are somewhat greater than corresponding errors of maximum likelihood estimates. However, higher-moment estimates are consistent and they are functionly explicit. Nothing more complex than the inversion of a 5×5 matrix is involved in their calculation. Accordingly, they are suitable for use in large samples. Even in moderate-sized or small samples, they might be employed to provide first approximations to be used in iterating to maximum likelihood estimates.

11.2 THE PEARSON DISTRIBUTIONS

The Pearson system of frequency curves has its genesis in the differential equation

$$\frac{1}{f(x)} \frac{df(x)}{dx} = \frac{a - x}{b_0 + b_1 x + b_2 x^2}, \quad r \le x \le s, \qquad (11.2.1)$$

Higher-Moment Estimators

where the origin is arbitrarily taken. For truncated distributions, as we explain later, it is convenient to take our origin at the left truncation point.

In the derivations that follow, we regard the coefficients of (11.2.1), a, b_0, b_1, and b_2, as primary characterizing parameters. Distribution means, standard deviations, α_3, and α_4 are expressed as functions of these coefficients. A moment recursion formula in which moments are expressed as functions of the coefficients of (11.2.1) can be obtained from the differential equation. We separate the variables, multiply both sides by x^k, and integrate over the full range of permissible values, that is, $r \leq x \leq s$. Thereby, we obtain

$$\int_r^s (b_0 + b_1 x + b_2 x^2) x^k \, df(x) = \int_r^s (a - x) x^k f(x) \, dx. \qquad (11.2.2)$$

Let the kth moment of the complete (untruncated) distribution about the left terminal be defined as

$$\mu_k' = \int_r^s x^k f(x) \, dx, \qquad (11.2.3)$$

and the right member of (11.2.2) becomes $a\mu_k' - \mu_{k+1}'$. We integrate the left member by parts to obtain

$$[(b_0 + b_1 x + b_2 x^2) x^k f(x)]_r^s - k b_0 \mu_{k-1}'$$
$$- (k - 1) b_1 \mu_k' - (k + 2) b_2 \mu_{k+1}' \qquad (11.2.4)$$

Only solutions of (11.2.1) for which $f(r) = f(s) = 0$ are included as members of the Pearson system. Moreover, only solutions for which the left side of (11.2.4) vanishes at both limits are included. As a consequence of these restrictions, we may combine the two sides of (11.2.2) to yield the following recursion formula for moments of the complete distribution about the origin:

$$h\mu_k' + b_0 k \mu_{k-1}' + b_1 k \mu_k' + b_2 (k + 2) \mu_{k+1}' = \mu_{k+1}', \qquad (11.2.5)$$

where in order to simplify our notation, we have written

$$h = a + b_1. \qquad (11.2.6)$$

We normalize $f(x)$ so that $\mu_0' = 1$, and let $k = 0, 1, 2,$ and 3 in succession in (11.2.5) to obtain the following system of equations:

$$\begin{aligned}
(2b_2 - 1)\mu_1' &= -h, \\
(b_1 + h)\mu_1' + (3b_2 - 1)\mu_2' &= -b_0, \\
2b_0 \mu_1' + (2b_1 + h)\mu_2' + (4b_2 - 1)\mu_3' &= 0, \\
3b_0 \mu_2' + (3b_1 + h)\mu_3' + (5b_2 - 1)\mu_4' &= 0.
\end{aligned} \qquad (11.2.7)$$

As a simultaneous solution of these equations, we find

$$\mu'_1 = h/(1 - 2b_2),$$
$$\mu'_2 = [b_0 + (b_1 + h)\mu'_1]/(1 - 3b_2),$$
$$\mu'_3 = [2b_0\mu'_1 + (2b_1 + h)\mu'_2]/(1 - 4b_2), \quad (11.2.8)$$
$$\mu'_4 = [3b_0\mu'_2 + (3b_1 + h)\mu'_3]/(1 - 5b_2),$$

With h, b_0, b_1, and b_2 known, the moments μ'_1, μ'_2, μ'_3, and μ'_4 can be easily calculated from (11.2.8) in the order named. Corresponding central moments follow from the translation formula

$$\mu_k = \sum_{i=0}^{k} \binom{k}{i} \mu'_{k-i}(\mu'_1)^i, \quad (11.2.9)$$

and the standard moments follow from

$$\alpha_k = \frac{\mu_k}{\sigma^k}, \quad (11.2.10)$$

where $\sigma^2 = \mu_2$. The second central moment now becomes

$$\mu_2 = \frac{1}{1 - 3b_2}\left[b_0 + \frac{h}{1 - 2b_2}\left(b_1 + \frac{hb_2}{1 - 2b_2}\right)\right]. \quad (11.2.11)$$

Similar formulas could also be written for μ_3 and μ_4, but they are too unwieldy to be useful. For each practical application it seems preferable to compute non-central moments about the left terminal from (11.2.8). Central moments as required can then be obtained from (11.2.9).

As previously mentioned, we select the left truncation point as the origin from which measurements are made. Let ξ designate this truncation point in standard units; that is, $\xi = (0 - \mu'_1)/\sigma$, and it follows that

$$\mu'_1 = -\sigma\xi. \quad (11.2.12)$$

Formulas for explicitly expressing the mean, standard deviation, α_3, and α_4 as functions of h, b_0, b_1, and b_2 for the four-parameter distributions are too complex to be useful, but as shown below, they are relatively simple for the normal and the Type III distributions.

Type III distributions: In this case, $b_2 = 0$, and we have

$$\mu'_1 = h, \qquad h = -\sigma\xi,$$
$$\sigma = \sqrt{b_0 + b_1 h}, \qquad b_1 = \sigma\alpha_3/2, \quad (11.2.13)$$
$$\alpha_3 = 2b_1/\sqrt{b_0 + b_1 h}, \qquad b_0 = \sigma^2[1 + \xi\alpha_3/2].$$

Higher-Moment Estimators

Normal distribution: In this case, $b_1 = b_2 = 0$, and we find

$$\mu_1' = h, \qquad h = -\sigma\xi,$$
$$\sigma = \sqrt{b_0}, \qquad b_0 = \sigma^2. \qquad (11.2.14)$$

11.3 RECURSION FORMULA FOR MOMENTS OF A TRUNCATED DISTRIBUTION

Let Y designate the original random variable, and let T_1 and T_2 designate truncation points in a doubly truncated distribution. We translate the origin to the left point of truncation; that is, $X = Y - T_1$. Now let $d = T_2 - T_1$, and the truncation points of X as measured from the translated origin become 0 and d, where $r \leq 0 \leq X \leq d \leq s$. We reduce the limits of integration in (11.2.2) to obtain

$$\int_0^d (b_0 + b_1 x + b_2 x^2) x^k df(x) = \int_0^d (a - x) x^k f(x)\, dx. \qquad (11.3.1)$$

Define the kth moment of the truncated distribution about the left terminal as

$$m_k = \int_0^d x^k f(x)\, dx, \qquad (11.3.2)$$

with $m_0 = 1$ and the right member of (11.3.1) becomes $am_k - m_{k+1}$. On integrating the left member by parts, we obtain

$$[(b_0 + b_1 x + b_2 x^2) x^k f(x)]_0^d - k b_0 m_{k-1} - (k+1) b_1 m_k. \qquad (11.3.3)$$

Since we are not integrating over the full range of X, the first term in the preceding expression does not vanish as it did for the complete distribution. However, if we define

$$H = f(0) b_0 \quad \text{and} \quad J = f(d)[b_0 + d b_1 + d^2 b_2], \qquad (11.3.4)$$

and then combine left and right members of (11.3.1), we obtain a recursion formula for moments of the truncated distribution. Thus we have

$$h m_k + b_0 k m_{k-1} + b_1 k m_k + b_2 (k+2) m_{k+1} - d^k J$$
$$= m_{k+1}, \qquad k \geq 1. \qquad (11.3.5)$$

If we let $k = 0$ in (11.3.1) prior to integrating, and then proceed as outlined above, we obtain the following companion equation to (11.3.5) for the case of $k = 0$:

$$h + 2 m_1 b_2 + H - J = m_1. \qquad (11.3.6)$$

11.4 PARAMETER ESTIMATION FROM DOUBLY TRUNCATED SAMPLES

The method of higher moments differs from the usual method of moments only in that it involves moments of an order that exceeds the number of parameters to be estimated by one for each tail that is removed. To estimate parameters of a four-parameter distribution from a doubly truncated sample, we equate the first six population moments to corresponding sample moments, whereas the usual method of moments would involve only the first four moments. A major advantage resulting from the use of higher-order moments is that estimating equations are linear and thus can be more easily solved. We let $k = 1, 2, 3, 4, 5$ in (11.3.5), replace parameters with corresponding estimates, and equate $m_k = v_k$, where v_k is the kth sample moment about the left point of truncation. We thereby obtain the following five linear estimating equations:

$$\begin{aligned} v_1 h^* + b_0^* + v_1 b_1^* + 3v_2 b_2^* - dJ^* &= v_2, \\ v_2 h^* + 2v_1 b_0^* + 2v_2 b_1^* + 4v_3 b_2^* - d^2 J^* &= v_3, \\ v_3 h^* + 3v_2 b_0^* + 3v_3 b_1^* + 5v_4 b_2^* - d^3 J^* &= v_4, \\ v_4 h^* + 4v_3 b_0^* + 4v_4 b_1^* + 6v_5 b_2^* - d^4 J^* &= v_5, \\ v_5 h^* + 5v_4 b_0^* + 5v_5 b_1^* + 7v_6 b_2^* - d^5 J^* &= v_6. \end{aligned} \quad (11.4.1)$$

Estimates h^*, b_0^*, b_1^*, b_2^*, and J^* are obtained as the simultaneous solution of these equations. The estimate H^* can be obtained from (11.3.6) as

$$H^* = v_1 + J^* - h^* - 2v_1 b_2^* \quad (11.4.2)$$

Estimates of the noncentral moments of the complete distribution are calculated by substituting estimates h^*, b_0^*, b_1^*, b_2^* into equations (11.2.8). Estimates of central moments are then calculated by substitution in (11.2.9), and estimates of standard moments follow from (11.2.10). The resulting estimates are summarized as

$$E(X) = T_1 + \mu_1', \qquad V(X) = \mu_2 = \sigma^2, \quad (11.4.3)$$
$$\alpha_3^2(X) = \beta_1 = \frac{\mu_3}{\sigma^3}, \qquad \alpha_4(X) = \beta_2 = \frac{\mu_4}{\sigma^4},$$

where β_1 and β_2 are Pearson's notation for the shape parameters. Note that estimates are distinguished from parameters throughout this presentation by starring (*) the estimates.

A chart of Pearson's β_1–β_2 curves which serve to classify the different distributions of the Pearson system is included as Figure 11.1.

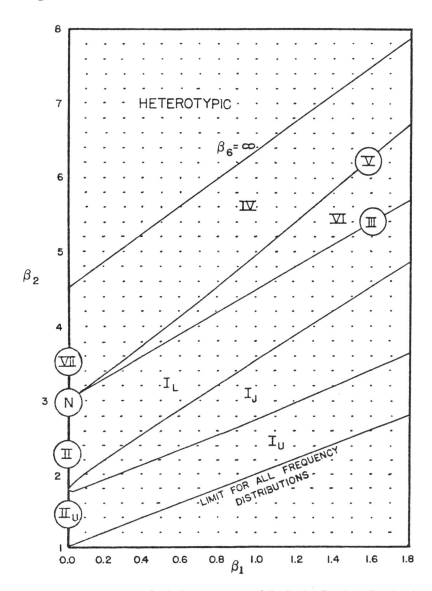

Figure 11.1 β_1–β_2 curves for the Pearson system of distribution functions. Reprinted from Whitten (1972), p. 101, by courtesy of the author.

Although neither H nor J is required in estimating moments of the complete distribution, a comparison of these estimates with corresponding values calculated from finally fitted curves provides a check on agreement between fitted curves and observed sample data.

11.5 DETERMINING THE DISTRIBUTION TYPE

With estimates of μ'_1, σ, α_3, and α_4 calculated as described in Section 11.4, the type of distribution can be established from the original Pearson criteria, or from the Carver–Craig criteria as presented by Craig (1936). Excellent descriptions of the original Pearson criteria are contained in volumes by Elderton (1938) and by Elderton and Johnson (1969). In the present instance, however, since estimates b_0, b_1, and b_2 must of necessity be computed before estimates of the population parameters can be obtained, it seems more appropriate to determine the type directly from the quadratic equation

$$b_0 + b_1 x + b_2 x^2 = 0. \qquad (11.5.1)$$

The general solution of the differential equation (11.2.1) can be written as

$$f(x) = C(x - r_1)^{m_1}(r_2 - x)^{m_2}, \qquad (11.5.2)$$

where r_1 and r_2 are roots of (11.5.1) (cf., for example Craig (1936). The nature of these roots determines the type of the distribution. Let D designate the discriminant, $D = b_1^2 - 4b_0 b_2$, and the principal Pearson distributions may be classified as follows

Type I: $r_1 - \mu'_1 < 0 < r_2 - \mu'_1$, $D > 0$;
Type II: $(r_1 - \mu'_1) = -(r_2 - \mu'_1)$, $b_1 = 2b_2\mu'_1$, $D > 0$;
Type III: $b_2 = 0$;
Type IV: r_1 and r_2 imaginary, $b_1 \neq 2b_2\mu'_1$, $D < 0$;
Type V: $(r_1 - \mu'_1)$, $(r_2 - \mu'_1)$ of the same sign, $D > 0$;
Type VI: r_1 and r_2 imaginary, $b_1 = 2b_2\mu'_1$, $D < 0$;
Normal: $b_1 = b_2 = 0$.

With $\beta_1 = \alpha_3^2$ and $\beta_2 = \alpha_4$ estimated from sample data, the graphs of Figure 10.1 can also be employed in determining distribution types. The numbering of types used here is that of Craig (1936).

A necessary condition for the odd central moments to equal zero [i.e., for $f(x)$ to be symmetrical about the mean] is that

$$b_1 = 2b_2\mu'_1. \qquad (11.5.3)$$

11.6 SINGLY TRUNCATED SAMPLES

If only the left tail is truncated, then J vanishes and we can drop the last equation from (11.4.1) and set $J = 0$ in the four equations that remain. If only the right tail is truncated, then $H = 0$, and by changing the variable from x to $d - x$,

Higher-Moment Estimators

we can translate the origin to the truncation point on the right, and again drop the last equation from (11.4.1)

In practical applications, finding either J or H equal or nearly equal to zero in a sample that is represented as being doubly truncated suggests that perhaps the sample was in fact only singly truncated. If both J and H are zero or nearly zero, this suggests that the sample is in fact from a complete rather than from a truncated distribution. In this case, parameter estimates would be calculated from complete sample estimators.

When samples are sufficiently large, occasions might arise when a sample minimum or maximum observation might be considered as a satisfactory estimate of the population terminal (threshold parameter). In this case, we may drop an additional estimating equation from (11.4.1) since one less parameter is being estimated from the sample data. To illustrate, consider a Type III distribution for which the first-order statistic x_1 is deemed to be an appropriate estimate of the distribution lower limit. In this case, we have

$$h = \frac{2\sigma}{\alpha_3} \qquad (11.6.1)$$

and from (11.2.13),

$$h = \frac{b_0 + b_1 h}{b_1}. \qquad (11.6.2)$$

It therefore follows that $b_0 = 0$, and the system of estimating equations to be solved consists of the first two equations of (11.4.1) plus (11.4.2) with $b_0 = b_2 = 0$. Parameters in (11.4.2) are, of course, replaced by their estimates.

11.7 TYPE III AND NORMAL DISTRIBUTIONS

As mentioned previously, Pearson's Type III distribution is a three-parameter gamma distribution. Maximum likelihood and modified maximum likelihood estimators for parameters of this distribution were considered in Chapter 7. In this chapter we are concerned only with higher-moment estimators. Since $b_2 = 0$, we need only the first five moments and the first four equations of (11.4.1) in order to calculate higher-moment estimates of Type III parameters from a doubly truncated sample. With singly truncated samples from which the left tail has been eliminated, we need only the first four sample moments, and it is necessary only that we solve the first three equations of (11.4.1) after setting $J = 0$.

To estimate parameters of a normal distribution from a doubly truncated sample, we calculate the first four moments and solve the first three equations of (11.4.1) with $b_1 = b_2 = 0$. With singly truncated samples from which the left tail is missing, we require only the first three sample moments, and we need to solve only the first two equations of (10.4.1) after setting $J = 0$. A more complete account of these three-moment estimators of normal distribution parameters from singly truncated samples was presented in Section 2.4.

11.8 A NUMERICAL EXAMPLE

To illustrate the practical application of estimators presented in this chapter, we consider an example of grouped data given by Lilian Shook (1930) on weights of 1000 women students. Shook considered these data as a complete sample from a Pearson Type III population. She used the method of moments to calculate estimates of population parameters and then to graduate the observed sample data.

For the purpose of this illustration, we truncate Shook's data on the left with $T_1 = 79.95$ pounds and on the right at $T_2 = 159.95$ pounds, and thereby eliminate the first and the last six cells of the grouped data. The retained (truncated) sample then consists of 981 observations within the range 79.95 to 159.95 pounds, with all knowledge concerning the truncated observations disregarded. We likewise disregard the prior knowledge that this sample might be from a Type III distribution. We proceed as though we have a doubly truncated sample from a general four-parameter Pearson distribution and calculate the first six sample moments about the left point of truncation. In order to compensate for moment errors due to grouping, we apply Sheppard's corrections for noncentral moments as given by Cramér (1946). Both corrected and uncorrected moments are tabulated as follows:

Uncorrected moments	Corrected moments
$n_1 = (7.56676860)\ 5$	$v_1 = (7.56676860)\ 5$
$n_2 = (66.4026504)\ 5^2$	$v_2 = (66.0693171)\ 5^2$
$n_3 = (649.817533)\ 5^3$	$v_3 = (642.250764)\ 5^3$
$n_4 = (6913.71764)\ 5^4$	$v_4 = (6781.37901)\ 5^4$
$n_5 = (78479.9827)\ 5^5$	$v_5 = (76331.5834)\ 5^5$
$n_6 = (937015.638)\ 5^6$	$v_6 = (902910.393)\ 5^6$

The corrected moments were substituted into (10.4.1), and a simultaneous solution of these equations yielded $h^* = 44.973178$, $b_0^* = -53.5929$, $b_1^* = 12.339508$, $b_2^* = -0.084107$, and $J^* = 0.578321$. From (10.4.2), we find $H^* = -0.196817$. The small negative value of b_2^* suggests that this sample is

Higher-Moment Estimators

indeed from a Type III distribution, and the small negative value of H^* suggests that the sample was singly truncated on the right with no truncation on the left.

With the sample now considered as a singly right truncated sample from a Type III distribution, we set $b_2 = 0$ and $H = 0$ and solve the system consisting of the first three equations of (11.4.1) plus (11.4.2) to obtain $h^* = 38.600670$, $b_0^* = 54.1194$, $b_1^* = 5.247727$, and $J^* = 0.766827$. When these estimates are substituted into (11.2.13), we calculate $\mu_1'^* = 38.60$, $\sigma^* = 16.027$, and $\alpha_3^* = 0.655$. The mean referred to zero becomes $E(X) = 38.60 + 79.95 = 118.55$, and the lower limit is 69.61. A graduation of the sample data based on these estimates along with the original sample data and Shook's original graduation based on complete sample estimates is given in Table 11.1.

Table 11.1 Weight Distribution of 1000 Women Students

Weights lbs.	Observed frequencies	Graduated (Expected) Frequencies Based on Type III Distribution	
		Shook's Graduation For Complete Sample	For Right Singly Truncated Sample T = 159.95
70 - 79.9	2	0	0.2
80 - 89.9	16	4	12.8
90 - 99.9	82	102	94.0
100 - 109.9	231	238	213.7
110 - 119.9	248	250	254.1
120 - 129.9	196	184	200.9
130 - 139.9	122	111	120.7
140 - 149.9	63	59	59.5
150 - 159.9	23	29	25.2
160 - 169.9	5	13	9.5
170 - 179.9	7	6	3.3
180 - 189.9	1	3	1.0
190 - 199.9	2	1	0.3
200 - 209.9	1	0	0.1
210 - 219.9	1	0	0.0
TOTALS	1000	1000	995.3
ESTIMATES Mean, $E^*(X)$		118.74	118.55
Std. Dev., σ_x^*		16.92	16.03
Skewness, α_3^*		0.976	0.655
Threshold, γ^*		84.09	69.61

Source: Adapted from Cohen (1951), Table 1, p. 262.

Agreement between observed and graduated frequencies is found to be considerably better for estimates calculated from the truncated sample than for estimates calculated from the complete sample. The improvement noted with truncation suggests that some of the extreme observations might have been outliers that unduly influenced the complete sample estimates.

12
Truncated and Censored Samples from Bivariate and Multivariate Normal Distributions

12.1 INTRODUCTION

In this chapter we consider parameter estimation in multivariate normal distributions from samples that may be truncated or censored with respect to one of the variables. Samples of this type arise when acceptance or screening procedures imposed on one variable eliminate certain sample specimens from further observation with respect to other variables. In regression or correlation studies, we may be concerned with predictions of success as measured by one or more variables from information provided by an acceptance examination score. Low-scoring candidates are eliminated and do not receive success scores. A manufacturer might wish to correlate physical characteristics such as weight, hardness, density, size, etc., with one or more performance characteristics such as life-span, operating costs, sales volume, or other characteristic for which observations are available only on accepted items. The determination of means, variances, and correlation coefficients from samples of these types poses a broad class of estimation problems, some of which are quite complex. Here our concern is limited to samples from a p-dimensional multivariate normal distribution where truncation and/or censoring applies to only a single screening (acceptance) variable.

In its most general form, the pdf of the complete unrestricted distribution may be written as

$$f(x_1, x_2, \ldots, x_p) = (2\pi)^{-p/2} |\sigma^{ij}|^{1/2}$$
$$\times \exp\left[-\frac{1}{2} \sum_{i=1}^{p} \sum_{j=1}^{p} \sigma^{ij} (x_i - \mu_i)(x_j - \mu_j)\right], \quad (12.1.1)$$

where the symmetric matrix $\|\sigma^{ij}\|$ of the quadratic form in the exponent is the inverse of the variance–covariance matrix $\|\sigma_{ij}\|$ and has the positive determinant $|\sigma^{ij}|$. In this notation, $\sigma_{ij} = \sigma_i \sigma_j \rho_{ij}$ and $\sigma_{ii} = \sigma_i^2$, where ρ_{ij} is the coefficient of correlation between x_i and x_j.

We consider samples in which N is the total number of sample specimens, n is the number of accepted specimens, and $c = N - n$ is the number of rejected (censored) items. In truncated samples, N and c are unknown. Only n, the number of acceptances, is known. In selected samples, full measurement of the screening variable is made on both accepted and rejected items, but measurement of the remaining variables is made only on the n accepted sample specimens.

In the two-dimensional case, where X is an independent screening variable and Y is a dependent performance variable, the pdf of (12.2.1) becomes

$$f(x, y) = \frac{\exp \frac{-1}{2(1-\rho^2)} \left[\left(\frac{x-\mu_x}{\sigma_x}\right)^2 - 2\rho \left(\frac{x-\mu_x}{\sigma_x}\right)\left(\frac{y-\mu_y}{\sigma_y}\right) + \left(\frac{y-\mu_y}{\sigma_y}\right)^2\right]}{2\pi \sigma_x \sigma_y \sqrt{1-\rho^2}}, \quad (12.1.2)$$

where μ_x, μ_y, σ_x, σ_y, and ρ are parameters to be estimated from sample data.

The presentation that follows is, to a major extent, based on previous results of Cohen (1953, 1955, 1957). Various aspects of some of the basic problems involved in this presentation have previously been considered by Pearson (1902), Aitkin (1934), Wilks (1932), Lawly (1943), Birnbaum (1950), Birnbaum et al. (1950), Campbell (1945), Des Raj (1953), Votaw et al. (1950), and perhaps by others.

12.2 ESTIMATION IN THE BIVARIATE NORMAL DISTRIBUTION

In many applications it is more meaningful to write the pdf of (12.1.2) as the product of the marginal density $f(x)$ and the conditional (array) density $f(y|x)$. We thereby obtain

$$f(x, y) = \left\{\frac{1}{\sigma_x \sqrt{2\pi}} \exp\left[-\frac{1}{2}\left(\frac{x-\mu_x}{\sigma_x}\right)^2\right]\right\}$$
$$\times \left\{\frac{1}{\sigma \sqrt{2\pi}} \exp\left[-\frac{1}{2}\left(\frac{y - \alpha - \beta(x-\bar{x})}{\sigma}\right)^2\right]\right\}, \quad (12.2.1)$$

Bivariate and Multivariate Normal Distributions

where

$$\beta = \frac{\rho \sigma_y}{\sigma_x}, \quad \alpha = \mu_y - \beta(\mu_x - \bar{x}), \quad \sigma^2 = \sigma_y^2 (1 - \rho^2), \quad (12.2.2)$$

and for a given sample, $\bar{x}$ is the mean of the screening variable as measured on the n accepted sample items. In making the change from (12.1.2) to (12.2.1), we have in effect reparameterized from $\mu_x, \mu_y, \sigma_x, \sigma_y, \rho_{xy}$, to $\mu_x, \sigma_x, \alpha, \beta, \sigma$.

Results obtained by Hotelling (1948), Tukey (1949), Pittman (1936), and Chapman (1952) guarantee that both the method of moments and the method of maximum likelihood lead to identical estimators in the case of truncated samples from multinormal distributions. Hence, we might have employed the method of moments for truncated samples. However, we elected to employ the method of maximum likelihood since it permits a uniform treatment of the various cases under consideration and it introduces no unusual algebraic difficulties.

12.2.1 Truncated Samples

Let a doubly truncated random sample be selected from a population with pdf (12.2.1) where T_1 and T_2 are known truncation points. Let n designate the number of observations for which $T_1 \leq x \leq T_2$. Let ξ_1 and ξ_2 designate the corresponding standardized truncation points

$$\xi_1 = \frac{(T_1 - \mu_x)}{\sigma_x} \quad \text{and} \quad \xi_2 = \frac{T_2 - \mu_x}{\sigma_x}. \quad (12.2.3)$$

The loglikelihood function for a sample as thus described is

$$\ln L = -n \ln \sigma_x - n \ln \sigma - \frac{1}{2\sigma_x^2} \sum_{i=1}^{n} (x_i - \mu_x)^2 - \frac{1}{2\sigma^2} \sum_{i=1}^{n} [y_i - \alpha - \beta(x_i - \bar{x})]^2 - n \ln [\Phi(\xi_2) - \Phi(\xi_1)] + \text{const}, \quad (12.2.4)$$

where $\Phi(\ )$ is the cdf of the standard normal distribution $(0, 1)$.

Maximum likelihood estimating equations follow as

$$\frac{\partial \ln L}{\partial \mu_x} = \frac{1}{\sigma_x^2} \sum_{i=1}^{n} (x_i - \mu_x) + \frac{n}{\sigma_x} \left[\frac{\phi(\xi_2) - \phi(\xi_1)}{\Phi(\xi_2) - \Phi(\xi_1)} \right] = 0,$$

$$\frac{\partial \ln L}{\partial \sigma_x} = \frac{1}{\sigma_x^3} \sum_{i=1}^{n} (x_i - \mu_x)^2 - \frac{n}{\sigma_x} \left[1 - \left(\frac{\xi_2 \phi(\xi_2) - \xi_1 \phi(\xi_1)}{\Phi(\xi_2) - \Phi(\xi_1)} \right) \right] = 0,$$

$$\frac{\partial \ln L}{\partial \alpha} = \frac{1}{\sigma^2} \sum_{i=1}^{n} [y_i - \alpha - \beta(x_i - \bar{x})] = 0, \quad (12.2.5)$$

$$\frac{\partial \ln L}{\partial \beta} = \frac{1}{\sigma^2} \sum_{i=1}^{n} [y_i - \alpha - \beta(x_i - \bar{x})][x_i - \bar{x}] = 0,$$

$$\frac{\partial \ln L}{\partial \sigma} = \frac{1}{\sigma^3} \sum_{i=1}^{n} [y_i - \alpha - \beta(x_i - \bar{x})]^2 \frac{n}{\sigma} = 0,$$

where $\phi(\ )$ and $\Phi(\ )$ are the pdf and the cdf of the standard normal distribution.

The first two of the preceding estimating equations are the same as equations (3.2.5), obtained in Chapter 3 for univariate normal distribution parameters from doubly truncated samples. Hence, the calculation of estimates $\hat{\mu}_x$ and $\hat{\sigma}_x$ is carried out as described in Chapter 3.

The last three equations of (12.2.5) are independent of μ_x and σ_x, and are therefore not affected by the truncation on the screening variable X. Consequently, regardless of the acceptance criteria or the type of restriction imposed on observations of X, these are the familiar "normal" equations, and they have the well-known solutions

$$\hat{\beta} = \frac{\bar{r}\bar{s}_y}{\bar{s}_x}, \qquad \hat{\alpha} = \bar{y}, \qquad \hat{\sigma} = \bar{s}_y \sqrt{1 - \bar{r}^2}, \qquad (12.2.6)$$

where (^) serves to distinguish maximum likelihood estimates from parameters, and the bars (¯) indicate that statistics thus designated are computed solely from observations recorded from the n accepted sample specimens. Accordingly

$$\bar{x} = \frac{1}{n} \sum_{i=1}^{n} x_i, \qquad \bar{s}_x^2 = \frac{1}{n} \sum_{i=1}^{n} (x_i - \bar{x})^2, \qquad \bar{y} = \frac{1}{n} \sum_{i=1}^{n} y_i,$$

$$\bar{s}_y^2 = \frac{1}{n} \sum_{i=1}^{n} (y_i - \bar{y})^2. \qquad (12.2.7)$$

$$\bar{r} = \sum_{i=1}^{n} \frac{(x_i - \bar{x})(y_i - \bar{y})}{n \bar{s}_x \bar{s}_y}.$$

Estimates $\hat{\mu}_y$, $\hat{\sigma}_y$, and $\hat{\rho}_{xy}$ follow from (12.2.6) and (12.2.2) as

$$\hat{\mu}_y = \hat{\alpha} + \hat{\beta}(\hat{\mu}_x - \bar{x}),$$

$$\hat{\sigma}_y = \sqrt{\hat{\sigma}^2 + \hat{\sigma}_x^2 \hat{\beta}^2}, \qquad (12.2.8)$$

$$\hat{\rho} = \hat{\sigma}_x \hat{\beta} / \sqrt{\hat{\sigma}^2 + \hat{\sigma}_x^2 \hat{\beta}^2},$$

and in equivalent forms as

Bivariate and Multivariate Normal Distributions

$$\hat{\mu}_y = \bar{y} - \tilde{r}\left(\frac{\tilde{s}_y}{\tilde{s}_x}\right)(\bar{x} - \hat{\mu}_x),$$

$$\hat{\sigma}_y = \tilde{s}_y \sqrt{\frac{1 - \hat{\lambda}(1 - \tilde{r}^2)}{1 - \hat{\lambda}}}, \qquad (12.2.9)$$

$$\hat{\rho} = \tilde{r}/\sqrt{1 - \hat{\lambda}(1 - \tilde{r}^2)},$$

where

$$\hat{\lambda} = 1 - \frac{\tilde{s}_x^2}{\hat{\sigma}_x^2}. \qquad (12.2.10)$$

Readers are cautioned not to confuse λ as defined here with the auxiliary estimating function $\lambda(h, \alpha)$, defined in Chapter 2 for use in calculating estimates of univariate normal distribution parameters from singly censored samples.

The estimators given in (12.2.9) were originally derived by Cohen (1953) from the likelihood function based on the pdf or (12.1.2) rather than on (12.2.1). The likelihood function was then maximized with respect to μ_x, σ_x, μ_y, σ_y, and ρ_{xy} without the introduction of α, β, and σ. Similar results were obtained independently at about the same time by Des Raj (1953).

Singly truncated samples: In the case of single truncation on the screening variable X, the first two estimating equations of (12.2.5) are identical to corresponding equations derived in Chapter 2 for the univariate distribution, and estimates $\hat{\mu}_x$ and $\hat{\sigma}_x$ can be calculated as described there.

Doubly censored samples: In this case the first two equations of (12.2.5) are identical to corresponding univariate estimating equations that were derived in Chapter 3.

Singly censored samples: The first two equations of (12.2.5) in this case are identical to the univariate singly censored normal distribution estimators of Chapter 2.

Selected samples: When the sampling procedure is such that scores (measurements) x_i, $i = 1, 2, \ldots, N$, are available from both accepted and rejected sample items, whereas values of y_i are available only on the n accepted items, then the first two equations of (12.2.5) yield

$$\hat{\mu}_x = \sum_{i=1}^{n} \frac{x_i}{N} \quad \text{and} \quad \sigma_x^2 = s_x^2 = \sum_{i=1}^{n} \frac{(x_i - \bar{x}_N)^2}{N}, \qquad (12.2.11)$$

which are recognized as the univariate estimators from an unrestricted sample of size N.

In each of these special cases, the estimators of (12.2.6), (12.2.8), and (12.2.9) are applicable for the calculation of estimates $\hat{\beta}$, $\hat{\alpha}$, $\hat{\sigma}$, and subsequently for the calculation of $\hat{\mu}_y$, $\hat{\sigma}_y$, and $\hat{\rho}$.

12.3 RELIABILITY OF ESTIMATES

The variance–covariance matrix of $(\hat{\alpha}, \hat{\beta}, \hat{\sigma})$ can be derived from the second partials of $\ln L$ with respect to these parameters. We therefore differentiate the last three equations of (12.2.5) to obtain

$$\frac{\partial^2 \ln L}{\partial \alpha^2} = -\frac{n}{\sigma^2}, \quad \frac{\partial^2 \ln L}{\partial \beta^2} = -\frac{1}{\sigma^2} \sum_{i=1}^{n} (x_i - \bar{x})^2,$$

$$\frac{\partial^2 \ln L}{\partial \sigma^2} = -\frac{3}{\sigma^4} \sum_{i=1}^{n} [y_i - \alpha - \beta(x_i - \bar{x})^2] + \frac{n}{\sigma^2},$$

$$\frac{\partial^2 \ln L}{\partial \alpha \, \partial \beta} = -\frac{1}{\sigma^2} \sum_{i=1}^{n} (x_i - \bar{x}), \qquad (12.3.1)$$

$$\frac{\partial^2 \ln L}{\partial \alpha \, \partial \sigma} = -\frac{2}{\sigma^3} \sum_{i=1}^{n} [y_i - \alpha - \beta(x_i - \bar{x})],$$

$$\frac{\partial^2 \ln L}{\partial \beta \, \partial \sigma} = -\frac{2}{\sigma^3} \sum_{i=1}^{n} [y_i - \alpha - \beta(x_i - \bar{x})].$$

To determine the expected values of the preceding partials, we require the following results

$$E\left[\frac{y - \alpha - \beta(x - \bar{x})}{\sigma}\right] = \frac{1}{E(n/N)} \int_R \phi(u) \, du \int_{-\infty}^{\infty} v\phi(v) \, dv = 0,$$

$$E\left[\frac{y - \alpha - \beta(x - \bar{x})}{\sigma}\right]^2 = \frac{1}{E(n/N)} \int_R \phi(u) \, du \int_{-\infty}^{\infty} v^2\phi(v) \, dv = 1,$$

$$E\left[\left(\frac{y - \alpha - \beta(x - \bar{x})}{\sigma}\right)(x - \bar{x})\right]$$
$$= \frac{1}{E(n/N)} \int_R (x - \bar{x})\phi(u) \, du \int_{-\infty}^{\infty} v\phi(v) \, dv = 0, \quad (12.3.2)$$

where $u = (x - \mu_x)/\sigma_x$, $v = [y - \alpha - \beta(x - \bar{x})]/\sigma$, and R designates the acceptance interval with respect to x.

By using these results, expected values of the partials of (12.3.1) become

Bivariate and Multivariate Normal Distributions

$$-E\left(\frac{\partial^2 \ln L}{\partial \alpha^2}\right) = \frac{E(n)}{\sigma^2}, \qquad E\left(\frac{\partial^2 \ln L}{\partial \alpha\, \partial \beta}\right) = 0,$$

$$-E\left(\frac{\partial^2 \ln L}{\partial \beta^2}\right) = \frac{E(n)\,\bar{\sigma}_x^2}{\sigma^2}, \qquad E\left(\frac{\partial^2 \ln L}{\partial \alpha\, \partial \sigma}\right) = 0, \qquad (12.3.3)$$

$$-E\left(\frac{\partial^2 \ln L}{\partial \sigma^2}\right) = \frac{2E(n)}{\sigma^2}, \qquad E\left(\frac{\partial^2 \ln L}{\partial \beta\, \partial \sigma}\right) = 0.$$

For the asymptotic variance–covariance matrix of $(\hat{\alpha}, \hat{\beta}, \hat{\sigma})$, we now write

$$V(\hat{\alpha}, \hat{\beta}, \hat{\sigma}) = \begin{vmatrix} -E\left(\frac{\partial^2 \ln L}{\partial \alpha^2}\right) & -E\left(\frac{\partial^2 \ln L}{\partial \alpha\, \partial \beta}\right) & -E\left(\frac{\partial^2 \ln L}{\partial \alpha\, \partial \sigma}\right) \\ -E\left(\frac{\partial^2 \ln L}{\partial \beta\, \partial \alpha}\right) & -E\left(\frac{\partial^2 \ln L}{\partial \beta^2}\right) & -E\left(\frac{\partial^2 \ln L}{\partial \beta\, \partial \sigma}\right) \\ -E\left(\frac{\partial^2 \ln L}{\partial \sigma\, \partial \alpha}\right) & -E\left(\frac{\partial^2 \ln L}{\partial \sigma\, \partial \beta}\right) & -E\left(\frac{\partial^2 \ln L}{\partial \sigma^2}\right) \end{vmatrix}^{-1} = 0$$

$$= \begin{vmatrix} \frac{E(n)}{\sigma^2} & 0 & 0 \\ 0 & \frac{E(n)\,\bar{\sigma}_x^2}{\sigma^2} & 0 \\ 0 & 0 & \frac{2E(n)}{\sigma^2} \end{vmatrix} \qquad (12.3.4)$$

From this result, we have

$$V(\hat{\alpha}) = \frac{\sigma^2}{E(n)}, \qquad V(\hat{\beta}) = \frac{\sigma^2}{\sigma_x^2 E(n)}, \qquad V(\hat{\sigma}) = \frac{\sigma^2}{2E(n)}. \qquad (12.3.5)$$

For the asymptotic variances and covariance of $\hat{\mu}_x$ and $\hat{\sigma}_x$, results given in Chapters 2 and 3 are likewise applicable here.

12.4 AN ILLUSTRATIVE EXAMPLE

Example 12.1. To illustrate the practical application of results obtained here for parameter estimates of the bivariate normal distribution, we have selected a sample from Goedicke (1953) consisting of entrance examination scores, x, and subsequent course averages, y, achieved by 529 students at an unnamed university.

The complete sample data are displayed in the following two-way frequency table.

x \ y	158	161	164	167	170	173
65	2	5		1		1
70	5	41	34	5		
75	4	46	81	46	5	1
80		22	84	49	18	1
85	1		23	23	7	2
90		3	4	8	4	
95			1	1		1

For the purpose of this illustration we assume that the minimum qualifying score on the entrance examination is $x = T = 159.5$. Under this requirement $n = 517$ applicants were admitted. By making appropriate further assumptions, we use the same basic data to illustrate truncated, censored, and selected samples. For comparison, estimates are also calculated from the complete sample. Except for the truncated sample, we consider N as being fixed. In the truncated sample, we consider N to be unknown and n to be fixed. Complete sample data with $N = 529$ are summarized as: $\bar{x}_N = 164.441$, $s_x = 2.847$, $r = 0.409$, $\bar{y}_N = 77.380$, and $s_y = 5.405$. Restricted sample data with $n = 517$ and $c = 12$ are summarized as: $\bar{x} = 164.592$, $\bar{s}_x = 2.7036$, $\bar{r} = 0.3892$, $\bar{y} = 77.505$, and $\bar{s}_y = 5.347$.

For the first stage of calculations, we turn to Chapter 2 and calculate estimates of μ_x and σ_x from the truncated and censored samples. For the truncated sample, we calculate $\hat{\alpha} = \bar{s}_x^2/(T - \bar{x})^2 = (2.7036)^2/(159.5 - 164.592)^2 = 0.2819$. Interpolation in Table 2.1 gives $\hat{\theta} = 0.078946$, and on substituting in (2.3.13) we calculate $\hat{\mu}_x = 164.190$. From (2.3.9), we calculate $\hat{\sigma}_x = 3.059$. For the censored sample, we calculate $h = c/N = 12/529 = 0.0227$, and $\hat{\alpha} = 0.2819$ (which is the same value calculated for the truncated sample). Two-way interpolation in Table 2.3 gives $\lambda(h, \hat{\alpha}) = 0.02737$, and from (2.5.11), we calculate $\hat{\mu}_x = 164.453$ and $\hat{\sigma}_x = 2.832$. Note that $\hat{\alpha}$ used in this paragraph is not to be confused with α introduced in Eq. (12.2.1).

The remaining estimates were calculated from (12.2.6) and (12.2.8). Asymptotic variances and covariances were calculated from (12.3.4) and (12.3.5) for estimates $\hat{\alpha}$, $\hat{\beta}$, and $\hat{\sigma}$, and from applicable variance formulas given in Chapter 2 for estimates $\hat{\mu}_x$ and $\hat{\sigma}_x$. All of these results are listed in Table 12.1.

With due regard for the degree of approximation involved in their calculation, the tabulated variances reflect the varying amounts of information provided by

Bivariate and Multivariate Normal Distributions

Table 12.1 Summary of Estimates and their Variances for Example 12.1

Parameters	Sample Type			
	Truncated	Censored	Selected	Complete
μ_x	164.190	164.453	164.441	164.441
σ_x	3.059	2.832	2.847	2.847
β	0.770	0.770	0.770	0.776
α	77.505	77.505	77.505	77.380
σ	4.925	4.925	4.925	4.932
μ_y	77.195	77.398	77.505	77.380
σ_y	5.459	5.386	5.391	5.405
ρ	0.431	0.405	0.407	0.409
ξ	-1,533	-1.749	-1.736	-1.736
No. Unrestricted Observations	517	517	517	529
Asymptotic Variances and Covariance				
$V(\hat{\mu}_x)$	0.0295	0.0153	0.0153	0.0153
$V(\hat{\sigma}_x)$	0.0180	0.0081	0.0077	0.0077
$Cov(\hat{\mu}_x, \hat{\sigma}_x)$	-0.0107	-0.0002	0	0
$V(\hat{\alpha})$	0.0469	0.0478	0.0478	0.0460
$V(\hat{\beta})$	0.0064	0.0071	0.0071	0.0057
$V(\hat{\sigma})$	0.0235	0.239	0.239	0.0230

the different samples. Minimum information is provided by the truncated sample and maximum information is provided by the complete sample. The calculated variances are approximate, because of not only the use of asymptotic formulas, but also their dependence on sample values. When comparing truncated sample variances with those for other samples, allowances need to be made for differences in sample sizes. The truncated sample with $n = 517$ (fixed) corresponds to a censored sample with $N = 551$.

12.5 PARAMETER ESTIMATION IN THE MULTIVARIATE NORMAL DISTRIBUTION

Estimation procedures for parameters of the multivariate normal distribution from truncated and censored samples involve a modest extension of procedures that apply in the bivariate normal distribution. When the restriction (truncation, censoring, or selection) occurs on only one of the p variates, say X_1 of the multivariate distribution with pdf (12.1.1), estimates $\hat{\mu}_1$ and $\hat{\sigma}_1$ are calculated solely from marginal data of X_1 without regard for any of the associated variates. Accordingly, the univariate estimators of Chapters 2 and 3 are applicable here just as they were in the bivariate distribution. In a selected sample, $\hat{\mu}_1 = \bar{x}_1$ and $\hat{\sigma}_1 = s_1$, where $\bar{x}_1$ and s_1 are calculated from the complete sample of size N.

Estimation of parameters of the remaining $(p - 1)$ variates and their correlation coefficients follows essentially the same pattern as that involved in the estimation of parameters of the variate y in the bivariate distribution. A complete derivation of maximum likelihood estimators for parameters of the multivariate normal distribution was given by Cohen (1957). Applicable estimators in this case are

$$\hat{\mu}_j = \bar{x}_j - \bar{r}_{1j}\left(\frac{\bar{s}_j}{\bar{s}_1}\right)(\bar{x}_1 - \hat{\mu}_1),$$

$$\hat{\sigma}_j = \bar{s}_j \sqrt{\frac{1 - \hat{\lambda}(1 - \bar{r}_{1j}^2)}{1 - \hat{\lambda}}},$$

$$\hat{\rho}_{ij} = \frac{\bar{r}_{ij} - \hat{\lambda}(\bar{r}_{ij} - \bar{r}_{1i}\bar{r}_{1j})}{\sqrt{[1 - \hat{\lambda}(1 - \bar{r}_{1i}^2)][1 - \hat{\lambda}(1 - \bar{r}_{1j}^2)]}},$$

$$i = 1, 2, \ldots, p - 1, j = 2, 3, \ldots, p, i < j,$$

(12.5.1)

where, in agreement with (12.2.10),

$$\hat{\lambda} = 1 - \frac{\bar{s}_1^2}{\hat{\sigma}_1^2}.$$

(12.5.2)

Since by definition $\bar{r}_{ii} \equiv 1$, the last equation of (12.5.1) for $i = 1$ becomes

$$\hat{\rho}_{1j} = \bar{r}_{1j}/\sqrt{1 - \hat{\lambda}(1 - \bar{r}_{1j}^2)}.$$

(12.5.3)

Note that the bivariate estimators of (12.2.9) can be obtained as special cases of (12.5.1) with $p = 2$.

12.5.1 Estimate Reliability

Asymptotic variances and covariances of estimates given in the preceding subsection can, of course, be obtained by inverting information matrices with elements that are negatives of expected values of second partials of the loglikelihood functions. These are of the order of $1/n$, but their exact expressions are too unwieldy to be of much practical value. For parameters of the restricted variable, asymptotic variances and covariances given in Chapters 2 and 3 for the univariate normal distribution are applicable. When a selection is based on X_1, but does not restrict observation of X_1 itself, then the complete sample variances and covariance

Bivariate and Multivariate Normal Distributions

$$V(\hat{\mu}_1) = \frac{\sigma_1^2}{N},$$

$$V(\hat{\sigma}_1) = \frac{\sigma_1^2}{2N}, \quad (12.5.4)$$

$$\text{Cov}(\hat{\mu}_1, \hat{\sigma}_1) = 0,$$

are applicable as are various exact small sample results based on the X_1 marginal distribution alone.

12.5.2 A Multivariate Illustrative Example

Example 12.2. The practical application of the multivariate estimators presented here is illustrated with a sample given by Baten (1938) and attributed by him to Harry C. Carver. The basic sample data consists of weight, height, shoulder, chest, waist, and hip measurements on 119 individuals. We designate these variates in the order listed as x_1, x_2, x_3, x_4, x_5, and x_6, respectively. Baten's data included 120 sets of measurements, but it was necessary to eliminate the last set because of a typographical error. As given by Baten, the sample was considered to be complete, but for the purpose of this illustration, it is arbitrarily left truncated with respect to weight at $x_{11} = T = 119.5$. Thereby $c = 11$ sets of observations are eliminated. Estimates are then calculated with the sample considered as truncated of size $n = 108$ and censored of size $N = 119$, with $n = 108$ and $c = 11$. A summary of the sample data is listed in Table 12.2, and a summary of the sample estimates for these two cases along with estimates calculated from the complete sample are displayed in Table 12.3.

Table 12.2 Summary of Sample Data for Example 12.2

$n = 108$	Truncation at $x_1 = T = 119.5$ lb			
$\bar{x}_1 = 140.8056$	$\bar{s}_1 = 11.8419$	$\bar{r}_{12} = 0.4701$	$\bar{r}_{25} = -0.1389$	
$\bar{x}_2 = 67.9241$	$\bar{s}_2 = 2.4008$	$\bar{r}_{13} = 0.4326$	$\bar{r}_{26} = 0.3019$	
$\bar{x}_3 = 16.4500$	$\bar{s}_3 = 0.7103$	$\bar{r}_{14} = 0.6501$	$\bar{r}_{34} = 0.5904$	
$\bar{x}_4 = 35.4537$	$\bar{s}_4 = 1.5373$	$\bar{r}_{15} = 0.4415$	$\bar{r}_{35} = 0.1852$	
$\bar{x}_5 = 28.1574$	$\bar{s}_5 = 1.6375$	$\bar{r}_{16} = 0.7873$	$\bar{r}_{36} = 0.4059$	
$\bar{x}_6 = 35.5898$	$\bar{s}_6 = 1.3746$	$\bar{r}_{23} = 0.2361$	$\bar{r}_{45} = 0.4931$	
		$\bar{r}_{24} = 0.1194$	$\bar{r}_{46} = 0.5491$	
			$\bar{r}_{56} = 0.4310$	

Table 12.3 Summary of Estimates for Example 12.2

Parameters	Complete Sample Estimates	Truncated Sample Estimates	Censored Sample Estimates
ξ		-1.379	-1.342
μ_1	138.2353	138.4883	138.2376
μ_2	67.6664	67.7033	67.6794
μ_3	16.3672	16.3899	16.3834
μ_4	35.1899	35.2581	35.2370
μ_5	27.9252	28.0159	28.0007
μ_6	35.3513	35.3780	35.3549
σ_1	13.9421	13.7696	13.9622
σ_2	2.5230	2.4923	2.5021
σ_3	0.7417	0.7333	0.7350
σ_4	1.7280	1.6477	1.6557
σ_5	1.7857	1.6927	1.6987
σ_6	1.5235	1.5172	1.5318
ρ_{12}	0.5239	0.5265	0.5318
ρ_{13}	0.5446	0.4872	0.4924
ρ_{14}	0.7339	0.7053	0.7037
ρ_{15}	0.5566	0.4966	0.5018
ρ_{16}	0.8369	0.8294	0.8330
ρ_{23}	0.2996	0.2872	0.2992
ρ_{24}	0.2406	0.2040	0.2114
ρ_{25}	-0.0120	-0.0613	-0.0536
ρ_{26}	0.3732	0.3772	0.3842
ρ_{34}	0.6193	0.6229	0.6265
ρ_{35}	0.3193	0.2365	0.2416
ρ_{36}	0.4908	0.4615	0.4667
ρ_{45}	0.5943	0.5362	0.5404
ρ_{46}	0.6569	0.6166	0.6220
ρ_{56}	0.5344	0.4849	0.4902
λ		0.2604	0.2806
Sample size	N = 119	n = 108	n = 108 c = 11
Asymptotic Variances*			
$V(\hat{\mu}_1)$	1.389	3.377	1.836
$V(\hat{\sigma}_1)$	0.695	1.959	1.041

* Calculated from (12.5.4) and from (2.6.1).

Bivariate and Multivariate Normal Distributions

Computing procedures are illustrated in the following paragraphs.

Truncated Sample

To estimate parameters of X_1, we employ estimators from Chapter 2 for a singly truncated sample from a univariate normal distribution. With $\bar{x}_1 = 140.8056$, $\bar{s}_1 = 11.8419$, and $T = 119.5$, we calculate

$$\hat{\alpha} = \frac{\bar{s}_1^2}{(\bar{x}_1 - T)^2} = \left[\frac{11.8419}{(140.8056 - 119.5)}\right]^2 = 0.30893.$$

We interpolate in Table 2.1 to obtain $\hat{\theta} = 0.108763$, and from (2.3.9) and (2.3.13) we calculate

$$\hat{\mu}_1 = 140.8056 - 0.108763(140.8056 - 119.5) = 138.4883,$$

$$\hat{\sigma}_1^2 = (11.8419)^2 + 0.108763(140.8056 - 119.5)^2 = 189.6012,$$

$$\hat{\sigma}_1 = 13.7696.$$

As an estimate of the truncation point in standard units of the complete distribution, we calculate $\xi = (119.5 - 138.4883)/13.7696 = -1.379$.

Censored Sample

For the censored sample, we require $\hat{\alpha} = 0.30893$ as calculated for the truncated sample, and $h = c/N = 11/119 = 0.09244$. Two-way interpolation in Table 2.3 gives $\lambda(\hat{\alpha}, h) = 0.12053$, and from (2.5.11), we calculate

$$\hat{\mu}_1 = 140.8056 - 0.12053(140.8056 - 119.5) = 138.2376,$$

$$\hat{\sigma}_1^2 = (11.8419)^2 + 0.12053(140.8056 - 119.5)^2 = 194.9426,$$

$$\hat{\sigma}_1 = 13.9622.$$

As an estimate of the censoring point in standard units of the complete distribution, we calculate $\xi = (119.5 - 138.2376)/13.9622 = -1.342$.

As a specimen calculation of estimates of the remaining parameters, based on the truncated sample, we compute

$$\hat{\lambda} = 1 - \frac{s_1^2}{\hat{\sigma}_1^2} = 1 - \left(\frac{11.8419}{13.7696}\right)^2 = 0.2604,$$

and from (12.5.1), we calculate

$$\hat{\mu}_2 = 67.9241 - 0.4701\left(\frac{2.4008}{11.8419}\right)(140.8046 - 138.4884) = 67.7033,$$

$$\hat{\sigma}_2 = 2.4008 \sqrt{\frac{1 - 0.2604(1 - 0.4701^2)}{1 - 0.2604}} = 2.4923,$$

$$\hat{\rho}_{12} = \frac{0.4701}{\sqrt{1 - 0.2604(1 - 0.4701^2)}} = 0.5265,$$

$$\hat{\rho}_{23} = \frac{0.2361 - 0.2604[0.2361 - 0.4701(0.4326)]}{\sqrt{[1 - 0.2604(1 - 0.4701^2)][1 - 0.2604(1 - 0.4326^2)]}} = 0.2872.$$

13
Truncated and Censored Samples from Discrete Distributions

13.1 INTRODUCTION

Truncated samples from discrete distributions arise in numerous situations where counts of zero are not observed. As an example, consider the distribution of the number of children per family in developing nations, where records are maintained only if there is at least one child in the family. The number of childless families remains unknown. The resulting sample is thus truncated with the zero class missing. In a continuous distribution, a sample of this type would be described as singly left truncated. In other situations, samples from discrete distributions might be censored on the right. This occurs when for larger values of the random variable, sample data include only the information that a certain number of observations exceed a specified cutoff point. Similarly, in some situations, samples might be censored on the left, and on some occasions they might even be doubly censored or doubly truncated. In this chapter we consider truncated and censored samples from the Poisson, negative binomial, binomial, and hypergeometric distributions.

13.2 THE POISSON DISTRIBUTION

The Poisson distribution is an appropriate mathematical model for studying such diverse classes of discrete data as haemocytometer counts of blood cells per square, the number of noxious weed seed per unit of field seed, and the number of defects per unit of a manufactured product. It is thus of interest to biologists, agronomists, and quality control practitioners as well as research workers in various other fields of scientific endeavor. It was first derived by Poisson (1837) and was later used by Bortkiewicz (1898) to explain the occurrence of events in which the probability of each occurrence was small. Various aspects of estimation involving singly truncated and singly censored Poisson samples with known terminals have previously been considered by numerous investigators. Among these are Tippett (1932), Bliss (1948), Plackett (1953), Rider (1953), David and Johnson (1952), Moore (1954), Cohen (1954), and others. The presentation in this section is primarily based on the author's 1954 paper, in which maximum likelihood estimators of the Poisson parameter were derived for singly and doubly truncated samples as well as for singly and doubly censored samples, all with known terminals. In each case, estimators are expressed in simple algebraic forms for easy application to practical problems.

13.2.1 Some Fundamentals

A discrete random variable X is said to have a Poisson distribution with parameter λ if its probability function is

$$\Pr[X = x] = \frac{e^{-\lambda}\lambda^x}{x!}, \quad x = 0, 1, 2, \ldots \quad (13.2.1)$$

This distribution arises as the limit of a sequence of binomial distributions with

$$\Pr[X = x] = \begin{cases} \binom{N}{x} p^x (1-p)^{N-x}, & x = 0, 1, \ldots, N, \\ 0 & \text{otherwise.} \end{cases} \quad (13.2.2)$$

If we let $N \to \infty$ and $p \to 0$, while $Np = \lambda$ is held constant, it can be shown that

$$\lim \Pr[X = x] = \frac{e^{-\lambda}\lambda^x}{x!}. \quad (13.2.3)$$

An excellent discussion of the Poisson distribution, its genesis, properties, and applications is included in Volume 3 of the three-volume set by Johnson and Kotz (1969).

The kth factorial moment of this distribution is

$$\mu_{[k]} = E[X(X-1)(X-2)\cdots(X-k+1)] = \lambda^k. \quad (13.2.4)$$

Let $k = 1, 2, 3$, and 4 in this equation and the noncentral moments about zero follow as

$$\begin{aligned} \mu_1'(X) &= E(X) = \lambda, \\ \mu_2'(X) &= \lambda + \lambda^2, \\ \mu_3'(X) &= \lambda + 3\lambda^2 + \lambda^3, \\ \mu_4'(X) &= \lambda + 7\lambda^2 + 6\lambda^3 + \lambda^4. \end{aligned} \quad (13.2.5)$$

Samples from Discrete Distributions

It subsequently follows that the central and standard moments become

$$\mu_2(X) = V(X) = \lambda, \qquad \mu_3(X) = \lambda, \qquad \mu_4(X) = \lambda + 3\lambda^2, \qquad (13.2.6)$$

$$\alpha_3(X) = \frac{1}{\sqrt{\lambda}}, \qquad \alpha_4(X) = 3 + \frac{1}{\lambda}.$$

From these results, we have

$$E(X) = V(X) \quad \text{and} \quad \alpha_4 - \alpha_3^2 - 3 = 0. \qquad (13.2.7)$$

The moments obey the recurrence formula

$$\mu_{k+1} = k\lambda\mu_{k-1} + \lambda \frac{\partial \mu_k}{\partial \lambda}. \qquad (13.2.8)$$

The moment-generating function (MGF) is

$$\text{MGF} = E(e^{tx}) = e^{-\lambda} \sum_{x=0}^{\infty} \frac{(\lambda e^t)^x}{x!} = \exp[\lambda(e^t - 1)]. \qquad (13.2.9)$$

The cumulant-generating function (CGF) is

$$\text{CGF} = \ln(\text{MGF}) = \lambda(e^t - 1), \qquad \kappa_k(X) = \lambda; \qquad k \geq 2. \qquad (13.2.10)$$

The cumulative probability function is

$$P(a, \lambda) = \Pr[X \leq a] = \sum_{x=0}^{a} \frac{e^{-\lambda}\lambda^x}{x!} = 1 - \sum_{x=a+1}^{\infty} \frac{e^{-\lambda}\lambda^x}{x!}, \qquad (13.2.11)$$

and the complementary cumulative function is

$$\bar{P}(a+1, \lambda) = \sum_{x=a+1}^{\infty} \frac{e^{-\lambda}\lambda^x}{x!}. \qquad (13.2.12)$$

Accordingly, it follows that

$$P(a, \lambda) = 1 - \bar{P}(a+1, \lambda) \quad \text{and} \quad \bar{P}(a, \lambda) = 1 - P(a-1, \lambda). \qquad (13.2.13)$$

The derivative of $\bar{P}(a, \lambda)$ with respect to λ is

$$\frac{d\bar{P}(a, \lambda)}{d\lambda} = \sum_{x=a}^{\infty} d\left(\frac{e^{-\lambda}\lambda^x}{x!}\right) \bigg/ d\lambda = \sum_{x=a}^{\infty} \left[\frac{e^{-\lambda}x\lambda^{x-1} - e^{-\lambda}\lambda^x}{x!}\right],$$

$$= \sum_{x=a}^{\infty} \frac{e^{-\lambda}x\lambda^{x-1}}{(x-1)!} - \sum_{x=a}^{\infty} \frac{e^{-\lambda}\lambda^x}{x!}$$

$$= \sum_{x=a-1}^{\infty} \frac{e^{-\lambda}\lambda^x}{x!} - \bar{P}(a; \lambda) \qquad (13.2.14)$$

$$= \bar{P}(a-1; \lambda) - \bar{P}(a; \lambda) = f(a-1)$$

$$= \Pr[x = (a-0)].$$

13.2.2 Singly Left Truncated Samples

We consider samples of size n that are truncated on the left at $X = a$. Thereby all observations for which $X = 0, 1, 2, \ldots, (a - 1)$ are eliminated. The likelihood function for a sample of this type is

$$L = [\bar{P}(a, \lambda)]^{-n} e^{-n\lambda} \lambda^{n\bar{x}} \left[\prod_{i=1}^{n} x_i! \right]^{-1}, \qquad (13.2.15)$$

and the loglikelihood function follows as

$$\ln L = -n\lambda + n\bar{x} \ln \lambda - n \ln [\bar{P}(a)] - \sum_{i=1}^{n} \ln x_i, \qquad (13.2.16)$$

where $\bar{x}$ is the sample mean and the notation $\bar{P}(a, \lambda)$ has been shortened to $\bar{P}(a)$.

The maximum likelihood estimating equation, $d \ln L/d\lambda = 0$, follows from (13.2.14) and (13.2.16) as

$$\frac{d \ln L}{d\lambda} = \frac{n\bar{x}}{\lambda} - n - \frac{nf(a - 1)}{\bar{P}(a)} = 0, \qquad (13.2.17)$$

where the probability function has been written as $f(x)$ rather than as $\Pr[X = x]$. This equation can be simplified to

$$\bar{x} = \lambda \left[1 + \frac{f(a - 1)}{\bar{P}(a)} \right]. \qquad (13.2.18)$$

Standard iterative procedures can be employed to solve this equation for $\hat{\lambda}$. However, in most applications the simple trial-and-error procedure will be satisfactory.

Zero Class Missing

In the special case where only the zero class has been truncated, the estimating equation (13.2.18) becomes

$$\bar{x} = \frac{\hat{\lambda}}{1 - e^{-\hat{\lambda}}}. \qquad (13.2.19)$$

Here as in (13.2.18) the trial-and-error procedure can be employed to solve for $\hat{\lambda}$. Tables of the Poisson distribution such as those of Molina (1942) can be helpful in solving estimating equations, but when tables are unavailable, it is a relatively simply task to calculate individual terms of the Poisson distribution as needed.

Samples from Discrete Distributions

13.2.3 Singly Right Censored Samples

Samples of the singly right censored type occur when observations in excess of some value $x = d$ are pooled such that in a sample of total size N, there are c censored observations greater than d and $n = N - c$ fully measured observations for which $x \leq d$. Tippett (1932) was concerned with a sample of this type.

The loglikelihood function may be written as

$$\ln L = -n\lambda + n\bar{x} \ln \lambda + c \ln \bar{P}(d+1) - \sum_{i=1}^{n} \ln x_i + \ln K, \quad (13.2.20)$$

where K is an ordering constant and $\bar{x}$ is the mean of the n uncensored sample observations. The maximum likelihood estimating equation follows as

$$\frac{d \ln L}{d\lambda} = \frac{n\bar{x}}{\lambda} - n + c \left[\frac{f(d)}{\bar{P}(d+1)}\right] = 0, \quad (13.2.21)$$

which we subsequently simplify to

$$\bar{x} = \lambda \left[1 - \frac{c}{n}\left(\frac{f(d)}{\bar{P}(d+1)}\right)\right] \quad (13.2.22)$$

The estimate $\hat{\lambda}$ can be readily calculated as the solution of this equation.

13.2.4 Doubly Truncated Samples

The probability function of a Poisson distribution truncated at $x = a$ on the left, and at $x = d$ on the right, may be written as

$$f(x) = \begin{cases} 0, & x < a, \\ [\bar{P}(a) - \bar{P}(d+1)]^{-1} \dfrac{e^{-\lambda}\lambda^x}{x!}, & a \leq x \leq d, \\ 0, & x > d. \end{cases} \quad (13.2.23)$$

The likelihood function of a sample of size n from this distribution becomes

$$L = [\bar{P}(a) - \bar{P}(d+1)]^{-n} e^{-n\lambda} \lambda^{n\bar{x}} \left[\prod_{i=1}^{n} x_i!\right]^{-1}. \quad (13.2.24)$$

where $\bar{x}$ is the sample mean.

We obtain this same likelihood function when we consider the population as being complete with probability function (13.2.1), and consider the sample as being truncated at a and d, respectively.

The loglikelihood function follows from (13.2.24) as

$$\ln L = -n\lambda + n\bar{x} \ln \lambda - \sum_{i=1}^{n} \ln x_i - n \ln[\bar{P}(a) - \bar{P}(d+1)], \quad (13.2.25)$$

and the maximum likelihood estimating function becomes

$$\frac{d \ln L}{d\lambda} = \frac{n\bar{x}}{\lambda} - n - n \left[\frac{f(a-1) - f(d)}{\bar{P}(a) - \bar{P}(d+1)} \right] = 0. \qquad (13.2.26)$$

We subsequently simplify this equation to

$$\bar{x} = \lambda \left[1 + \frac{f(a-1) - f(d)}{\bar{P}(a) - \bar{P}(d+1)} \right], \qquad (13.2.27)$$

and we obtain the estimate $\hat{\lambda}$ as the solution.

We note that the singly left truncated estimating equation (13.2.18) becomes a special case of (13.2.27) when we let $d \to \infty$ and thus $f(d) = 0$. In a similar manner, the estimating equation for a right singly truncated sample is obtained from (13.2.27) when we set $a = 0$, $f(a - 1) = 0$, and $\bar{P}(a) = 1$.

13.2.5 Doubly Censored Samples—Known Number of Censored Observations in Each Tail

In a sample of this type, we let c_1 and c_2 designate the number of censored observations in the left and right tails respectively and n is the number of complete observations for which $a \le x \le d$. The likelihood function of a sample as thus described from a Poisson distribution with probability function (13.2.1) is

$$L = K[1 - \bar{P}(a)]^{c_1} e^{-n\lambda} \lambda^{n\bar{x}} \left[\prod_{i=1}^{n} x_i! \right]^{-1} [\bar{P}(d+1)]^{c_2}, \qquad (13.2.28)$$

where K is an ordering constant and other symbols are as previously defined. We take logarithms of (13.2.28), differentiate with respect to λ, and equate to zero to obtain the estimating equation

$$\frac{d \ln L}{d\lambda} = \frac{n\bar{x}}{\lambda} - n - c_1 \left[\frac{f(a-1)}{1 - \bar{P}(a)} \right] + c_2 \left[\frac{f(d)}{\bar{P}(d+1)} \right] = 0. \qquad (13.2.29)$$

Subsequent simplification yields

$$\bar{x} = \lambda \left[1 + \frac{c_1}{n} \left(\frac{f(a-1)}{1 - \bar{P}(a)} \right) - \frac{c_2}{n} \left(\frac{f(d)}{\bar{P}(d+1)} \right) \right], \qquad (13.2.30)$$

and the estimate $\hat{\lambda}$ follows as the solution of this equation.

Note that the singly right censored estimating equation (13.2.22) can be obtained as a special case of (13.2.30) when we let $c_1 = 0$.

Singly Left Censored Sample

In this special case, we let $c_2 = 0$, and the estimating equation (13.2.30) becomes

Samples from Discrete Distributions

$$\bar{x} = \lambda \left[1 + \frac{c_1}{n} \left(\frac{f(a-1)}{1 - \bar{P}(a)} \right) \right]. \quad (13.2.31)$$

In all censored samples, $\bar{x}$ is the mean of the n uncensored observations.

13.2.6 Doubly Censored Samples—Total Number of Censored Observations Known, But Not the Number in Each Tail Separately

It occasionally may happen in practical applications that the total number of censored observations is known, but the number in each tail separately is not known. In these samples, we let a and d designate the terminals and let c designate the combined number of censored observations in the two tails. The total sample size is N and $n = N - c$ is the number of complete observations. The likelihood function of a sample of this type from a Poisson distribution with probability function (13.2.1) is

$$L = K e^{-n\lambda} \lambda^{n\bar{x}} \left[\prod_{i=1}^{n} x_i! \right]^{-1} [1 - \bar{P}(a) + \bar{P}(d+1)]^c. \quad (13.2.32)$$

On taking logarithms, differentiating, and equating to zero, we obtain the estimating equation

$$\frac{d \ln L}{d\lambda} = \frac{n\bar{x}}{\lambda} - n + c \left[\frac{f(d) - f(a-1)}{1 - \bar{P}(a) + \bar{P}(d+1)} \right] = 0. \quad (13.2.33)$$

We simplify this equation to obtain

$$\bar{x} = \lambda \left[1 + \frac{c_1}{n} \left(\frac{f(a-1) - f(d)}{1 - \bar{P}(a) + \bar{P}(d+1)} \right) \right]. \quad (13.2.34)$$

and the estimate $\hat{\lambda}$ is obtained as the solution.

13.2.7 Asymptotic Variance of Estimates

The asymptotic variance of estimates of λ can be obtained as the inverse of the negative of the expected value of the second derivative of $\ln L$. In large samples, a close approximation can be obtained as

$$V(\hat{\lambda}) \doteq - \left[\frac{d^2 \ln L}{d\lambda^2} \right]_{\lambda = \hat{\lambda}}^{-1}. \quad (13.2.35)$$

Second derivatives of $\ln L$ for the various cases considered here are as follows:

Doubly truncated samples:

$$\frac{d^2 \ln L}{d\lambda^2} = -\frac{n\bar{x}}{\lambda^2} - n\left[\frac{f(a-2) - f(a-1) - f(d-1) + f(d)}{\bar{P}(a) - \bar{P}(d+1)}\right]$$
$$+ n\left[\frac{f(a-1) - f(d)}{\bar{P}(a) - \bar{P}(d+1)}\right]^2. \quad (13.2.36)$$

Singly left truncated samples:

$$\frac{d^2 \ln L}{d\lambda^2} = -\frac{n\bar{x}}{\lambda^2} - n\left[\frac{f(a-2) - f(a-1)}{\bar{P}(a)}\right] + n\left[\frac{f(a-1)}{\bar{P}(a)}\right]^2. \quad (13.2.37)$$

Singly right truncated samples:

$$\frac{d^2 \ln L}{d\lambda^2} = -\frac{n\bar{x}}{\lambda^2} + n\left[\frac{f(d-1) - f(d)}{1 - \bar{P}(d+1)}\right] + n\left[\frac{f(d)}{1 - \bar{P}(d+1)}\right]^2. \quad (13.2.38)$$

Doubly censored samples—number of censored observations in each tail known:

$$\frac{d^2 \ln L}{d\lambda^2} = -\frac{n\bar{x}}{\lambda^2} - c_1\left[\frac{f(a-2) - f(a-1)}{1 - \bar{P}(a)} + \left(\frac{f(a-1)}{1 - \bar{P}(a)}\right)^2\right]$$
$$+ c_2\left[\frac{f(d-1) - f(d)}{\bar{P}(d+1)} - \left(\frac{f(d)}{\bar{P}(d+1)}\right)^2\right]. \quad (13.2.39)$$

Doubly censored samples—combined number of censored observations known but not number in each tail separately:

$$\frac{d^2 \ln L}{d\lambda^2} = -\frac{n\bar{x}}{\lambda^2} + c\left[\frac{f(d-1) - f(d) - f(a-2) + f(a-1)}{1 - \bar{P}(a) + \bar{P}(d+1)}\right]$$
$$- c\left[\frac{f(d) - f(a-1)}{1 - \bar{P}(a) + \bar{P}(d+1)}\right]^2. \quad (13.2.40)$$

Singly left censored samples:

$$\frac{d^2 \ln L}{d\lambda^2} = -\frac{n\bar{x}}{\lambda^2} - c_1\left[\frac{f(a-2) - f(a-1)}{1 - \bar{P}(a)} + \left(\frac{f(a-1)}{1 - \bar{P}(a)}\right)^2\right]. \quad (13.2.41)$$

Singly right censored samples:

$$\frac{d^2 \ln L}{d\lambda^2} = -\frac{n\bar{x}}{\lambda^2} + c_2\left[\frac{f(d-1) - f(d)}{\bar{P}(d+1)} - \left(\frac{f(d)}{\bar{P}(d+1)}\right)^2\right]. \quad (13.2.42)$$

13.2.8 Illustrative Examples

As previously mentioned, Poisson tables, such as those of Molina (1942), enable us to solve the various estimating equations given here with relative ease. Even

Samples from Discrete Distributions

when tables are not available, we can calculate individual terms of the Poisson distribution as they are needed. As a first approximation to $\hat{\lambda}$, the sample mean $\bar{x}$ will prove satisfactory in many practical applications. Other first approximations have been suggested by Moore (1954) and Rider (1953). Computational procedures are illustrated with the following examples.

Example 13.1. Sample data for this example consist of the number of α particles, observed by Rutherford and Geiger (1910) in one-eighth minute time intervals. Observations recorded for $N = 2608$ intervals are tabulated as follows with x designating the number of particles observed during an interval and O designating the observed sample frequencies.

x	0	1	2	3	4	5	6	7	8	9+	Total
O	57	203	383	525	532	408	273	139	45	43	2608

As tabulated, this is a singly right censored sample with $N = 2608$, $n = 2565$, $c = 43$, and $d = 8$. The mean of the censored observations is $\bar{x} = 3.7750$. The applicable estimating equation is (13.2.22), which we solve with $\lambda_1 = \bar{x} = 3.8$ (rounded off) as a first approximation. We find that $\hat{\lambda}$ lies between 3.8 and 3.9, and linear interpolation as shown below yields $\hat{\lambda} = 3.870$.

λ	$H(\lambda)$
3.900	3.8052
3.870	3.7750 = $\bar{x}$
3.800	3.7039

Calculations for $H(\lambda)$, which is the right side of equation (13.2.22), were made with the aid of Molina's tables. The standard deviation of $\hat{\lambda}$ as computed from (13.2.35) and (13.2.42) is $\sigma_{\hat{\lambda}} = \sqrt{V(\hat{\lambda})} = 0.04$.

Example 13.2. The basic data are the same as in Example 13.1, but for this example, we assume the sample to be singly right truncated at $d = 8$ with $n = 2565$ and $\bar{x} = 3.7750$. We assume that no information is available concerning the number of missing observations in excess of $d = 8$. In this case, the applicable estimating equation is (13.2.27) with $a = 0$, $f(a - 1) = 0$, and $\bar{P}(a) = 1$.

Calculations which follow the same pattern as those in Example 13.1 yield $\hat{\lambda} = 3.879$ and $\sigma_{\hat{\lambda}} = 0.04$.

We dispense with illustrations of the remaining sample types since solution of the applicable estimating equations can be accomplished in the same manner as in the two illustrations presented here. Standard iterative procedures such as Newton's method could be used to solve estimating equations for any of the sample types considered in this chapter. However, for simplicity and ease of computation, the trial-and-error method used here is preferred in most applications.

13.3 THE NEGATIVE BINOMIAL DISTRIBUTION

The negative binomial is a two-parameter discrete distribution that is used extensively for the description of data that are too heterogeneous to be fitted by the Poisson distribution. Since much of the data collected in studies of atmospheric phenomena exhibit marked heterogeneity, this distribution is of particular interest to aerospace and weather scientists. It is also of interest to social and medical scientists. It has attracted the attention of numerous investigators, among whom are Greenwood and Yule (1920), Fisher (1941), Haldane (1941), Anscombe (1950), Bliss and Fisher (1953), Bowman and Shenton (1965, 1966), and numerous others. Samples from this distribution when zero observations are missing have been studied by David and Johnson (1952), Sampford (1955), Rider (1955), Hartley (1958), Brass (1958), and Cohen (1965). In this section we are primarily concerned with the negative binomial as an alternative to the Poisson distribution when the zero class has been truncated.

13.3.1 The Probability Function

Whereas the terms of the binomial distribution are generated by the expansion of the binomial $(q + p)^n$, the terms of the negative binomial distribution are generated by the expansion of $(q - p)^{-n}$ where $q = p + 1$. Accordingly, the probability function of this distribution may be written as

$$\Pr[X = x] = f(x; p, k) = \frac{\Gamma(x + k)}{x!\, \Gamma(k)} p^k (1 - p)^x,$$

$$x = 0, 1, 2, \ldots \quad (13.3.1)$$

A special case of this distribution, which is sometimes referred to as the Pascal, sometimes as the Polya, and sometimes as the binomial waiting time distribution, results when k is limited to integer values. It can be obtained as the distribution of the number of failures in a sequence of independent Bernoulli trials that is

Samples from Discrete Distributions

terminated with the occurrence of the kth success, the total number of trials in any given sequence being $x + k$.

In its more general form where k may assume either integer or noninteger values, the negative binomial is perhaps the most widely used of all two-parameter discrete distributions, and its applications are many and varied. Different authors have written this function in various different forms. Fisher (1941) preferred a form that follows from (13.3.1) when we make the transformation $q = (1 - p)/p$. Anscombe (1950) used a form that results from the transformation $m = k(1 - p)/p = kq$. In this presentation we adhere to the form given in (13.3.1).

The Truncated Negative Binomial with Zero Class Missing

We note from (13.3.1) that $f(0) = p^k$, and the probability function in this case becomes

$$f_T(x; p, k) = \frac{\Gamma(k + x)}{x:!\,\Gamma(k)} \frac{p^k(1 - p)^x}{(1 - p^k)} \qquad x = 1, 2, 3, \ldots \qquad (13.3.2)$$

The factorial moments of the truncated distribution are

$$\mu_{[j]} = \frac{\Gamma(k + j)}{\Gamma(k)(1 - p^k)} \left(\frac{1 - p}{p}\right)^j. \qquad (13.3.3)$$

It follows that

$$\mu_{[1]} = \mu_1' = \mu = \left(\frac{k}{1 - p^k}\right)\left(\frac{1 - p}{p}\right),$$

$$\mu_{[2]} = \frac{k(k + 1)}{1 - p^k}\left(\frac{1 - p}{p}\right)^2 = \mu(k + 1)\left(\frac{1 - p}{p}\right), \qquad (13.3.4)$$

$$f_T(1) = \frac{kp^{k+1}}{1 - p^k}\left(\frac{1 - p}{p}\right)^2 = \mu p^{k+1}.$$

The second moment about zero and the second central moment follow from (13.3.4) as

$$\mu_2' = \mu_{[2]} + \mu = \frac{\mu}{p}[k(1 - p) + 1], \qquad (13.3.5)$$

$$\mu_2 = \mu_2' - \mu^2 = \mu\left[\frac{1}{p} - \mu p^k\right].$$

13.3.1 Parameter Estimation in the Truncated Distribution

Although maximum likelihood estimators in the truncated negative binomial distribution enjoy the usual optimal properties, they are not explicit, and their calculation often involves lengthy, time-consuming iterative procedures. The solution of moment-estimating equations based on the sample mean and variance is almost as time-consuming as calculation of the MLE. Explicit estimators based on the first three sample moments were proposed by David and Johnson (1952), but they were found to be rather inefficient. Sampford (1955) developed a reasonably rapid iterative technique for solving the two-moment estimating equations, but ultimately concluded that resulting estimates might only be suitable for use as first approximations in an iterative solution of maximum likelihood estimating equations. Brass (1958) derived explicit estimators based on the first two sample moments and the sample proportion of ones. These estimates are easy to calculate and are quite efficient for most combinations of parameter values. In most applications the Brass estimates provide close first approximations which can be used in an iterative calculation of the MLE, and they are often found to be satisfactory as final estimates. In this section we present derivations of the Brass estimators and of the maximum likelihood estimators.

Maximum Likelihood Estimation

The likelihood function of a random sample of size n from the truncated negative binomial with probability function (13.3.2) is

$$L = \left(\frac{p^k}{1-p^k}\right)^n \prod_{i=1}^{n} \frac{\Gamma(x_i + k)}{x_i!\,\Gamma(k)} (1-p)^{x_i}. \tag{13.3.6}$$

On taking logarithms of L, differentiating with respect to p and k, and equating to zero, we obtain the estimating equations

$$\frac{\partial \ln L}{\partial p} = \frac{nk}{p} + \frac{nkp^{k-1}}{1-p^k} - \frac{n\bar{x}}{1-p} = 0,$$

$$\frac{\partial \ln L}{\partial k} = n \ln p + \frac{np^k \ln p}{1-p^k} + \sum_{x=1}^{R} n_x \sum_{j=1}^{x} (k+j-1)^{-1} = 0, \tag{13.3.7}$$

where R is the largest sample value, and n_x is the sample frequency of x.

As given by Haldane (1941) and Sampford (1955), these equations can be simplified as follows

$$\frac{k(1-p)}{p(1-p^k)} = \bar{x},$$

$$\frac{-p \ln p}{1-p} = \frac{k}{n\bar{x}} \sum_{x=1}^{R} (k+x-1)^{-1} \sum_{x=1}^{R} n_x. \tag{13.3.8}$$

Note that the first equation of (13.3.8) equates the distribution mean as given in (13.3.4) to the sample mean. Standard iterative procedures can be employed for the calculation of maximum likelihood estimates $\hat{p}$ and $\hat{k}$. Components of the information matrix are

Samples from Discrete Distributions

$$-\frac{\partial^2 \ln L}{\partial p^2} = \frac{nk[1 - (k + 1)p^k]}{p^2(1 - p^k)^2} + \frac{n\bar{x}}{(1 - p)^2},$$

$$-\frac{\partial^2 \ln L}{\partial p\, \partial k} = \frac{n[1 - (1 - k \ln p)\, p^k]}{p(1 - p^k)^2}, \qquad (13.3.9)$$

$$-\frac{\partial^2 \ln L}{\partial k^2} = \sum_{j=1}^{R} (k + j - 1)^{-2} \sum_{i=1}^{R} n_i - \frac{n(\ln p)^2 p^k}{(1 - p^k)^2}.$$

Brass Estimators

As estimating equations, Brass (1958) employed $\mu = \bar{x}$, $\mu_2 = s^2$, and $f_T(1) = n_1/n$. Estimators of the parameters p and k follow from these equations. From the third equation of (13.3.4), we write

$$p^k = \frac{f_T(1)}{\mu p} \quad \text{and} \quad (1 - p^k) = \frac{\mu p - f_T(1)}{\mu p}. \qquad (13.3.10)$$

When these values are substituted into the first equation of (13.3.4) and the second equation of (13.3.5), we obtain

$$p = \frac{\mu}{\mu_2}[1 - f_T(1)] \quad \text{and} \quad k = \frac{p\mu - f_T(1)}{1 - p}. \qquad (13.3.11)$$

On substituting $\mu = \bar{x}$, $\mu_2 = s^2$, and $f_T(1) = n_1/n$, into these equations, the resulting estimators become

$$p^* = \frac{\bar{x}}{s^2}\left[1 - \frac{n_1}{n}\right] \quad \text{and} \quad k^* = \frac{p^*\bar{x} - (n_1/n)}{1 - p^*}. \qquad (13.3.12)$$

Brass demonstrated that these estimators are consistent, although they are not unbiased. However, when n is large, the effect of bias is slight. The efficiency for most combinations of parameter values is above 90%. Accordingly, these easily calculated estimators are satisfactory as final estimates in many applications, and when the situation calls for maximum likelihood estimates, they provide excellent first approximations from which to begin iterations to the MLE. Variances and covariances of estimates were given by Brass as

$$V(\mu^*) = \frac{\mu_2}{n}, \quad V(s^2) = \frac{\mu_4 - \mu_2^2}{n},$$

$$V\!\left(\frac{n_1}{n}\right) = \frac{f_T(1)[1 - f_T(1)]}{n},$$

$$\text{Cov}(\mu^*, s^2) = \frac{\mu_3}{n}, \quad \text{Cov}\!\left(\mu^*, \frac{n_1}{n}\right) = \frac{f_T(1)(1 - \mu)}{n}, \qquad (13.3.13)$$

$$\text{Cov}\!\left(s^2, \frac{n_1}{n}\right) = \frac{f_T(1)[1 - \mu_2 + \mu^2 - 2\mu]}{n}.$$

13.3.2 An Illustrative Example

Example 13.2. To illustrate estimation in the truncated negative binomial distribution, we consider a sample of chromosome breakage that was originally given by Sampford (1955). Sample data are as follows.

Number breaks	x	1	2	3	4	5	6	7	8	9	10	11	12	13
Frequency	n_x	11	6	4	5	0	1	0	2	1	0	1	0	1

In summary, $n = 32$, $n_1 = 11$, $n\bar{x} = 110$, $\bar{x} = 3.4375$, and $s^2 = 9.9315$. Brass estimates follow from (13.3.11) as

$$p^* = (3.4375/9.9315)[1 - 11/32)] = 0.2345,$$

$$k^* = [0.2345(3.4375) - (11/32)]/(1 - 0.2345) = 0.6040.$$

Maximum likelihood estimates for this example were calculated by Sampford as $\hat{p} = 0.2113$ and $\hat{k} = 0.493$, and their asymptotic variances and covariance as $V(\hat{p}) = 0.0098628$, $V(\hat{k}) = 0.276265$, and $\text{Cov}(\hat{p},\hat{k}) = 0.047185$.

13.4 THE BINOMIAL DISTRIBUTION

The binomial distribution with parameters n, p is defined as the distribution of the random variable X for which

$$f(x; n, p) = \Pr[X = x] = \binom{n}{x} p^x q^{n-x}, \quad x = 0, 1, 2, \ldots, n, \quad (13.4.1)$$

where $q = 1 - p$.

The first four noncentral moments about zero are

$$\mu_1'(X) = np,$$
$$\mu_2'(X) = np + n(n - 1)p^2,$$
$$\mu_3'(X) = np + 3n(n - 1)p^2 + n(n - 1)(n - 2)p^3, \quad (13.4.2)$$
$$\mu_4'(X) = np + 7n(n - 1)p^2 + 6n(n - 1)(n - 2)p^3$$
$$+ n(n - 1)(n - 2)(n - 3)p^4.$$

The central moments are

$$V(X) = \mu_2(X) = npq,$$
$$\mu_3(X) = npq(q - p), \quad (13.4.3)$$
$$\mu_4(X) = 3(npq)^2 + npq(1 - 6pq).$$

Samples from Discrete Distributions

The third and fourth standard moments are

$$\alpha_3(X) = \frac{q - p}{\sqrt{npq}},$$

$$\alpha_4(X) = 3 + \frac{1 - 6pq}{npq}. \tag{13.4.4}$$

13.4.1 The Truncated Binomial Distribution with Missing Zero Class

The probability function of the truncated binomial distribution with missing zero class, which is the distribution of primary interest in this section, is

$$f_T(x; n, p) = P_1[X = x] = \frac{\binom{n}{x} p^x q^{n-x}}{1 - q^n}, \qquad x = 1, 2, \ldots, n, \tag{13.4.5}$$

where, from (13.4.1), $q^n = f(0; n, p)$. Noncentral moments of this distribution about zero are obtained by dividing corresponding moments of the complete binomial distribution as given in (13.4.2) by $(1 - q^n)$. Accordingly, the mean and variance of the truncated binomial become

$$E(X) = \frac{np}{(1 - q^n)} \quad \text{and} \quad V(X) = \frac{npq}{1 - q^n} - \frac{n^2 p^2 q^n}{(1 - q^n)^2} \tag{13.4.6}$$

Parameter Estimation

Consider a sample consisting of k observations of the number of successes, x, obtained in repeated sequences of Bernoulli trials for which (13.4.5) is the probability function. The likelihood function of a sample as thus described is

$$K(x_1, x_2, \ldots, x_k) = (1 - q^n)^{-k} \prod_{i=1}^{k} \binom{n}{x_i} p^{x_i} q^{n - x_i}. \tag{13.4.7}$$

The loglikelihood function follows as

$$\ln L = \sum_{i=1}^{k} \ln \binom{n}{x_i} + \ln p \sum_{i=1}^{k} x_i + \ln q \sum_{i=1}^{n} (n - x_i) - k \ln (1 - q^n). \tag{13.4.8}$$

The maximum likelihood estimating equation follows when we differentiate $\ln L$ with respect to p, and equate this result to zero. We thereby obtain

$$\frac{d \ln L}{dp} = \frac{k\bar{x}}{p} - \frac{1}{q} \sum_{i=1}^{k} (n - x_i) - \frac{knq^{n-1}}{1 - q^n} = 0, \tag{13.4.9}$$

which we subsequently express in the simpler algebraic form

$$\bar{x} = np\left[1 + \frac{q^n}{1-q^n}\right] = \frac{np}{1-q^n}, \qquad (13.4.10)$$

where $\bar{x} = \sum_{i=1}^{k} x_i/k$.

Note that the maximum likelihood estimator of (13.4.10) is also the moment estimator, obtained from the estimating equation $E(X) = \bar{x}$. The estimate $\hat{p}$ follows as the solution of (13.4.10). Various standard iterative procedures can be used for making this calculation, but the trial-and-error procedure is likely to be most convenient. For a first approximation, $p_1 = \bar{x}/n$ will usually be satisfactory.

It is to be noted that $\lim_{i=\infty} [q^n/(1 - q^n)] = 0$ for all $q > 0$. Thus, for large values of n, $\hat{p} \to \bar{x}/n$, which is the estimate in the complete binomial distribution.

The asymptotic variance of $\hat{p}$ (for large k) is

$$V(\hat{p}) = \frac{pq(1-q^n)}{nk(1-q^n-npq^{n-1})}. \qquad (13.4.11)$$

An alternate estimator to the MLE, given by Mantel (1951), is

$$\tilde{p} = \frac{\bar{x} - (k_1/k)}{n - (k_1/k)}, \qquad (13.4.12)$$

where k_1 is the number of x's equal to 1, and k is the total number of observations of X. This estimator is quite easy to evaluate and Johnson and Kotz (1969) have shown it to be highly efficient (above 97% when $k \leq 5$). It can, of course, be used as an improved first approximation in lieu of $p = \bar{x}/n$ in iterative calculations of the MLE.

13.4.2 An Illustrative Example

Example 13.3. To illustrate the calculation of estimates in practical applications involving the truncated binomial distribution, we consider the following sample data:

Number successes x (each observation)	1	2	3	4	5	6	7	8	9	10
Number observations k_x	3	4	2	1	1	0	1	0	0	0

In summary, $n = 10$, $k = 12$, $\bar{x} = 33/12 = 2.75$, and $k_1 = 3$. As first approximations, we calculate $\bar{p} = \bar{x}/n = 2.75/10 = 0.275$, and from (13.4.12), the Mantel approximation is $\tilde{p} = [2.75 - (3/12)]/[10 - (3/12)] = 0.2564$. We

Samples from Discrete Distributions

subsequently try $p = 0.26$ and 0.27 in (13.4.10) and interpolate for the MLE as shown below.

p	$p[1 + q^n]/(1 - q^n)]$
0.2700	0.2821
0.2617	$\bar{x}/n = 0.2750$
0.2600	0.2735

Accordingly, the final estimate is $\hat{p} = 0.2617$.

13.5 THE HYPERGEOMETRIC DISTRIBUTION

In sampling from a binomial population, it is assumed that the trials (i.e., observations) are independent and that p remains constant from trial to trial. This assumption is justified if populations are sufficiently large, but in small finite populations, trials are not independent. Consider, for example, the probability of choosing x defective items in a sample of $n = 10$ from a population of size $N = 100$ if the population consists of 5 defective items and 95 nondefective items. For this example, $p = 0.05$, and

$$\Pr[X = x] = \frac{\binom{5}{x}\binom{95}{10-x}}{\binom{100}{10}}, \quad x = 0, 1, 2, 3, 4, 5. \quad (13.5.1)$$

For large finite populations, the error resulting from assuming that p is constant and that trials are independent is small enough to be ignored, and the binomial distribution can be used to calculate $\Pr[X = x]$. However, in small populations, when sampling is done without replacement, use of the binomial distribution will result in erroneous estimates of p. In these cases, the hypergeometric as exemplified in (13.5.1) is the appropriate distribution. In a more general context, we write the probability of the hypergeometric distribution as

$$f(x; N, n, p) = \Pr[X = x] = \frac{\binom{Np}{x}\binom{N-Np}{n-x}}{\binom{N}{n}},$$

$$x = 0, 1, 2, \ldots, n \text{ (or } Np), \quad (13.5.2)$$

where N is the size of the population, n is the sample size, p is the proportion, and Np is the number of members in this finite population that are characterized by, say, property A. Calculations with this formula will show that when n is only a small proportion of N, the value of N must be relatively small before any appreciable difference is noted between probabilities calculated from this formula and from the binomial formula (13.4.1). As an illustration, consider a population consisting of 100 manufactured items, of which 10 are defective ($p = 0.1$). The probability of selecting at most two defective items in a random sample of size 10 without replacement is

$$\Pr[X \leq 2] = \sum_{x=0}^{2} \frac{\binom{10}{x}\binom{90}{10-x}}{\binom{100}{10}} = 0.94.$$

whereas from the binomial formula, we calculate

$$\Pr[X \leq 2] = \sum_{x=0}^{2} \frac{10!}{x!(10-x)!}(0.1)^x(0.9)^{10-x} = 0.93.$$

The mean and variance of the hypergeometric distribution with pf (13.5.2) is

$$E(X) = np \quad \text{and} \quad V(X) = npq\left(\frac{N-n}{N-1}\right). \tag{13.5.3}$$

13.5.1 The Truncated Hypergeometric Distribution

We are primarily concerned here with the truncated distribution from which the zero class is missing. In this case the probability function can be written as

$$f_T(x) = \frac{\binom{Np}{x}\binom{Nq}{n-x}}{\binom{N}{n}[1-f(0)]}, \quad x = 1, 2, \ldots, n \text{ (or } Np\text{)}, \tag{13.5.4}$$

where $f(0)$ can be calculated from (13.5.2) as

$$f(0; n, n, p) = \frac{\binom{Np}{0}\binom{Nq}{n}}{\binom{N}{n}} = \frac{Nq(Nq-1)\ldots(Nq-n+1)}{N(N-1)\ldots(N-n+1)}. \tag{13.5.5}$$

From this it follows that

$$\lim_{N \to \infty} f(0) = q^n, \tag{13.5.6}$$

Samples from Discrete Distributions

which is equal to $\Pr[X = 0]$ for the binomial distribution. Accordingly, when N is large and n is small, we can use this approximation for $f(0)$ without fear of introducing appreciable error. The following illustrations provide an indication of actual discrepancies that occur between exact and approximate calculations.

Let $N = 100$, $n = 10$, $p = 0.1$, $q = 0.9$, and from (13.5.5)

$$f(0) = \frac{90(89)(88)(87)(86)(85)(84)(83)(82)(81)}{100(99)(98)(97)(96)(95)(94)(93)(92)(91)} = 0.3304,$$

whereas $q^n = (0.9)^{10} = 0.3487$.

Let $N = 100$, $n = 5$, $p = 0.05$, $q = 0.95$, and

$$f(0) = \frac{95(94)(93)(92)(91)}{100(99)(98)(97)(96)} = 0.7696 \quad \text{and} \quad q^n = (0.95)^5 = 0.7738.$$

Let us now examine the mean and the variance of the truncated distribution. The first two moments about zero are

$$\mu'_1 = E(X) = \frac{np}{1 - f(0)}, \qquad (13.5.7)$$

$$\mu'_2 = npq\left(\frac{N-n}{N-1}\right) + \frac{(np)^2}{1 - f(0)}.$$

It follows that

$$V(X) = \mu_2 = np\left[\frac{np + q[(N-n)/(N-1)]}{1 - f(0)} - \frac{np}{[1 - f(0)]^2}\right]. \qquad (13.5.8)$$

13.5.2 Parameter Estimation in the Truncated Hypergeometric Distribution

Moment Estimation

Although the moment estimate of p from a random sample with a missing zero class is not explicit, the calculations are relatively simple. We equate the expected value of X as given in (13.5.7) to the sample mean $\bar{x}$, and thereby obtain

$$\frac{np}{1 - f(0)} = \bar{x}, \qquad (13.5.9)$$

where $f(0)$ is given by (13.5.5). As a first approximation to $f(0)$, we might use $f_1(0) = q^n$, or we might use Mantel's approximation to p, given in (13.4.12) for the truncated binomial. These calculations can best be illustrated with an example.

Example 13.4. For this illustration, we use the sample data from example 13.3, which we now assume has been randomly selected from a finite population of size $N = 100$. In summary, $N = 100$, $n = 10$, $k = 12$, $k_1 = 3$, and $\bar{x} = 33/12 = 2.75$. If we use the binomial estimator of (13.4.10) as a first approximation to p, thereby using q^n as an approximation to $f(0)$, we obtain $p_1 = 0.2617$, and if we use the Mantel estimator of (13.4.12), we have $p_1 = 0.2564$. These calculations are the same as those illustrated in Example 13.3 for the binomial distribution. In order to solve (13.5.9), which applies to the hypergeometric distribution, we select $p_1 = 0.26$ as a trial value, which is substituted into (13.5.5) to calculate

$$f(0) = \frac{74(73)(72)(71)(70)(69)(68)(67)(66)(65)}{100(99)(98)(97)(96)(95)(94)(93)(92)(91)} = 0.0415.$$

When this value is substituted into the left side of (13.5.9), we have $0.26/(1 - 0.0415) = 0.2713$. As a second trial value, we select $p_2 = 0.27$, and this time we calculate

$$f(0) = \frac{73(72)(71)(70)(69)(68)(67)(66)(65)(64)}{100(99)(98)(97)(96)(95)(94)(93)(92)(91)} = 0.0359.$$

This value is now substituted into (13.5.9), and we calculate $0.27/(1 - 0.0359) = 0.2801$. For our final estimate, we interpolate as summarized below to obtain $\hat{p} = 0.2642$.

p	$f(0)$	$p/[1 - f(0)]$
0.2700	0.0359	0.2801
0.2642		$0.2750 = \bar{x}/n$
0.2600	0.0415	0.2713

We note that the hypergeometric estimate $\hat{p} = 0.2642$ differs very little from the binomial estimate $\hat{p} = 0.2617$ for the same data and thus confirms our earlier conclusion that when N is large and n is small the binomial is an excellent approximation for the hypergeometric distribution.

Maximum likelihood estimation in the truncated hypergeometric distribution becomes quite unwieldy. However, since the maximum likelihood and the moment estimators are identical in the truncated binomial distribution and since the binomial is a limiting form of the hypergeometric as $N \to \infty$, it is conjectured that the two estimators might also be identical or at least almost identical in the truncated hypergeometric distribution. Certainly, the ease with which moment

Samples from Discrete Distributions

estimates can be calculated in this case makes them attractive for most practical applications.

Further information concerning both theory and applications of the hypergeometric distribution can be found in the following references: Freeman (1973), Guenther (1975, 1977, 1983), Johnson and Kotz (1969), Lieberman and Owen (1961), Lund (1980), Mood et al. (1974), Patel and Read (1982), and Patil and Joshi (1968). Additional references are contained in the volume by Johnson and Kotz (1969).

14
Truncated Attribute Sampling and Related Topics

This chapter is concerned with truncated attribute acceptance sampling, with parameter estimation in discrete distributions when sample data are misclassified, and with inflated zero class discrete distributions. These topics are, of course, closely related to results presented in Chapter 13.

14.1 TRUNCATED ATTRIBUTE ACCEPTANCE SAMPLING

Sampling plans in which inspection is terminated as soon as a decision can be reached to either accept or reject are called truncated (or curtailed) plans. For specified risks, sizes of these samples are variable, and on the average they are smaller than corresponding fixed-size samples. Phatak and Bhatt (1967) derived maximum likelihood estimators for the fraction defective and for the average sample number (ASN) in single-stage curtailed samples. Craig (1968a, b) simplified these estimators and extended them for use in two-stage curtailed samples. Cohen (1970a, b) further simplified some of the earlier results of Phatak and Bhatt.

Our concern in this section is with attribute acceptance plans in which randomly selected individual items from a lot are inspected in sequence until either

1. An accumulated total of k defectives are found, in which case inspection is terminated and the lot is rejected, or

Truncated Attributes and Other Topics

2. An accumulated total of K nondefectives are found, in which case inspection is terminated and the lot is accepted.

We let Y designate the number of items inspected in order to reach a decision with respect to acceptance or rejection of a given lot. Thus Y is a discrete random variable that may assume the values $k, k + 1, \ldots, n$, where

$$n = k + K - 1 \quad \text{and} \quad k < K. \tag{14.1.1}$$

Sampling may thus be terminated with rejection after inspecting as few as k items from a lot, but in no case will the number inspected exceed n as given by (14.1.1). As described here, sampling plans are fully specified by the rejection and acceptance numbers k and K. However, this description is equivalent to that of the usual single-stage acceptance sampling plan in which n and k are specified. In this latter case, $K = n - k + 1$. As a result of this equivalency relationship, the operating characteristics of the truncated samples are the same as those of corresponding fixed-size single stage samples.

14.1.1 Probability Function of Random Variable Y

Let p designate the probability of selecting a defective item in a single trial and let this probability remain constant from trial to trial. Furthermore, let trials be stochastically independent. Accordingly, p may be interpreted as the process average proportion of defectives. For sufficiently large lots, we consider p as being the proportion of defectives in the lot. The probability function of Y, the number of items inspected from a given lot, may be expressed as

$$f(y; p) = f(y \cap R; p) + f(y \cap A; p),$$
$$y = k, k + 1, k + 2, \ldots, n, \tag{14.1.2}$$

where $f(y \cap R; p)$ is the joint probability that $Y = y$ and that the lot will be rejected, whereas $f(y \cap A; p)$ is the joint probability that $Y = y$ and that the lot will be accepted. It follows that

$$f(y \cap R; p) = \binom{y-1}{k-1} p^k q^{y-k}, \quad y = k, k+1, \ldots, n,$$
$$f(y \cap A; p) = \binom{y-1}{K-1} q^K p^{y-K}, \quad y = K, K+1, \ldots, n, \tag{14.1.3}$$

where $q = 1 - p$.

The two probability functions of (14.1.3) are recognized as negative binomial distributions. The probability function of (14.1.2) may be written as

$$f(y; p) = \begin{cases} f(y \cap R; p); & y = k, k+1, \ldots, K-1, \\ f(y \cap R; p) + f(y \cap A; p); & y = K, K+1, \ldots, n. \end{cases} \tag{14.1.4}$$

We substitute the two functions given in (14.1.3) into (14.1.4) and thereby we have

$$f(y; p) = \begin{cases} \binom{y-1}{k-1} p^k q^{y-k}; & y = k, k+1, \ldots, K-1, \\ \binom{y-1}{k-1} p^k q^{y-k} + \binom{y-1}{K-1} q^K p^{y-K}; & y = K, K+1, \ldots, n. \\ 0 & \text{elsewhere.} \end{cases} \quad (14.1.5)$$

The probability of rejecting a lot follows from the first equation of (14.1.3) when we sum on y from k to n. We thus obtain

$$P(R) = \sum_{y=k}^{n} \binom{y-1}{k-1} p^k q^{y-k}. \quad (14.1.6)$$

The following identity between the binomial and the negative binomial distribution has been demonstrated by Patil (1960, 1963), by Morris (1963) and perhaps by others:

$$\sum_{s=k}^{n} \binom{n}{z} p^s q^{n-s} = \sum_{y=k}^{n} \binom{y-1}{k-1} p^k q^{y-k}. \quad (14.1.7)$$

Therefore, the probability of rejecting a lot can be expressed in the following alternate form:

$$P(R) = \sum_{s=k}^{n} \binom{n}{z} p^s q^{n-s}, \quad (14.1.8)$$

where Z is the number of defectives present in a sample of size $n = k + K - 1$. It is noted that this is merely the probability of rejection from an ordinary fixed-size single-stage acceptance sample of size n.

It consequently follows that the operating characteristic curve of a curtailed sampling plan as specified by k and K is identical with that of an ordinary single sampling plan specified by k and n where $n = k + K - 1$. Curtailment, of course, permits a decision to be reached with smaller samples than when fixed-size samples are employed.

14.1.2 Parameter Estimation from Curtailed Samples

Suppose that m lots have been subjected to acceptance inspections in accordance with a curtailed plan as described in the preceding section. Let a designate the number of lots that were accepted and let r designate the number that were rejected so that

$$m = a + r. \quad (14.1.9)$$

Truncated Attributes and Other Topics

Let the number of defectives found and the number of items inspected be recorded for each lot. The sample data then consists of the paired values (z_1, y_1), (z_2, y_2), ..., (z_a, y_a), (k, y_{a+1}), (k, y_{a+2}), ..., (k, y_{a+n}), where $z_i(i = 1, 2, \ldots, a)$ is the number of defectives found in the ith accepted lot ($z_i < k$) and k is the number of defectives found in each rejected lot. This sample could be described more concisely as consisting of the paired values (z_i, y_i), $i = 1, 2, \ldots, m$ with $z_i < k$ in accepted lots and $z_i = k$ in rejected lots. It is assumed that no further inspections are made from a rejected lot after the decision to reject has been reached. It follows from (14.1.3) that the likelihood function of a sample as described is

$$L = \prod_{i=1}^{r} \binom{y_i - 1}{k - 1} p^k q^{y_i - k} \prod_{i=1}^{a} \binom{y_i - 1}{K - 1} q^K p^{y_i - K}. \tag{14.1.10}$$

We take logarithms of L, differentiate with respect to p, equate to zero, and solve for p to obtain

$$\hat{p} = \frac{\left[\sum_{i=1}^{a} y_i - aK\right] + rk}{\sum_{i=1}^{m} y_i} \tag{14.1.11}$$

where

rk = number defectives found in r rejected lots
aK = number nondefectives found in a accepted lots
$\sum_{j=1}^{a} y_j$ = total number items inspected in a accepted lots
$\left[\sum_{j=1}^{a} y_j - aK\right]$ = number defectives in a accepted lots
$\sum_{i=1}^{m} y_i$ = total number inspections from $m = a + r$ lots

It subsequently follows from (14.1.11) that

$$\hat{p} = \frac{\text{total number of defectives found}}{\text{total number of units inspected}}. \tag{14.1.12}$$

The asymptotic variance of $\hat{p}$ can be obtained as

$$V(\hat{p}) = -1/E(\partial^2 \ln L/\partial p^2), \tag{14.1.13}$$

which after certain algebraic reduction can be written as

$$V(\hat{p}) \sim \frac{\hat{p}\hat{q}}{mE(Y)}. \tag{14.1.14}$$

For a sufficiently large number of lots, the mean of the observed values of Y should provide a reasonable approximation to $E(Y)$, and the preceding variance might thereby be approximated as

$$V(\hat{p}) \doteq \frac{\hat{p}\hat{q}}{\sum_{i=1}^{m} y_i}. \qquad (14.1.15)$$

In the notation employed here, ASN = $E(Y)$; that is, the average sample number is merely the expected value of the random variable Y. From the probability function of (14.1.5), it follows that

$$E(Y) = \sum_{y=k}^{n} y \binom{y-1}{k-1} p^k q^{y-k} + \sum_{y=K}^{n} y \binom{y-1}{K-1} q^K p^{y-K}, \qquad (14.1.16)$$

Use of the identity of (14.1.7) enables us to write

$$E(Y) = \frac{k}{p}\{1 - B[p, n+1, k]\} + \frac{K}{q} B[p, n+1, k-1], \qquad (14.1.17)$$

where $B[p, n, k]$ is the cumulative binomial function

$$B[p, n, k] = \sum_{s=0}^{k} \binom{n}{x} p^s q^{n-s} \quad \text{and} \quad n = k + K - 1. \qquad (14.1.18)$$

The expression given in (14.1.17) for $E(Y)$ was given by Phatak and Bhatt (1967), and an equivalent result was given by Craig (1968b).

Entries in Table 14.1 indicate the reduction in sample sizes that can be realized through truncation in selected sample plans. As expected, each of these plans shows the greatest reduction in sample size when rejection is made as a result of large values of p.

14.2 ESTIMATION FROM MISCLASSIFIED INSPECTION DATA

In attribute inspections it sometimes happens that inspectors are reluctant to reject product when the number of defects per item or the number of defective items per sample exceeds an allowable limit by only one. Accordingly it may happen that observations of $k + 1$, which would call for rejection, are erroneously reported as being only k. In particular, when no defects or defectives are allowed, the inclination to report ones as zeroes is often a factor to be reckoned with. In these cases any estimator of the overall average percentage (or fraction) of defects or defective items in production presented for inspection that fails to properly

Table 14.1 Characteristics of Curtailed Sampling Plans

n = 10, k = 4, K = 7

p	P(R)	ASN
0	0.0000	7
0.10	0.0128	7.74
0.15	0.0450	8.09
0.20	0.1209	8.35
0.25	0.2241	8.49
0.30	0.3504	8.50
0.35	0.4862	8.37
0.40	0.6177	8.13
0.45	0.7340	7.79
0.50	0.8281	7.39

n = 20, k = 5, K = 16

p	P(R)	ASN
0	0.0000	16
0.10	0.0432	17.57
0.15	0.1702	17.86
0.20	0.3704	17.49
0.25	0.5852	16.51
0.30	0.7625	15.16
0.35	0.8818	13.69
0.40	0.9490	12.29
0.45	0.9811	11.05
0.50	0.9941	9.98

n = 40, k = 1, K = 40

p	P(R)	ASN
0	0.0000	40
0.01	0.0607	39.18
0.02	0.1905	37.24
0.03	0.3385	34.76
0.04	0.4790	32.09
0.05	0.6009	29.45
0.06	0.7010	26.95
0.07	0.7799	24.64
0.08	0.8406	22.56
0.09	0.8860	20.70
0.10	0.9195	10.05
0.11	0.9438	17.59
0.12	0.9612	16.29

n = 50, k = 1, K = 50

p	P(R)	ASN
0	0.0000	50
0.01	0.0894	48.44
0.02	0.2642	45.00
0.03	0.4447	40.89
0.04	0.5995	36.74
0.05	0.7206	32.87
0.06	0.8100	29.41
0.07	0.8735	26.39
0.08	0.9173	23.77
0.09	0.9468	21.53
0.10	0.9662	19.61

n = 80, k = 2, K = 79

p	P(R)	ASN
0	0.0000	79
0.01	0.0466	78.99
0.02	0.2156	74.70
0.03	0.4319	68.02
0.04	0.6252	60.54
0.05	0.7694	53.34
0.06	0.8656	46.92
0.07	0.9250	41.44
0.08	0.9596	36.85
0.09	0.9789	33.04

n = 60, k = 2, K = 59

p	P(R)	ASN
0	0.0000	60
0.01	0.0224	59.65
0.02	0.1187	57.96
0.03	0.2685	54.95
0.04	0.4324	51.10
0.05	0.5826	46.90
0.06	0.7060	42.73
0.07	0.8002	38.81
0.08	0.8683	35.25
0.09	0.9154	32.09
0.10	0.9470	29.31

Table 14.1, *continued*

n = 100, k = 2, K = 99			n = 120, k = 3, K = 118		
p	P(R)	ASN	p	P(R)	ASN
0	0.0000	99	0	0.0000	118
0.01	0.0794	97.76	0.01	0.0330	119.1
0.02	0.3233	89.37	0.02	0.2200	112.9
0.03	0.5802	77.94	0.03	0.4867	101.4
0.04	0.7679	66.60	0.04	0.7113	88.0
0.05	0.8817	56.77	0.05	0.8356	75.6
0.06	0.9434	48.77	0.06	0.9340	65.1
0.07	0.9742	42.39	0.07	0.9719	56.6

n = 140, k = 3, K = 138			n = 160, k = 3, K = 158		
p	P(R)	ASN	p	P(R)	ASN
0	0.0000	138	0	0.0000	158
0.01	0.0528	138.3	0.01	0.0778	157.0
0.02	0.3076	127.6	0.02	0.3979	140.6
0.03	0.6080	110.4	0.03	0.7099	117.3
0.04	0.8151	92.8	0.04	0.8861	95.8
0.05	0.9235	77.8	0.05	0.9612	78.9
0.06	0.9713	66.0			

n = 180, k = 3, K = 178			n = 160, k = 4, K = 157		
p	P(R)	ASN	p	P(R)	ASN
			0	0.0000	157
			0.01	0.0230	159.3
0	0.0000	178	0.02	0.2179	151.5
0.01	0.1077	175.1	0.03	0.5260	134.8
0.02	0.4861	151.8	0.04	0.7706	115.0
0.03	0.7910	122.3	0.05	0.9061	97.0
0.04	0.9320	97.6			
0.05	0.9810	79.5	0.06	0.9662	82.5

take these misclassifications into account will result in erroneous estimates. In this section we employ the method of maximum likelihood to estimate the process or population fraction p when samples are subject to misclassification as thus described.

Truncated Attributes and Other Topics

14.2.1 Misclassification in the Poisson Distribution

Misclassification in the Poisson distribution in which observations of $x = k + 1$ were reported as $x = k$ with probability θ were previously considered by Cohen (1960b, f). In this case, the probability function of X, the reported number of defects per item, becomes

$$f(x; \lambda, \theta) = \begin{cases} e^{-\lambda}\lambda^x/x!, & x = 0, 1, \ldots, (k-1), (k+2) \ldots, \\ [e^{-\lambda}\lambda^k/k!][1 + \lambda\theta/(k+1)], & x = k, \\ (1 - \theta)e^{-\lambda}\lambda^{k+1}/(k+1)!, & x = k+1, \end{cases} \quad (14.2.1)$$

where $\lambda > 0$, $0 \le \theta \le 1$, and $\Pr[X = x] = f(x; \lambda, \theta)$.

The likelihood function of a random sample consisting of observations $\{x_i\}$, $i = 1, 2, \ldots, N$ of the reported number of defects per item is

$$L = \frac{e^{-N\lambda}\lambda^{N\bar{x}}}{\prod_{i=1}^{N} x_i!} \left(1 + \frac{\lambda\theta}{k+1}\right)^{n_k} [1 - \theta]^{n_{k+1}}, \quad (14.2.2)$$

where n_k and n_{k+1} are the sample frequencies of k and $k+1$, respectively. We take logarithms of L, differentiate with respect to λ and θ, and equate to zero, to obtain the estimating equations

$$\frac{\partial \ln L}{\partial \lambda} = -N + \frac{N\bar{x}}{\lambda} + \frac{n_k \theta}{k + 1 + \lambda\theta} = 0,$$
$$\frac{\partial \ln L}{\partial \theta} = \frac{n_k \lambda}{k + 1 + \lambda\theta} - \frac{n_{k+1}}{1 - \theta} = 0. \quad (14.2.3)$$

When they exist, estimates $\hat{\lambda}$ and $\hat{\theta}$ can be found as the simultaneous solution of these two equations. However, no solution exists when (i) all sample observations are k's, in which case $n_k = N$, $n_{k+1} = 0$, and $\bar{x} = k$, and (ii) all sample observations are $k + 1$, in which case $n_k = 0$, $n_{k+1} = N$, and $\bar{x} = k + 1$. In the event that $n_k = n_{k+1} = 0$, no estimate exists for θ, but the first equation of (14.2.3) leads to the estimate $\hat{\lambda} = \bar{x}$. Although these exceptional cases are of theoretical interest, they are not likely to be of any practical importance unless samples are small, and it is not expected that these estimators will be employed unless samples are reasonably large. The occurrence of samples for which acceptable estimates of λ and θ fail to exist should be viewed as a suggestion that perhaps the probability function of (14.2.1) is not applicable to the random variable actually being observed.

To facilitate their solution in cases where solutions exist, the two equations of (14.2.3) are simplified to

$$\lambda^2 - \left[\bar{x} - (k+1) + \frac{n_k}{N}\right]\lambda - (k+1)\left(\bar{x} - \frac{n_{k+1}}{N}\right) = 0,$$

$$\theta = \left[\frac{n_k - (k+1)n_{k+1}/\lambda}{n_k + n_{k+1}}\right],$$

(14.2.4)

where $\bar{x}$ is the sample mean ($\bar{x} = \sum_{i=1}^{N} x_i/N$). The first of these equations is quadratic in λ, and in cases where estimates exist, this equation has one positive and one negative root. The positive root is the required estimate, which the quadratic formula enables us to write as

$$\hat{\lambda} = \frac{[\bar{x} - (k+1) + n_k/N] + \sqrt{[\bar{x} - (k+1) + n_k/N]^2 + 4(k+1)(\bar{x} - n_{k+1}/N)}}{2}.$$

(14.2.5)

The estimate $\hat{\theta}$ then follows from the second equation of (14.2.4) when we replace λ with $\hat{\lambda}$.

The Special Case in Which k = 0

In this special case the estimating equations are somewhat simpler in form. Equations (14.2.4) become

$$\lambda^2 - \left(\bar{x} - 1 + \frac{n_0}{N}\right)\lambda - \left(\bar{x} - \frac{n_1}{N}\right) = 0,$$

$$\theta = \frac{n_0 - n_1/\lambda}{n_0 + n_1}.$$

(14.2.6)

and the estimator (14.2.5) becomes

$$\hat{\lambda} = \frac{1}{2}\left[\left(\bar{x} - 1 + \frac{n_0}{N}\right) + \sqrt{\left(\bar{x} - 1 + \frac{n_0}{N}\right)^2 + 4\left(\bar{x} - \frac{n_1}{N}\right)}\right].$$

(14.2.7)

With $\hat{\lambda}$ thus calculated, $\hat{\theta}$ follows from the second equation of (14.2.6).

14.2.2 Sampling Errors of Estimates from Misclassified Poisson Data

The asymptotic variance–covariance matrix of $(\hat{\lambda}, \hat{\theta})$ is obtained in the usual manner by inverting the information matrix with elements that are negatives of expected values of second partials of $\ln L$.

Truncated Attributes and Other Topics

Further differentiation of (14.2.3) gives

$$\frac{\partial^2 \ln L}{\partial \lambda^2} = -\frac{N\bar{x}}{\lambda^2} - \frac{n_k \theta^2}{(k+1+\lambda\theta)^2},$$

$$\frac{\partial^2 \ln L}{\partial \lambda \, \partial \theta} = \frac{n_k(k+1)}{(k+1+\lambda\theta)^2} = \frac{\partial^2 \ln L}{\partial \theta \, \partial \lambda}, \quad (14.2.8)$$

$$\frac{\partial^2 \ln L}{\partial \theta^2} = -\frac{n_k \lambda^2}{(k+1+\lambda\theta)^2} - \frac{n_{k+1}}{(1-\theta)^2}.$$

Since expected values of n_k, n_{k+1}, and X are

$$E(n_k) = N\left(\frac{e^{-\lambda}\lambda^k}{k!}\right)\left(\frac{1+\lambda\theta}{k+1}\right),$$

$$E(n_{k+1}) = N\left[\frac{(1-\theta)e^{-\lambda}\lambda^{k+1}}{(k+1)!}\right], \quad (14.2.9)$$

$$E(X) = \lambda\left[1 - \left(\frac{\theta}{k+1}\right)\frac{e^{-\lambda}\lambda^k}{k!}\right],$$

it follows that

$$-E\left(\frac{\partial^2 \ln L}{\partial \lambda^2}\right) = N\phi_{11} = N\left(\frac{1}{k+1}\right)\left[\frac{k+1-\theta\psi}{\lambda} + \frac{\theta^2\psi}{k+1+\lambda\theta}\right],$$

$$-E\left(\frac{\partial^2 \ln L}{\partial \lambda \, \partial \theta}\right) = N\phi_{12} = -\frac{N}{k+1+\lambda\theta} = N\phi_{21} = -E\left(\frac{\partial^2 \ln L}{\partial \theta \, \partial \lambda}\right) \quad (14.2.10)$$

$$-E\left(\frac{\partial^2 \ln L}{\partial \theta^2}\right) = N\phi_{22} = N\left(\frac{\lambda\psi}{k+1}\right)\left[\frac{\lambda}{k+1+\lambda\theta} + \frac{1}{1-\theta}\right],$$

where $\psi = e^{-\lambda}\lambda^k/k!$ and ϕ_{ij} is written for $\phi_{ij}/(\lambda, \theta)$. The asymptotic variances and covariance follow as

$$V(\hat{\lambda}) \sim \frac{\phi_{22}}{N(\phi_{11}\phi_{22} - \phi_{12}^2)},$$

$$V(\hat{\theta}) \sim \frac{\phi_{11}}{N(\phi_{11}\phi_{22} - \phi_{12}^2)}, \quad (14.2.11)$$

$$\text{Cov}(\hat{\lambda}, \hat{\theta}) \sim \frac{-\phi_{12}}{N(\phi_{11}\phi_{22} - \phi_{12}^2)},$$

and the coefficient of correlation between estimates is

$$\rho_{\hat{\lambda},\hat{\theta}} \sim \frac{-\phi_{12}}{\sqrt{\phi_{11}\phi_{22}}}. \quad (14.2.12)$$

14.2.3 Illustrative Examples—Misclassified Poisson Data

To illustrate the practical application of estimators presented here, we employ two examples given by Cohen (1960b, f).

Example 14.1 Data for this example were generated from a Poisson population for which $\lambda = 2.1$. The sample consists of assumed inspections of defects per unit on $N = 1000$ units, the acceptance number being $k = 3$. It was assumed that the fraction $\theta = 0.394$ of observations for which $x = 4$ was misclassified as being $x = 3$. Following is a tabulation of the reported inspection results.

Number defects per unit x	Frequency n_x
0	122
1	258
2	270
3	228
4	60
5	42
6	15
7	4
8	1

In summary, for this example, $\bar{x} - 2.058$, $k = 3$, $n_3 = 228$, $n_4 = 60$, and $N = 1000$. When these values are substituted into (14.2.5), we calculate

$$\hat{\lambda} = \frac{-1.714 + \sqrt{1.714^2 - 4(-7.992)}}{2} = 2.092.$$

From the second equation of (14.2.4), we calculate

$$\hat{\theta} = \frac{228 - 4(60)/2.092}{228 + 60} = 0.393.$$

These estimates are to be compared with population parameters $\lambda = 2.1$ and $\theta = 0.394$. In order to calculate estimate variances and covariance, we employ (14.2.10):

$$\hat{\phi}_{11} = \frac{1}{4} \left[\frac{4 - (0.393)(0.189)}{2.092} + \frac{(0.393)^2(0.189)}{4 + (2.092)(0.393)} \right] = 0.4702,$$

Truncated Attributes and Other Topics

$$\hat{\phi}_{13} = \frac{-0.189}{4 + (2.092)(0.393)} = -0.0392,$$

$$\hat{\phi}_{23} = \left[\frac{(2.092)(0.189)}{4}\right]\left[\frac{2.092}{4 + (2.092)(0.393)} + \frac{1}{1 - 0.393}\right] = 0.2057,$$

where $\hat{\phi}_{ij}$ is written for $\phi_{ij}(\hat{\lambda}, \hat{\theta})$. When these values are substituted into (14.2.11) and (14.2.12), we have $V(\hat{\lambda}) = 0.0022$, $V(\hat{\theta}) = 0.0049$, $\text{Cov}(\hat{\lambda}, \hat{\theta}) = 0.0004$, and $\rho_{\hat{\lambda},\hat{\theta}} = 0.13$.

Example 14.2 This example is designed to illustrate estimation in the special case where $k = 0$. Sample data have been obtained by altering Bortkiewicz's classical example on deaths from the kick of a horse from records of a certain group of 10 Prussian Army Corps over the 20-year period 1875–1894. The original data consisted of 200 annual reports of deaths from the stated cause. For the purpose of this illustration, it is assumed that 20 of the reports which should have shown one death each were misclassified to report zero deaths. Both the original and the altered data follow:

Number deaths per army corps per year	Number observations	
	Original data	Altered data
0	109	129
1	65	45
2	22	22
3	3	3
4	1	1
5	0	0

In summary, for the altered data, we have $N = 200$, $\bar{x} = 0.51$, $n_0/N = 0.645$, $n_1/N = 0.225$, $(\bar{x} - 1 + n_0/N) = 0.155$, and $(\bar{x} - n_1/N) = 0.285$. On substituting these values into (14.2.5), we calculate

$$\hat{\lambda} = \frac{0.155 + \sqrt{0.155^2 + 4(0.285)}}{2} = 0.617.$$

Subsequent substitution into the second equation of (14.2.6) yields

$$\hat{\theta} = \frac{129 - 45/0.617}{129 + 45} = 0.322.$$

The estimate $\hat{\lambda} = 0.617$, obtained above, is to be compared with $\hat{\lambda} = 0.610$, which follows from the original unaltered data. The estimate $\hat{\theta} = 0.322$ is to be compared with $20/65 = 0.308$, which is the actual proportion of ones that were misclassified in the process of altering the original data for this illustration. With λ and θ replaced by their estimates, the asymptotic variances and covariance were calculated from (14.2.11) and (14.2.12) as $V(\hat{\lambda}) = 0.0046$, $V(\hat{\theta}) = 0.0097$, $\text{Cov}(\hat{\lambda},\hat{\theta}) = 0.0031$, and $\rho_{\hat{\lambda},\hat{\theta}} = 0.47$. The variance $V(\hat{\lambda}) = 0.0046$ as calculated above from the altered sample is to be compared with $V(\hat{\lambda}) = \hat{\lambda}/N = .610/200 = 0.00305$ from the original unaltered sample.

14.2.4 Misclassification in the Binomial Distribution

Let a record be made of the number of defective items in random samples from a binomial distribution with parameters n and p, where n is known and p is to be estimated from the sample data. Suppose that samples actually containing $k + 1$ defective items are incorrectly reported to contain only k defectives with probability θ. Let X designate the number of defectives reported as being present in samples. The probability function of X, $f(x; n, p, \theta) = \Pr[X = x]$, may be written as

$$f(x; n, p, \theta) = \begin{cases} \binom{n}{x} p^x (1-p)^{n-x}, & x = 0, 1, \ldots, k-1, k+2, \ldots, n, \\ \binom{n}{k} p^k (1-p)^{n-k} \left[1 + \left(\frac{n-k}{k+1}\right)\left(\frac{\theta p}{1-p}\right)\right], & x = k, \\ (1-\theta)\binom{n}{k-1} p^{k+1} (1-p)^{n-k-1}, & x = k+1, \end{cases}$$

(14.2.13)

where $0 \leq p \leq 1$, $0 \leq \theta \leq 1$, and $k = 0, 1, 2, \ldots, (n-1)$.

Consider N observations of X. Let r_x be the sample frequency of $X = x$. It follows that $\sum_{x=0}^{n} r_x = N$, and the likelihood function of these N observations may be written as

$$L = \left[1 + \left(\frac{n-k}{k+1}\right)\left(\frac{\theta p}{1-p}\right)\right]^{r_k}$$

$$\times [1-\theta]^{r_{k+1}} \prod_{x=0}^{n} \left[\binom{n}{x} p^x (1-p)^{n-x}\right]^{r_x}. \quad (14.2.14)$$

We take logarithms of L, differentiate with respect to p and θ, and equate to zero to obtain the estimating equations

Truncated Attributes and Other Topics

$$\frac{\partial \ln L}{\partial p} = \left[\frac{1}{p} + \frac{1}{1-p}\right] \sum_{x=0}^{n} xr_x - \frac{nN}{1-p}$$

$$+ \frac{r_k(n-k)\theta}{(1-p)[(k+1)(1-p) + (n-k)\theta p]} = 0, \quad (14.2.15)$$

$$\frac{\partial \ln L}{\partial \theta} = \frac{r_k(n-k)p}{(k+1)(1-p) + (n-k)\theta p} - \left(\frac{r_{k+1}}{1-\theta}\right) = 0.$$

The required estimates $\hat{p}$ and $\hat{\theta}$, when they exist, can be found by simultaneously solving these two equations. As a first step toward a solution, the two equations of (14.2.15) are simplified to the equivalent equations

$$\frac{\theta r_{k+1}}{1-\theta} + \sum_{x=0}^{n} xr_x = Nnp, \quad (14.2.16)$$

$$\frac{r_k(n-k)}{[(k+1)(1-p) + (n-k)\theta p]} = \frac{r_{k+1}}{p(1-\theta)}.$$

The second equation of (14.2.16) is solved for θ and substituted into the first equation. We thereby obtain

$$p^2 nN[n - 2k - 1] + p\left[nN(k+1) - (n - 2k - 1)\sum_{x=0}^{n} xr_x - r_k(n-k)\right.$$

$$\left. - r_{k+1}(k+1)\right] - (k+1)\left[\sum_{x=0}^{n} xr_x - r_{k+1}\right] = 0 \quad (14.2.17)$$

$$\theta = \frac{1}{r_k + r_{k+1}}\left[r_k - \frac{r_{k+1}(k+1)(1-p)}{(n-k)p}\right].$$

The first of these equations is quadratic in p, and the required estimate $\hat{p}$ must then be a positive real root in the closed interval $(0, 1)$. With $\hat{p}$ thus calculated, $\hat{\theta}$ follows from the second equation of (14.2.17).

When no misclassification has occurred and when thus $\theta = 0$, the first equation of (14.2.17) yields the well-known estimate

$$\hat{p} = \sum_{x=0}^{n} \frac{xr_x}{nN}.$$

In the event that (i) all sample observations are k's, in which case $r_k = N$, $r_{k+1} = 0$, and $\bar{x} = k$, or (ii) all sample observations are $(k+1)$'s, in which case, $r_k = 0$, $r_{k+1} = N$, and $\bar{x} = k+1$, then estimates $\hat{p}$ and $\hat{\theta}$ do not exist. In the event that $r_k = r_{k+1} = 0$, no estimate is possible for θ, but from the first equation (14.2.15), we would have $\hat{p} = \sum_{x=0}^{n} xr_x/nN$. Although of theoretical interest, these exceptions seem unlikely to be important in practical applications for the same reasons that were noted in connection with estimation from misclassified Poisson data.

The Special Case in Which k = 0

When $k = 0$, the estimating equations of (14.2.17) become

$$p^2 nN(n-1) + p\left[nN - (n-1)\sum_{x=0}^{n} xr_x - nr_0 - r_1\right]$$

$$- \sum_{x=0}^{n} xr_x + r_1 = 0, \quad \theta = \frac{r_0 - r_1(1-p)/np}{n_0 + r_1}, \qquad (14.2.18)$$

and from the quadratic equation in p, we have

$$\hat{p} = \frac{1}{2Nn(n-1)}\left\{-\left[nN - (n-1)\sum_{x=0}^{n} xr_x - nr_0 - r_1\right]\right.$$

$$\left. + \sqrt{\left[nN - (n-1)\sum_{x=0}^{n} xr_x - nr_0 - r_1\right]^2 + 4Nn(n-1)\left[\sum_{x=0}^{n} xr_x - r_1\right]}\right\}.$$

(14.2.19)

With $\hat{p}$ thus determined, $\hat{\theta}$ follows from the second equation of (14.2.18).

14.2.5 Sampling Errors of Estimates from Misclassified Binomial Data

The asymptotic variance–covariance matrix of $(\hat{p}, \hat{\theta})$ is obtained in the usual manner by inverting the information matrix with elements that are negatives of expected values of the second partials of $\ln L$. Thus,

$$-\frac{1}{N}E\left(\frac{\partial^2 \ln L}{\partial p^2}\right) = \phi_{11} = \frac{1}{(1-p)^2}\left[\frac{n(1-p)}{p} - \frac{f_k}{A}\left(\frac{1}{p} - \frac{1}{A}\right)\right],$$

$$-\frac{1}{N}E\left(\frac{\partial^2 \ln L}{\partial p\, \partial \theta}\right) = \phi_{12} = \frac{f_k(k+1)}{(n-k)\theta^2 A^2} = \phi_{21} = -\frac{1}{N}E\left(\frac{\partial^2 \ln L}{\partial \theta\, \partial p}\right), \quad (14.2.20)$$

$$-\frac{1}{N}E\left(\frac{\partial^2 \ln L}{\partial \theta^2}\right) = \phi_{22} = \frac{f_k p^2}{\theta^2 A^2} + \frac{f_{k+1}}{(1-\theta)^2},$$

where $A = [(k+1)(1-p)/(n-k)\theta + p]$, $f_k = f(k)$, $f_{k+1} = f(k+1)$, and $\phi_{ij} = \phi_{ij}(\hat{p}, \hat{\theta})$ with $i, j = 1, 2$. The asymptotic variances and covariance follow as

Truncated Attributes and Other Topics

$$V(\hat{p}) = \frac{\phi_{22}}{N(\phi_{11}\phi_{22} - \phi_{12}^2)},$$

$$V(\hat{\theta}) = \frac{\phi_{11}}{N(\phi_{11}\phi_{22} - \phi_{12}^2)}, \quad (14.2.21)$$

$$\text{Cov}(\hat{p},\hat{\theta}) = \frac{-\phi_{12}}{N(\phi_{11}\phi_{22} - \phi_{12}^2)}.$$

The correlation coefficient between estimates is

$$\rho_{\hat{p},\hat{\theta}} \sim -\frac{\phi_{12}}{\sqrt{\phi_{11}\phi_{22}}}. \quad (14.2.22)$$

14.2.6 An Illustrative Example—Misclassified Binomial Data

Example 14.3 An example generated by Cohen (1960a) consisted of $N = 1000$ random observations of the number of defectives in samples of $n = 40$ from a binomial population in which $p = 0.3$ where sample values of $k + 1 = 3$ were misclassified as being $k = 2$ with probability $\theta = 0.605$. The reported observations for this example are as follows:

Number defective x	Frequency r
0	296
1	366
2	273
3	34
4	25
5	5
6	1

In summary, we have $N = 1000$, $n = 40$, $k = 2$, $\bar{x} = 1.145$, $r_2 = 273$, and $r_3 = 34$. On substituting these values into (14.2.17) and solving, we find $\hat{p} = 0.299$ and $\hat{\theta} = 0.61$, which are to be compared with population values of 0.3 and 0.605, respectively.

To calculate estimate variances and covariance, we first employ (14.2.20) to calculate $\phi_{11} = 13.39$, $\phi_{12} = -1.94$, and $\phi_{22} = 0.59$, where p and θ have been replaced by their estimates 0.299 and 0.61. From (14.2.21) and (14.2.22), we calculate $V(\hat{p}) = 0.00000075$, $V(\hat{\theta}) = 0.0017$, $\text{Cov}(\hat{p}, \hat{\theta}) = 0.0000025$, and $\rho_{\hat{p},\hat{\theta}} = 0.07$.

14.3 INFLATED ZERO DISTRIBUTIONS

An inflated zero distribution is the result of mixing a discrete distribution in which the random variable may assume the values, 0, 1, 2, . . . , with a degenerate distribution in which the random variable may only be zero. For an example, consider the counts of specified organisms present in blood samples from patients who may have been infected with a certain disease. The population sampled is composed of "infected" and "noninfected" individuals. The organism count is zero from "noninfected" individuals, whereas it may be 0, 1, 2, . . . from infected individuals. When sample observations are made at random from this mixed population without regard for or knowledge of whether a selected individual is infected or not, the observed distribution of organisms exhibits an inflated zero class. Although any discrete distribution may form the basis for an inflated zero distribution, we will limit our consideration in this section to the Poisson and the negative binomial distributions. Many of the results presented in this section are due to or related to previous results of Cohen (1960e, 1966), David and Johnson (1952), Hartley (1958), Muench (1938), Sampford (1955), and Singh (1962, 1963).

Let ω designate the mixing parameter; that is, the proportion of "infected" individuals in the population, and let $f(x; \lambda_1, \lambda_2)$ be the probability function of the number of organisms in samples from "infected" individuals. Now let $g(x; \omega, \lambda_1, \lambda_2)$ designate the probability function of the resulting inflated zero distribution, which we write as

$$g(x; \omega, \lambda_1, \lambda_2) = \begin{cases} (1 - \omega) + \omega f(0), & x = 0, \\ \omega f(x; \lambda_1, \lambda_2), & x = 1, 2, 3, \ldots, \end{cases} \quad (14.3.1)$$

where $0 \leq \omega \leq 1$.

14.3.1 Poisson Inflated Zero Distribution

In this case, the probability function of (14.3.1) becomes

$$g(x; \omega, \lambda) = \begin{cases} (1 - \omega) + \omega e^{-\lambda}, & x = 0, \\ \omega e^{-\lambda} \lambda^x / x!, & x = 1, 2, \ldots. \end{cases} \quad (14.3.2)$$

The likelihood function of a sample consisting of N observations of X from a distribution with probability function (14.3.2) is written as

$$L = [1 - \omega(1 - e^{-\lambda})]^{n_0} \omega^n \prod_{i=1}^{n} \frac{e^{-\lambda} \lambda^{x_i}}{x_i!}, \quad (14.3.3)$$

where n_0 is the number of zero observations and n is the number of nonzero observations such that $N = n_0 + n$. The loglikelihood function now becomes

Truncated Attributes and Other Topics

$$\ln L = n_0 \ln[1 - \omega(1 - e^{-\lambda})] + n \ln \omega$$
$$- n\lambda + n\bar{x}^* \ln \lambda - \sum_{i=1}^{n} \ln(x_i!), \quad (14.3.4)$$

where $\bar{x}^* = \sum_{i=1}^{n} x_i/n$, $\bar{x} = \sum_{i=1}^{N} x_i/N$, and thus $\bar{x}^* = (N/n)\bar{x}$.

To derive the maximum likelihood estimating equations, we take derivatives of $\ln L$, equate to zero, and thereby obtain

$$\frac{\partial \ln L}{\partial \omega} = \frac{-n_0(1 - e^{-\lambda})}{1 - \omega(1 - e^{-\lambda})} + \frac{n}{\omega} = 0,$$
$$\frac{\partial \ln L}{\partial \lambda} = \frac{-n_0 \omega e^{-\lambda}}{1 - \omega(1 - e^{-\lambda})} - n + \frac{1}{\lambda} n\bar{x}^* = 0. \quad (14.3.5)$$

The two equations of (14.3.5) are subsequently simplified to

$$\frac{\hat{\lambda}}{1 - e^{-\hat{\lambda}}} = \frac{N}{n}\bar{x},$$
$$\hat{\omega} = \frac{n}{N(1 - e^{-\hat{\lambda}})}. \quad (14.3.6)$$

The first of these equations can be solved for $\hat{\lambda}$, and $\hat{\omega}$ then follows from the second equation.

An Alternate Approach

As an alternate approach in the derivation of estimating equations, we let

$$\theta = \omega[1 - f(0)] \quad \text{and thus} \quad \omega = \frac{\theta}{1 - f(0)}. \quad (14.3.7)$$

With this change of parameter, the zero inflated probability function of (14.3.1) becomes

$$g(x; \theta, \lambda_1, \lambda_2) = \begin{cases} 1 - \theta, & x = 0, \\ \theta f(x; \lambda_1, \lambda_2)/[1 - f(0)], & x = 1, 2, \ldots \end{cases} \quad (14.3.8)$$

The likelihood function of a random sample of size N from this distribution can be expressed as

$$L = (1 - \theta)^{n_0}[1 - f(0)]^{-n} \sum_{i=1}^{n} f(x_i; \lambda_1, \lambda_2). \quad (14.3.9)$$

where n_x is the sample frequency of x and n is the number of nonzero observations ($n = N - n_0$).

We take logarithms of L, differentiate with respect to the parameters, and equate to zero to obtain

$$\frac{\partial \ln L}{\partial \theta} = -\frac{n_0}{1-\theta} + \frac{n}{\theta} = 0, \qquad (14.3.10)$$

$$\frac{\partial \ln L}{\partial \lambda_j} = \frac{n}{1-f(0)}\frac{\partial f(0)}{\partial \lambda_j} + \sum_{i=1}^{n}\frac{1}{f(x_i)}\frac{\partial f(x_i)}{\partial \lambda_j} = 0, \qquad j = 1, 2.$$

When they exist, maximum likelihood estimates $\hat{\theta}$, $\hat{\lambda}_1$, $\hat{\lambda}_2$, can be found by simultaneously solving the preceding system of equations. However, regardless of the form assumed by $f(x; \lambda_1, \lambda_2)$, the estimate of θ follows from the first equation of (14.3.10) as

$$\hat{\theta} = \frac{n}{N}. \qquad (14.3.11)$$

Estimates of the remaining parameters can be obtained as the simultaneous solution of $\partial \ln L/\partial \lambda_j = 0$, $j = 1, 2$, which derives information only from the nonzero sample observations and is the same system of estimating equations obtained in Chapter 13 for a truncated distribution with missing zero class. The estimate of ω follows from (14.3.7) as

$$\hat{\omega} = \frac{\hat{\theta}}{1-f(0)}. \qquad (14.3.12)$$

For the Poisson distribution, the alternate probability function (14.3.8) becomes

$$f(x; \theta, \lambda) = \begin{cases} (1-\theta), & x = 0, \\ \theta e^{-\lambda}\lambda^x/(1-e^{-\lambda})x!, & x = 1, 2, \ldots, \end{cases} \qquad (14.3.13)$$

Maximum likelihood estimating equations based on this probability function are

$$\frac{\partial \ln L}{\partial \theta} = \frac{n}{\theta} - \frac{n_0}{1-\theta} = 0,$$

$$\frac{\partial \ln L}{\partial \lambda} = -\frac{n}{1-e^{-\lambda}} + \frac{n\bar{x}*}{\lambda} = 0, \qquad (14.3.14)$$

and the resulting estimators are

$$\hat{\theta} = \frac{n}{N}, \qquad (14.3.15)$$

$$\bar{x}* = \frac{\lambda}{1-e^{-\lambda}},$$

in agreement with those given in (14.3.6).

Truncated Attributes and Other Topics

Estimate Variances

Asymptotic estimate variances of parameters of the inflated Poisson distribution are obtained by inverting the information matrix with elements that are negatives of expected values of the second partials of ln L. Further differentiation of (14.3.14) gives

$$\frac{\partial^2 \ln L}{\partial \theta^2} = -\frac{n}{\theta_2} - \frac{n_0}{(1-\theta)^2},$$

$$\frac{\partial^2 \ln L}{\partial \theta \, \partial \lambda} = 0 = \frac{\partial^2 \ln \lambda}{\partial \lambda \, \partial \theta}, \quad (14.3.16)$$

$$\frac{\partial^2 \ln L}{\partial \lambda^2} = -n\left[\frac{\bar{x}*}{\lambda^2} - \frac{e^{-\lambda}}{(1-e^{-\lambda})^2}\right].$$

It then follows that

$$V(\hat{\theta}) = \frac{\theta(1-\theta)}{N}, \quad \text{Cov}(\hat{\theta}, \hat{\lambda}) = 0,$$

$$\text{and} \quad V(\hat{\lambda}) = \left[\frac{\lambda}{E(n)}\right]\psi(\lambda), \quad (14.3.17)$$

where $E(n) = N\theta$, and $\psi(\lambda) = (1 - e^{-\lambda})^2/[1 - (\lambda + 1)e^{-\lambda}]$.

Cohen (1960d) demonstrated that $\lim_{\lambda \to 0} \psi(\lambda) = 2$, $\lim_{\lambda \to \infty} \psi(\lambda) = 1$, and therefore that regardless of the value of λ, the asymptotic variance satisfies the inequality $\lambda/N\theta \leq V(\hat{\lambda}) \leq 2\lambda/N\theta$. Tables and a graph of the function $\psi(\lambda)$ were given as aids to facilitate the calculation of $V(\hat{\lambda})$. In addition, a table of $\lambda/(1 - e^{-\lambda})$ was included to aid in the calculation of estimates of λ.

14.3.2 The Negative Binomial Inflated Zero Distribution

Let the probability function of the negative binomial distribution with parameters p and k be written as

$$f(x; p, k) = \frac{\Gamma(x+k)}{x!\Gamma(k)} p^k (1-p)^x, \quad x = 0, 1, 2, \ldots \quad (14.3.18)$$

The corresponding inflated zero distribution follows from (14.3.8) as

$$g(x; \theta, p, k) = \begin{cases} 1 - \theta, & x = 0, \\ \theta \frac{\Gamma(x+k) p^k (1-p)^x}{x! \, \Gamma(k)(1-p^k)}, & x = 1, 2, \ldots \end{cases} \quad (14.3.19)$$

Maximum likelihood estimating equations follow from (14.3.10) in agreement with the estimators of (13.3.8) from Chapter 13 as

$$\theta = \frac{n}{N},$$

$$\frac{k(1-p)}{p(1-p^k)} = \bar{x}, \qquad (14.3.20)$$

$$-\frac{p \ln p}{1-p} = \frac{k}{n\bar{x}} \sum_{x=1}^{R} (k+x-1)^{-1} \sum_{i=x}^{R} n_i,$$

where R is the largest sample observation, and, consistent with previously used notation, $\bar{x}$ is the mean of the nonzero observations. When they exist, estimates $\hat{p}$ and $\hat{k}$ can be calculated by simultaneously solving the last two equations of (14.3.20), as described in Chapter 13. As first approximations for use in an iterative solution of these equations, the estimators of Brass (1958), which were derived in Chapter 13, are recommended. These estimators are

$$p^* = \left(\frac{\bar{x}}{s^2}\right)\left(1 - \frac{n_1}{n}\right) \quad \text{and} \quad k^* = \frac{(p^*\bar{x} - n_1/n)}{1 - p^*}. \qquad (14.3.21)$$

To estimate ω, we employ (14.3.7) and it follows that

$$\hat{\omega} = \frac{n}{N(1-p^k)}. \qquad (14.3.22)$$

Estimate Variances

From (14.3.17), we have $V(\hat{\theta}) = \theta(1 - \theta)/N$, and $\text{Cov}(\hat{\theta}, \hat{p}) = 0$, $\text{Cov}(\hat{\theta}, \hat{k}) = 0$. Variances and covariances of $\hat{p}$ and $\hat{k}$, which are also applicable here, are given in Chapter 13.

14.3.3 Illustrative Examples

Example 14.4 To illustrate the calculation of estimates for parameters of the inflated zero Poisson distribution, we reexamine a sample given by Beall and Rescia (1953) of the number of European corn-borers, *Pyrausta nubilalis Hubn.* observed in small unit areas of a field in 1937. Corn stalks were selected at random and a count was made of the number of borers found on each stalk. The population sampled consisted of a mixture of infested and noninfested stalks. Following is a tabulation of the observed data along with expected frequencies calculated for the Poisson distribution with parameters estimated from (14.3.15).

Truncated Attributes and Other Topics 241

No. Insects, x	0	1	2	3	4	5+	Total
Observed frequency	33	12	6	3	1	1	56
Expected frequency	33	10.8	7.4	3.3	1.1	0.4	56

In summary, for this sample, we have $N = 56$, $n_0 = 33$, $n = 23$, and $\bar{x}^* = 1.8261$ (if we assume that the observation reported as 5+ was actually 5). From the first equation of (14.3.15), we calculate $\hat{\theta} = 23/56 = 0.4107$, and from the second equation, we have $1.8261 = \lambda/(1 - e^{-\lambda})$, and thus $\hat{\lambda} = 1.355$. The expected frequencies were calculated by substituting these estimates into the probability function (14.3.13). Asymptotic variances of $\hat{\theta}$ and $\hat{\lambda}$ were calculated from (14.3.17) with estimates 0.4107 and 1.355 substituted for θ and λ, respectively. We calculate $\psi(1.355) = 1.39$, $E(n) = N\hat{\theta} = n$, $V(\hat{\theta}) = 0.0043$, and $V(\hat{\lambda}) = 0.082$.

Beal and Rescia considered the possibility that these data might best be fitted by one of Neyman's contageous distributions and they calculated expected frequencies for those models, but the fit was not good. At least, a better fit is provided by the inflated zero Poisson model as presented here. A more complete discussion of comparisons between expected and observed data for this example has been given by Cohen (1960e).

Example 14.5 As an illustration of the inflated zero negative binomial distribution, we have chosen a sample consisting of counts of the number of eggs of an intestinal trematode, *Schistosoma mansoni*, on single slides collected from 926 inhabitants of an Egyptian village by Dr. J. Allen Scott, and studied by Muench (1938). The population sampled consisted of a mixture of an infected and a noninfected group. Maximum likelihood estimates of the parameters were calculated as described in this section, and expected frequencies were calculated for comparison with the observed frequencies. The observed data along with expected frequencies are tabulated below.

No eggs per slide x	Observed frequencies n_0	Expected frequencies n_E
0	603	603
1	112	121.3
2	93	78.1
3	53	48.7
4	19	29.8

No eggs per slide x	Observed frequencies n_O	Expected frequencies n_E
5	21	18.1
6	7	10.9
7	6	6.5
8	5	3.9
9	2	2.3
10	1	1.4
11	2	0.8
12	0	0.5
13	0	0.3
14	2	0.2
15+	0	0.2
Totals	926	926.0

In summary, for this example, we have $N = 926$, $n_0 = 603$, $n = 323$, $n_1 = 112$, $\bar{x} = 0.90713$, $\bar{x}^* = 2.60062$, $s^{*2} = 4.24683$, and $s^2 = 3.01623$. From (14.3.11), we calculate

$$\hat{\theta} = \frac{323}{926} = 0.34881.$$

As first approximations to $\hat{p}$ and $\hat{k}$, we employ the Brass estimators of (14.3.21) to calculate

$$p^* = 0.4000 \quad \text{and} \quad k^* = 1.156.$$

Maximum likelihood estimates were subsequently calculated as

$$\hat{p} = 0.4181 \quad \text{and} \quad \hat{k} = 1.213.$$

To estimate the proportion ω of infected inhabitants in the village, we employ (14.3.7) and calculate

$$\hat{\omega} = \frac{\hat{\theta}}{1 - \hat{p}^{\hat{k}}} = \frac{0.34881}{0.6527} = 0.5344.$$

To the nearest whole number, it is accordingly estimated that out of the sample of 926 inhabitants examined, 495 were infected and 431 were free of infection. Since 323 of those infected were identified by the presence of one or more trematode eggs on their slides, it is then estimated that $495 - 323 = 172$ infected members of the sample remain unidentified.

Truncated Attributes and Other Topics

For the asymptotic variances and covariances, we invert the information matrix with estimates substituted for parameters and with partials as given in Chapter 13 substituted for expected values, to calculate $V(\hat{p}) = 0.0030$, $V(\hat{k}) = 0.1382$, and $\text{Cov}(\hat{p}, \hat{k}) = -0.0193$. From (14.3.17), the variance of $\hat{\theta}$, which is applicable in all inflated zero distribution, is calculated as $V(\hat{\theta}) = \hat{\theta}(1 - \hat{\theta})/N = 0.000245$.

Appendix: Tables of Cumulative Standard Distribution Functions

Table A.1 Cumulative Distribution Function of the Standardized Weibull Distribution (α_3; 0, 1)

z	0.0	0.1	0.2	0.3	0.4	0.5	0.6	z
-3.0	.000061							-3.0
-2.9	.000210	.000000						-2.9
-2.8	.000528	.000033						-2.8
-2.7	.001099	.000202						-2.7
-2.6	.002020	.000636	.000021					-2.6
-2.5	.003397	.001480	.000252					-2.5
-2.4	.005349	.002886	.000949	.000028				-2.4
-2.3	.007999	.005016	.002348	.000476				-2.3
-2.2	.011480	.008031	.004675	.001807	.000127			-2.2
-2.1	.015930	.012099	.008146	.004361	.001296			-2.1
-2.0	.021487	.017384	.012966	.008421	.004127	.000848		-2.0
-1.9	.028296	.024049	.019326	.014228	.008981	.004055	.000501	-1.9
-1.8	.036496	.032250	.027404	.021991	.016115	.010027	.004283	-1.8
-1.7	.046226	.042133	.037359	.031882	.025722	.018975	.011890	-1.7
-1.6	.057615	.053835	.049330	.044044	.037938	.031011	.023329	-1.6
-1.5	.070786	.067474	.063431	.058580	.052848	.046176	.038529	-1.5
-1.4	.085847	.083151	.079751	.075559	.070486	.064446	.057355	-1.4
-1.3	.102888	.100944	.098348	.095011	.090843	.085746	.079620	-1.3
-1.2	.121983	.120906	.119249	.116930	.113859	.109946	.105091	-1.2
-1.1	.143179	.143060	.142447	.141265	.139435	.136874	.133495	-1.1
-1.0	.166497	.167398	.167898	.167931	.167427	.166317	.164525	-1.0
-.9	.191928	.193879	.195522	.196800	.197654	.198024	.197851	-.9
-.8	.219429	.222425	.225201	.227709	.229898	.231717	.233118	-.8
-.7	.248922	.252922	.256782	.260460	.263911	.267092	.269963	-.7
-.6	.280294	.285220	.290077	.294823	.299420	.303829	.308014	-.6
-.5	.313393	.319136	.324864	.330541	.336130	.341595	.346902	-.5
-.4	.348031	.354449	.360895	.367335	.373732	.380052	.386263	-.4
-.3	.383986	.390911	.397897	.404909	.411910	.418867	.425748	-.3
-.2	.421003	.428248	.435577	.442954	.450344	.457711	.465024	-.2
-.1	.458797	.466164	.473630	.481160	.488719	.496271	.503782	-.1
.0	.497064	.504347	.511743	.519215	.526729	.534249	.541742	.0
.1	.535480	.542479	.549604	.556818	.564086	.571374	.578648	.1
.2	.573712	.580239	.586906	.593678	.600521	.607400	.614281	.2
.3	.611426	.617313	.623358	.629528	.635790	.642109	.648452	.3
.4	.648295	.653399	.658685	.664122	.669677	.675317	.681008	.4
.5	.684008	.688219	.692641	.697246	.702000	.706871	.711825	.5
.6	.718274	.721518	.725008	.728714	.732606	.736651	.740817	.6
.7	.750836	.753078	.755601	.758378	.761379	.764572	.767925	.7
.8	.781473	.782716	.784274	.786123	.788235	.790578	.793120	.8
.9	.810008	.810289	.810918	.811871	.813123	.814643	.816399	.9
1.0	.836308	.835700	.835463	.835579	.836024	.836770	.837785	1.0
1.1	.860291	.858891	.857878	.857237	.856948	.856985	.857319	1.1
1.2	.881920	.879848	.878166	.876865	.875929	.875336	.875062	1.2
1.3	.901208	.898598	.896367	.894513	.893027	.891892	.891088	1.3
1.4	.918210	.915200	.912546	.910253	.908319	.906735	.905483	1.4

Source: From Cohen and Whitten (1988), Table A.3.1, pp. 292–299, by courtesy of Marcel Dekker, Inc.

Table A.1 *Continued*

z \ α_3	0.0	0.1	0.2	0.3	0.4	0.5	0.6	z
1.5	.933020	.929748	.926796	.924178	.921900	.919959	.918344	1.5
1.6	.945762	.942361	.939231	.936398	.933875	.931668	.929772	1.6
1.7	.956589	.953177	.949981	.947034	.944360	.941972	.939873	1.7
1.8	.965670	.962349	.959184	.956215	.953474	.950983	.948752	1.8
1.9	.973187	.970041	.966987	.964074	.961341	.958816	.956518	1.9
2.0	.979325	.976416	.973538	.970745	.968082	.965583	.963273	2.0
2.1	.984267	.981638	.978984	.976360	.973816	.971393	.969120	2.1
2.2	.988190	.985864	.983464	.981046	.978660	.976351	.974153	2.2
2.3	.991259	.989243	.987113	.984923	.982722	.980556	.978464	2.3
2.4	.993623	.991910	.990055	.988103	.986103	.984101	.982137	2.4
2.5	.995416	.993990	.992400	.990688	.988897	.987071	.985251	2.5
2.6	.996755	.995590	.994251	.992771	.991188	.989544	.987878	2.6
2.7	.997739	.996805	.995695	.994435	.993055	.991591	.990082	2.7
2.8	.998450	.997715	.996810	.995752	.994563	.993275	.991923	2.8
2.9	.998955	.998387	.997662	.996785	.995773	.994652	.993451	2.9
3.0	.999307	.998877	.998301	.997587	.996737	.995771	.994715	3.0
3.1	.999549	.999229	.998783	.998205	.997498	.996676	.995755	3.1
3.2	.999711	.999478	.999137	.998677	.998096	.997400	.996605	3.2
3.3	.999819	.999651	.999395	.999033	.998561	.997979	.997298	3.3
3.4	.999888	.999771	.999580	.999300	.998920	.998438	.997859	3.4
3.5	.999932	.999851	.999712	.999498	.999196	.998800	.998311	3.5
3.6	.999960	.999905	.999805	.999643	.999405	.999083	.998674	3.6
3.7	.999977	.999940	.999870	.999749	.999563	.999303	.998964	3.7
3.8	.999987	.999963	.999914	.999825	.999682	.999474	.999194	3.8
3.9	.999993	.999978	.999944	.999879	.999770	.999605	.999376	3.9
4.0	.999996	.999987	.999964	.999917	.999835	.999705	.999519	4.0
4.1	.999998	.999992	.999977	.999944	.999882	.999781	.999631	4.1
4.2	.999999	.999995	.999985	.999962	.999917	.999839	.999718	4.2
4.3	.999999	.999997	.999991	.999975	.999942	.999882	.999786	4.3
4.4	1.000000	.999999	.999994	.999984	.999959	.999914	.999838	4.4
4.5	1.000000	.999999	.999997	.999989	.999972	.999937	.999878	4.5
4.6	1.000000	1.000000	.999998	.999993	.999981	.999955	.999909	4.6
4.7	1.000000	1.000000	.999999	.999996	.999987	.999968	.999932	4.7
4.8	1.000000	1.000000	.999999	.999997	.999991	.999977	.999949	4.8
4.9	1.000000	1.000000	1.000000	.999998	.999994	.999984	.999963	4.9
5.0	1.000000	1.000000	1.000000	.999999	.999996	.999989	.999972	5.0
5.1	1.000000	1.000000	1.000000	.999999	.999997	.999992	.999980	5.1
5.2	1.000000	1.000000	1.000000	1.000000	.999998	.999994	.999985	5.2
5.3	1.000000	1.000000	1.000000	1.000000	.999999	.999996	.999989	5.3
5.4	1.000000	1.000000	1.000000	1.000000	.999999	.999997	.999992	5.4
5.5	1.000000	1.000000	1.000000	1.000000	1.000000	.999998	.999994	5.5
5.6	1.000000	1.000000	1.000000	1.000000	1.000000	.999999	.999996	5.6
5.7	1.000000	1.000000	1.000000	1.000000	1.000000	.999999	.999997	5.7
5.8	1.000000	1.000000	1.000000	1.000000	1.000000	.999999	.999998	5.8
5.9	1.000000	1.000000	1.000000	1.000000	1.000000	1.000000	.999999	5.9

(Continued)

Table A.1 *Continued*

z \ α_3	0.7	0.8	0.9	1.0	1.1	1.2	1.3	z
-3.0								-3.0
-2.9								-2.9
-2.8								-2.8
-2.7								-2.7
-2.6								-2.6
-2.5								-2.5
-2.4								-2.4
-2.3								-2.3
-2.2								-2.2
-2.1								-2.1
-2.0								-2.0
-1.9								-1.9
-1.8	.000284							-1.8
-1.7	.005068	.000226						-1.7
-1.6	.015105	.006899	.000485					-1.6
-1.5	.029927	.020502	.010673	.001843				-1.5
-1.4	.049143	.039769	.029260	.017819	.006227			-1.4
-1.3	.072365	.063878	.054062	.042843	.030214	.016390	.002651	-1.3
-1.2	.099191	.092134	.083800	.074059	.062769	.049787	.034993	-1.2
-1.1	.129205	.123905	.117488	.109834	.100807	.090248	.077968	-1.1
-1.0	.161977	.158593	.154287	.148968	.142533	.134865	.125826	-1.0
-.9	.197073	.195630	.193458	.190492	.186664	.181896	.176105	-.9
-.8	.234053	.234475	.234338	.233598	.232209	.230126	.227300	-.8
-.7	.272484	.274619	.276333	.277595	.278377	.278649	.278385	-.7
-.6	.311943	.315583	.318909	.321895	.324522	.326772	.328629	-.6
-.5	.352021	.356925	.361589	.365994	.370121	.373959	.377497	-.5
-.4	.392334	.398240	.403956	.409465	.414750	.419799	.424604	-.4
-.3	.432521	.439161	.445645	.451952	.458068	.463980	.469680	-.3
-.2	.472250	.479364	.486342	.493162	.499810	.506271	.512537	-.2
-.1	.511223	.518565	.525784	.532858	.539771	.546509	.553061	-.1
.0	.549175	.556521	.563756	.570856	.577804	.584587	.591193	.0
.1	.585876	.593030	.600084	.607017	.613809	.620445	.626916	.1
.2	.621134	.627928	.634639	.641244	.647724	.654063	.660249	.2
.3	.654789	.661089	.667326	.673478	.679524	.685450	.691240	.3
.4	.686719	.692420	.698085	.703690	.709214	.714641	.719956	.4
.5	.716832	.721862	.726886	.731880	.736822	.741695	.746482	.5
.6	.745071	.749383	.753725	.758070	.762396	.766682	.770913	.6
.7	.771405	.774980	.778622	.782303	.785999	.789689	.793353	.7
.8	.795828	.798670	.801616	.804637	.807709	.810807	.813912	.8
.9	.818359	.820491	.822762	.825145	.827611	.830137	.832700	.9
1.0	.839038	.840497	.842130	.843907	.845799	.847782	.849831	1.0
1.1	.857921	.858757	.859798	.861011	.862370	.863846	.865417	1.1
1.2	.875076	.875350	.875852	.876552	.877422	.878435	.879566	1.2
1.3	.890587	.890363	.890386	.890627	.891058	.891652	.892385	1.3
1.4	.904543	.903888	.903494	.903333	.903376	.903599	.903977	1.4

Table A.1 *Continued*

z \ α_3	0.7	0.8	0.9	1.0	1.1	1.2	1.3	z
1.5	.917039	.916024	.915275	.914767	.914475	.914373	.914439	1.5
1.6	.928178	.926869	.925825	.925026	.924449	.924069	.923865	1.6
1.7	.938059	.936520	.935241	.934204	.933390	.932776	.932342	1.7
1.8	.946785	.945076	.943617	.942392	.941385	.940578	.939953	1.8
1.9	.954456	.952632	.951041	.949674	.948518	.947557	.946775	1.9
2.0	.961170	.959279	.957603	.956135	.954867	.953787	.952881	2.0
2.1	.967019	.965105	.963382	.961850	.960505	.959338	.958337	2.1
2.2	.972094	.970191	.968457	.966894	.965501	.964275	.963206	2.2
2.3	.976476	.974617	.972899	.971332	.969919	.968658	.967543	2.3
2.4	.980246	.978452	.976776	.975229	.973817	.972542	.971402	2.4
2.5	.983473	.981765	.980149	.978641	.977249	.975979	.974831	2.5
2.6	.986225	.984616	.983076	.981621	.980265	.979014	.977873	2.6
2.7	.988562	.987062	.985607	.984218	.982910	.981691	.980568	2.7
2.8	.990538	.989152	.987791	.986476	.985225	.984048	.982953	2.8
2.9	.992201	.990931	.989668	.988434	.987247	.986120	.985062	2.9
3.0	.993596	.992442	.991279	.990129	.989011	.987938	.986923	3.0
3.1	.994761	.993720	.992657	.991592	.990545	.989532	.988564	3.1
3.2	.995730	.994798	.993832	.992852	.991879	.990927	.990009	3.2
3.3	.996533	.995704	.994831	.993935	.993035	.992145	.991280	3.3
3.4	.997195	.996462	.995679	.994864	.994036	.993209	.992397	3.4
3.5	.997739	.997095	.996396	.995659	.994900	.994135	.993377	3.5
3.6	.998184	.997622	.997002	.996338	.995646	.994941	.994236	3.6
3.7	.998547	.998059	.997511	.996916	.996289	.995642	.994988	3.7
3.8	.998842	.998420	.997939	.997408	.996841	.996250	.995647	3.8
3.9	.999080	.998718	.998297	.997826	.997315	.996777	.996222	3.9
4.0	.999272	.998963	.998596	.998179	.997721	.997233	.996724	4.0
4.1	.999426	.999163	.998846	.998478	.998069	.997627	.997162	4.1
4.2	.999549	.999327	.999053	.998730	.998366	.997967	.997543	4.2
4.3	.999647	.999460	.999225	.998943	.998619	.998261	.997876	4.3
4.4	.999725	.999568	.999367	.999121	.998835	.998514	.998164	4.4
4.5	.999786	.999656	.999484	.999270	.999018	.998731	.998415	4.5
4.6	.999834	.999726	.999580	.999396	.999174	.998918	.998633	4.6
4.7	.999872	.999783	.999659	.999500	.999305	.999078	.998822	4.7
4.8	.999902	.999828	.999724	.999587	.999417	.999215	.998985	4.8
4.9	.999925	.999864	.999777	.999660	.999511	.999333	.999127	4.9
5.0	.999942	.999893	.999820	.999720	.999591	.999433	.999249	5.0
5.1	.999956	.999916	.999856	.999770	.999658	.999519	.999355	5.1
5.2	.999967	.999935	.999884	.999812	.999715	.999593	.999446	5.2
5.3	.999975	.999949	.999907	.999846	.999762	.999655	.999525	5.3
5.4	.999981	.999960	.999926	.999874	.999802	.999708	.999593	5.4
5.5	.999986	.999969	.999941	.999897	.999835	.999753	.999651	5.5
5.6	.999989	.999976	.999953	.999916	.999863	.999792	.999702	5.6
5.7	.999992	.999982	.999963	.999932	.999886	.999824	.999745	5.7
5.8	.999994	.999986	.999970	.999945	.999906	.999852	.999782	5.8
5.9	.999996	.999989	.999977	.999955	.999922	.999876	.999814	5.9

(Continued)

Table A.1 *Continued*

z \ α_3	1.4	1.5	1.6	1.7	1.8	1.9	2.0	z
-1.5								-1.5
-1.4								-1.4
-1.3								-1.3
-1.2	.018423	.001341						-1.2
-1.1	.063738	.047278	.028267	.006598				-1.1
-1.0	.115251	.102933	.088600	.071890	.052275	.028907		-1.0
-.9	.169192	.161043	.151518	.140439	.127580	.112632	.095163	-.9
-.8	.223679	.219206	.213815	.207429	.199956	.191283	.181269	-.8
-.7	.277561	.276148	.274119	.271445	.268090	.264018	.259182	-.7
-.6	.330080	.331114	.331721	.331891	.331614	.330881	.329680	-.6
-.5	.380726	.383643	.386244	.388527	.390492	.392139	.393469	-.5
-.4	.429160	.433461	.437508	.441302	.444844	.448138	.451188	-.4
-.3	.475160	.480419	.485456	.490270	.494866	.499245	.503415	-.3
-.2	.518601	.524459	.530110	.535553	.540792	.545830	.550671	-.2
-.1	.559421	.565584	.571548	.577312	.582878	.588250	.593430	-.1
.0	.597614	.603845	.609883	.615729	.621381	.626844	.632121	.0
.1	.633211	.639325	.645255	.650999	.656558	.661934	.667129	.1
.2	.666273	.672129	.677813	.683322	.688656	.693816	.698806	.2
.3	.696886	.702380	.707718	.712896	.717913	.722770	.727468	.3
.4	.725149	.730210	.735133	.739916	.744555	.749051	.753403	.4
.5	.751172	.755755	.760222	.764570	.768795	.772895	.776870	.5
.6	.775074	.779155	.783145	.787039	.790832	.794521	.798103	.6
.7	.796978	.800549	.804057	.807493	.810851	.814127	.817316	.7
.8	.817006	.820076	.823109	.826094	.829025	.831896	.834701	.8
.9	.835283	.837868	.840442	.842994	.845514	.847995	.850431	.9
1.0	.851928	.854054	.856193	.858334	.860465	.862578	.864665	1.0
1.1	.867060	.868756	.870490	.872247	.874015	.875783	.877544	1.1
1.2	.880792	.882093	.883453	.884855	.886287	.887738	.889197	1.2
1.3	.893233	.894175	.895194	.896273	.897398	.898558	.899741	1.3
1.4	.904486	.905106	.905818	.906605	.907453	.908349	.909282	1.4
1.5	.914649	.914983	.915421	.915948	.916548	.917207	.917915	1.5
1.6	.923814	.923897	.924095	.924391	.924771	.925220	.925726	1.6
1.7	.932068	.931934	.931923	.932017	.932203	.932466	.932794	1.7
1.8	.939490	.939172	.938980	.938900	.938917	.939018	.939190	1.8
1.9	.946156	.945683	.945339	.945110	.944981	.944941	.944977	1.9
2.0	.952135	.951534	.951063	.950708	.950456	.950295	.950213	2.0
2.1	.957492	.956788	.956213	.955753	.955397	.955133	.954951	2.1
2.2	.962285	.961501	.960842	.960297	.959855	.959506	.959238	2.2
2.3	.966568	.965724	.965000	.964388	.963876	.963456	.963117	2.3
2.4	.970392	.969505	.968734	.968069	.967502	.967024	.966627	2.4
2.5	.973802	.972888	.972083	.971380	.970770	.970247	.969803	2.5
2.6	.976839	.975912	.975086	.974356	.973715	.973158	.972676	2.6
2.7	.979542	.978612	.977777	.977030	.976369	.975786	.975276	2.7
2.8	.981944	.981022	.980186	.979433	.978759	.978159	.977629	2.8
2.9	.984078	.983172	.982343	.981590	.980911	.980302	.979758	2.9

Table A.1 *Continued*

z\α_3	1.4	1.5	1.6	1.7	1.8	1.9	2.0	z
3.0	.985971	.985087	.984272	.983526	.982848	.982235	.981684	3.0
3.1	.987648	.986792	.985996	.985263	.984592	.983981	.983427	3.1
3.2	.989134	.988309	.987537	.986821	.986161	.985556	.985004	3.2
3.3	.990448	.989658	.988913	.988218	.987573	.986977	.986431	3.3
3.4	.991610	.990856	.990142	.989470	.988842	.988260	.987723	3.4
3.5	.992636	.991921	.991238	.990591	.989984	.989417	.988891	3.5
3.6	.993541	.992865	.992215	.991596	.991010	.990461	.989948	3.6
3.7	.994339	.993702	.993086	.992495	.991933	.991402	.990905	3.7
3.8	.995042	.994444	.993862	.993300	.992762	.992252	.991770	3.8
3.9	.995661	.995102	.994553	.994020	.993507	.993017	.992553	3.9
4.0	.996205	.995683	.995168	.994664	.994176	.993708	.993262	4.0
4.1	.996683	.996198	.995715	.995240	.994777	.994331	.993903	4.1
4.2	.997103	.996653	.996202	.995755	.995317	.994892	.994483	4.2
4.3	.997471	.997055	.996634	.996215	.995801	.995398	.995008	4.3
4.4	.997794	.997410	.997019	.996626	.996236	.995855	.995483	4.4
4.5	.998077	.997723	.997360	.996993	.996627	.996266	.995913	4.5
4.6	.998325	.998000	.997663	.997321	.996977	.996637	.996302	4.6
4.7	.998542	.998243	.997932	.997613	.997292	.996971	.996654	4.7
4.8	.998731	.998458	.998171	.997875	.997574	.997272	.996972	4.8
4.9	.998897	.998647	.998383	.998108	.997826	.997543	.997261	4.9
5.0	.999041	.998813	.998570	.998315	.998053	.997788	.997521	5.0
5.1	.999167	.998960	.998736	.998501	.998257	.998008	.997757	5.1
5.2	.999277	.999089	.998884	.998666	.998439	.998206	.997971	5.2
5.3	.999373	.999202	.999014	.998813	.998602	.998385	.998164	5.3
5.4	.999456	.999301	.999130	.998944	.998749	.998546	.998338	5.4
5.5	.999529	.999389	.999232	.999061	.998880	.998691	.998497	5.5
5.6	.999592	.999465	.999322	.999166	.998998	.998822	.998640	5.6
5.7	.999647	.999532	.999402	.999258	.999103	.998939	.998769	5.7
5.8	.999695	.999591	.999473	.999341	.999198	.999045	.998886	5.8
5.9	.999736	.999643	.999535	.999414	.999282	.999141	.998992	5.9
6.0	.999772	.999688	.999590	.999480	.999358	.999227	.999088	6.0
6.1	.999803	.999728	.999639	.999538	.999426	.999304	.999175	6.1
6.2	.999830	.999762	.999682	.999590	.999486	.999374	.999253	6.2
6.3	.999853	.999793	.999720	.999636	.999541	.999436	.999324	6.3
6.4	.999874	.999819	.999754	.999677	.999589	.999493	.999389	6.4
6.5	.999891	.999843	.999783	.999713	.999633	.999544	.999447	6.5
6.6	.999906	.999863	.999809	.999745	.999672	.999590	.999500	6.6
6.7	.999919	.999881	.999832	.999774	.999707	.999631	.999547	6.7
6.8	.999930	.999896	.999852	.999800	.999738	.999668	.999590	6.8
6.9	.999940	.999909	.999870	.999822	.999766	.999701	.999629	6.9
7.0	.999949	.999921	.999886	.999842	.999791	.999731	.999665	7.0
7.1	.999956	.999932	.999900	.999860	.999813	.999758	.999696	7.1
7.2	.999962	.999940	.999912	.999876	.999833	.999782	.999725	7.2
7.3	.999967	.999948	.999923	.999890	.999851	.999804	.999751	7.3
7.4	.999972	.999955	.999932	.999903	.999867	.999824	.999775	7.4

(Continued)

Table A.1 *Continued*

z\α_3	2.1	2.2	2.3	2.5	3.0	4.0	5.0	z
-1.5								-1.5
-1.4								-1.4
-1.3								-1.3
-1.2								-1.2
-1.1								-1.1
-1.0								-1.0
-.9	.074506	.049502	.017379					-.9
-.8	.169733	.156436	.141048	.101845				-.8
-.7	.253529	.246996	.239502	.221221	.147016			-.7
-.6	.328000	.325826	.323141	.316159	.287790	.139295		-.6
-.5	.394484	.395183	.395569	.395401	.389439	.350496	.247418	-.5
-.4	.454000	.456578	.458928	.462967	.469462	.468717	.450570	-.4
-.3	.507379	.511144	.514716	.521309	.534883	.552022	.559306	-.3
-.2	.555320	.559783	.564067	.572121	.589569	.615450	.633006	-.2
-.1	.598425	.603238	.607877	.616654	.635975	.665812	.687555	-.1
.0	.637215	.642132	.646877	.655876	.675789	.706898	.729950	.0
.1	.672148	.676995	.681675	.690558	.710241	.741067	.763966	.1
.2	.703628	.708285	.712784	.721324	.740255	.769896	.791883	.2
.3	.732011	.736400	.740641	.748694	.766552	.794501	.815192	.3
.4	.757614	.761686	.765622	.773101	.789702	.815695	.834918	.4
.5	.780720	.784446	.788052	.794911	.810164	.834092	.851794	.5
.6	.801580	.804950	.808215	.814436	.828315	.850167	.866366	.6
.7	.820419	.823432	.826358	.831946	.844467	.864292	.879044	.7
.8	.837438	.840104	.842699	.847672	.858879	.876765	.890149	.8
.9	.852818	.855151	.857430	.861815	.871773	.887827	.899932	.9
1.0	.866720	.868739	.870719	.874550	.883334	.897676	.908595	1.0
1.1	.879290	.881015	.882716	.886030	.893721	.906477	.916301	1.1
1.2	.890657	.892111	.893555	.896391	.903072	.914364	.923183	1.2
1.3	.900939	.902145	.903352	.905751	.911504	.921454	.929353	1.3
1.4	.910242	.911222	.912214	.914213	.919121	.927844	.934902	1.4
1.5	.918661	.919437	.920234	.921872	.926011	.933617	.939908	1.5
1.6	.926281	.926873	.927496	.928808	.932252	.938844	.944437	1.6
1.7	.933178	.933608	.934075	.935095	.937914	.943587	.948546	1.7
1.8	.939423	.939708	.940037	.940798	.943056	.947899	.952281	1.8
1.9	.945079	.945237	.945444	.945974	.947732	.951826	.955685	1.9
2.0	.950200	.950248	.950348	.950675	.951988	.955408	.958793	2.0
2.1	.954840	.954792	.954798	.954947	.955866	.958682	.961637	2.1
2.2	.959043	.958912	.958838	.958833	.959403	.961678	.964244	2.2
2.3	.962851	.962650	.962507	.962368	.962633	.964423	.966637	2.3
2.4	.966302	.966042	.965840	.965586	.965584	.966942	.968837	2.4
2.5	.969429	.969121	.968870	.968517	.968283	.969257	.970864	2.5
2.6	.972264	.971915	.971623	.971188	.970753	.971386	.972734	2.6
2.7	.974834	.974453	.974127	.973623	.973016	.973347	.974461	2.7
2.8	.977164	.976757	.976405	.975845	.975091	.975155	.976058	2.8
2.9	.979276	.978851	.978478	.977872	.976994	.976824	.977536	2.9

Table A.1 *Continued*

z\α₃	2.1	2.2	2.3	2.5	3.0	4.0	5.0	z
3.0	.981191	.980753	.980364	.979722	.978741	.978366	.978907	3.0
3.1	.982928	.982481	.982081	.981412	.980346	.979791	.980179	3.1
3.2	.984504	.984052	.983645	.982956	.981821	.981111	.981361	3.2
3.3	.985933	.985480	.985069	.984367	.983178	.982333	.982461	3.3
3.4	.987229	.986778	.986367	.985657	.984426	.983467	.983484	3.4
3.5	.988405	.987959	.987550	.986837	.985576	.984519	.984437	3.5
3.6	.989472	.989032	.988628	.987917	.986636	.985495	.985327	3.6
3.7	.990440	.990009	.989610	.988905	.987612	.986403	.986157	3.7
3.8	.991319	.990898	.990507	.989810	.988513	.987248	.986932	3.8
3.9	.992116	.991707	.991324	.990639	.989344	.988033	.987658	3.9
4.0	.992840	.992442	.992070	.991398	.990111	.988765	.988336	4.0
4.1	.993497	.993112	.992750	.992094	.990820	.989447	.988972	4.1
4.2	.994093	.993722	.993371	.992732	.991475	.990083	.989567	4.2
4.3	.994634	.994277	.993938	.993317	.992081	.990677	.990125	4.3
4.4	.995125	.994783	.994456	.993853	.992641	.991231	.990649	4.4
4.5	.995572	.995243	.994928	.994345	.993159	.991740	.991142	4.5
4.6	.995977	.995662	.995360	.994797	.993638	.992232	.991604	4.6
4.7	.996344	.996044	.995754	.995211	.994082	.992685	.992039	4.7
4.8	.996678	.996392	.996115	.995592	.994494	.993108	.992447	4.8
4.9	.996982	.996709	.996444	.995941	.994875	.993504	.992832	4.9
5.0	.997257	.996998	.996745	.996263	.995229	.993876	.993195	5.0
5.1	.997507	.997261	.997020	.996558	.995557	.994224	.993536	5.1
5.2	.997735	.997501	.997271	.996829	.995861	.994550	.993859	5.2
5.3	.997941	.997720	.997501	.997078	.996143	.994856	.994162	5.3
5.4	.998129	.997919	.997711	.997307	.996405	.995143	.994449	5.4
5.5	.998299	.998101	.997904	.997518	.996649	.995413	.994720	5.5
5.6	.998454	.998267	.998079	.997712	.996875	.995666	.994976	5.6
5.7	.998595	.998418	.998240	.997890	.997085	.995904	.995217	5.7
5.8	.998722	.998556	.998388	.998054	.997280	.996127	.995446	5.8
5.9	.998838	.998681	.998522	.998205	.997462	.996338	.995662	5.9
6.0	.998944	.998796	.998646	.998344	.997631	.996535	.995867	6.0
6.1	.999040	.998901	.998759	.998472	.997788	.996722	.996061	6.1
6.2	.999127	.998996	.998862	.998590	.997934	.996897	.996244	6.2
6.3	.999206	.999083	.998957	.998699	.998070	.997062	.996418	6.3
6.4	.999278	.999163	.999043	.998799	.998197	.997218	.996583	6.4
6.5	.999344	.999235	.999123	.998891	.998315	.997364	.996739	6.5
6.6	.999403	.999301	.999195	.998976	.998425	.997502	.996887	6.6
6.7	.999457	.999362	.999262	.999055	.998527	.997633	.997028	6.7
6.8	.999506	.999417	.999323	.999127	.998623	.997756	.997162	6.8
6.9	.999551	.999467	.999379	.999193	.998712	.997872	.997288	6.9
7.0	.999592	.999513	.999430	.999255	.998795	.997981	.997409	7.0
7.1	.999629	.999555	.999477	.999311	.998873	.998085	.997524	7.1
7.2	.999662	.999593	.999520	.999364	.998945	.998182	.997632	7.2
7.3	.999693	.999628	.999560	.999412	.999012	.998275	.997736	7.3
7.4	.999720	.999660	.999596	.999456	.999075	.998362	.997834	7.4

Table A.2 Cumulative Distribution Function of the Standardized Lognormal Distribution $(\alpha_3; 0, 1)$

z \ α_3	0.0	0.1	0.2	0.3	0.4	0.5	0.6	z
-3.0	.001350	.000831	.000455	.000213	.000080	.000022	.000004	-3.0
-2.9	.001866	.001207	.000705	.000360	.000153	.000050	.000011	-2.9
-2.8	.002555	.001731	.001074	.000595	.000282	.000107	.000029	-2.8
-2.7	.003467	.002451	.001608	.000957	.000500	.000218	.000073	-2.7
-2.6	.004661	.003429	.002366	.001503	.000857	.000422	.000169	-2.6
-2.5	.006210	.004738	.003423	.002308	.001422	.000777	.000360	-2.5
-2.4	.008198	.006469	.004875	.003466	.002288	.001371	.000721	-2.4
-2.3	.010724	.008729	.006833	.005097	.003578	.002323	.001360	-2.3
-2.2	.013903	.011641	.009433	.007345	.005442	.003788	.002431	-2.2
-2.1	.017864	.015346	.012830	.010379	.008066	.005963	.004138	-2.1
-2.0	.022750	.020003	.017198	.014395	.011665	.009086	.006738	-2.0
-1.9	.028717	.025785	.022731	.019607	.016481	.013429	.010534	-1.9
-1.8	.035930	.032877	.029635	.026250	.022779	.019292	.015870	-1.8
-1.7	.044565	.041470	.038125	.034564	.030833	.026991	.023109	-1.7
-1.6	.054799	.051759	.048416	.044791	.040918	.036841	.032618	-1.6
-1.5	.066807	.063935	.060717	.057163	.053294	.049139	.044740	-1.5
-1.4	.080757	.078176	.075221	.071890	.068193	.064145	.059774	-1.4
-1.3	.096800	.094640	.092093	.089148	.085804	.082066	.077947	-1.3
-1.2	.115070	.113459	.111464	.109067	.106259	.103037	.099403	-1.2
-1.1	.135666	.134727	.133420	.131724	.129625	.127113	.124182	-1.1
-1.0	.158655	.158494	.157994	.157133	.155892	.154258	.152220	-1.0
-.9	.184060	.184761	.185162	.185239	.184972	.184346	.183348	-.9
-.8	.211855	.213474	.214836	.215919	.216701	.217165	.217298	-.8
-.7	.241964	.244521	.246868	.248980	.250837	.252422	.253719	-.7
-.6	.274253	.277732	.281045	.284166	.287076	.289755	.292190	-.6
-.5	.308538	.312883	.317099	.321164	.325054	.328752	.332242	-.5
-.4	.344578	.349694	.354714	.359614	.364370	.368964	.373378	-.4
-.3	.382089	.387846	.393532	.399123	.404593	.409923	.415095	-.3
-.2	.420740	.426980	.433166	.439275	.445280	.451162	.456900	-.2
-.1	.460172	.466712	.473212	.479647	.485993	.492225	.498325	-.1
.0	.500000	.506644	.513261	.519824	.526307	.532688	.538945	.0
.1	.539828	.546379	.552914	.559406	.565830	.572162	.578382	.1
.2	.579260	.585525	.591788	.598024	.604205	.610310	.616315	.2
.3	.617911	.623714	.629534	.635346	.641124	.646844	.652484	.3
.4	.655422	.660608	.665839	.671087	.676326	.681533	.686685	.4
.5	.691462	.695910	.700435	.705008	.709606	.714202	.718774	.5
.6	.725747	.729368	.733103	.736927	.740811	.744730	.748660	.6
.7	.758036	.760778	.763678	.766708	.769839	.773046	.776302	.7
.8	.788145	.789993	.792044	.794269	.796638	.799124	.801699	.8
.9	.815940	.816915	.818136	.819574	.821198	.822980	.824890	.9
1.0	.841345	.841497	.841935	.842628	.843548	.844662	.845942	1.0
1.1	.864334	.863740	.863463	.863476	.863748	.864248	.864946	1.1
1.2	.884930	.883685	.882781	.882191	.881886	.881835	.882010	1.2
1.3	.903200	.901412	.899978	.898872	.898069	.897541	.897257	1.3
1.4	.919243	.917029	.915167	.913639	.912422	.911489	.910815	1.4

Source: From Cohen and Whitten (1988), Table A.3.2, pp. 300–307, by courtesy of Marcel Dekker, Inc.

Table A.2 *Continued*

z \ α_3	0.0	0.1	0.2	0.3	0.4	0.5	0.6	z
1.5	.933193	.930667	.928482	.926624	.925076	.923815	.922818	1.5
1.6	.945201	.942475	.940066	.937968	.936169	.934652	.933399	1.6
1.7	.955435	.952611	.950070	.947815	.945841	.944137	.942689	1.7
1.8	.964070	.961237	.958647	.956310	.954229	.952400	.950813	1.8
1.9	.971283	.968518	.965949	.963595	.961468	.959569	.957893	1.9
2.0	.977250	.974612	.972122	.969807	.967684	.965761	.964042	2.0
2.1	.982136	.979671	.977305	.975072	.972996	.971090	.969363	2.1
2.2	.986097	.983837	.981628	.979511	.977515	.975659	.973955	2.2
2.3	.989276	.987239	.985210	.983233	.981342	.979561	.977906	2.3
2.4	.991802	.989995	.988159	.986338	.984570	.982882	.981296	2.4
2.5	.993790	.992211	.990571	.988915	.987281	.985700	.984196	2.5
2.6	.995339	.993978	.992531	.991043	.989549	.988083	.986671	2.6
2.7	.996533	.995376	.994116	.992791	.991439	.990092	.988778	2.7
2.8	.997445	.996474	.995388	.994221	.993008	.991781	.990568	2.8
2.9	.998134	.997330	.996403	.995385	.994306	.993197	.992085	2.9
3.0	.998650	.997991	.997210	.996320	.995377	.994301	.993368	3.0
3.1	.999032	.998499	.997846	.997090	.996256	.995368	.994451	3.1
3.2	.999313	.998886	.998345	.997702	.996976	.996189	.995364	3.2
3.3	.999517	.999179	.998735	.998192	.997564	.996871	.996132	3.3
3.4	.999663	.999399	.999038	.998582	.998042	.997435	.996777	3.4
3.5	.999767	.999563	.999271	.998892	.998431	.997901	.997317	3.5
3.6	.999841	.999684	.999451	.999137	.998745	.998285	.997769	3.6
3.7	.999892	.999773	.999588	.999330	.998999	.998601	.998148	3.7
3.8	.999928	.999838	.999692	.999481	.999203	.998861	.998463	3.8
3.9	.999952	.999885	.999771	.999600	.999367	.999074	.998726	3.9
4.0	.999968	.999919	.999831	.999692	.999498	.999248	.998945	4.0
4.1	.999979	.999943	.999875	.999764	.999603	.999390	.999127	4.1
4.2	.999987	.999961	.999908	.999819	.999686	.999506	.999279	4.2
4.3	.999991	.999973	.999933	.999862	.999753	.999600	.999404	4.3
4.4	.999995	.999981	.999951	.999895	.999805	.999677	.999508	4.4
4.5	.999997	.999987	.999965	.999921	.999847	.999739	.999594	4.5
4.6	.999998	.999991	.999974	.999940	.999880	.999790	.999666	4.6
4.7	.999999	.999994	.999982	.999955	.999906	.999831	.999725	4.7
4.8	.999999	.999996	.999987	.999966	.999927	.999864	.999773	4.8
4.9	1.000000	.999997	.999991	.999974	.999943	.999891	.999813	4.9
5.0	1.000000	.999998	.999993	.999981	.999956	.999912	.999847	5.0
5.1	1.000000	.999999	.999995	.999986	.999965	.999930	.999874	5.1
5.2	1.000000	.999999	.999997	.999989	.999973	.999944	.999896	5.2
5.3	1.000000	1.000000	.999998	.999992	.999979	.999955	.999915	5.3
5.4	1.000000	1.000000	.999998	.999994	.999984	.999964	.999930	5.4
5.5	1.000000	1.000000	.999999	.999996	.999988	.999971	.999943	5.5
5.6	1.000000	1.000000	.999999	.999997	.999990	.999977	.999953	5.6
5.7	1.000000	1.000000	.999999	.999998	.999993	.999982	.999961	5.7
5.8	1.000000	1.000000	1.000000	.999998	.999994	.999985	.999968	5.8
5.9	1.000000	1.000000	1.000000	.999999	.999996	.999988	.999974	5.9

(Continued)

Table A.2 *Continued*

z \ α_3	0.7	0.8	0.9	1.0	1.1	1.2	1.3	z
-3.0	.000000	.000000	.000000	.000000				-3.0
-2.9	.000001	.000000	.000000	.000000				-2.9
-2.8	.000005	.000000	.000000	.000000	.000000			-2.8
-2.7	.000017	.000002	.000000	.000000	.000000			-2.7
-2.6	.000050	.000009	.000001	.000000	.000000	.000000		-2.6
-2.5	.000132	.000034	.000005	.000000	.000000	.000000		-2.5
-2.4	.000316	.000106	.000023	.000003	.000000	.000000	.000000	-2.4
-2.3	.000692	.000289	.000090	.000017	.000001	.000000	.000000	-2.3
-2.2	.001402	.000697	.000281	.000082	.000014	.000001	.000000	-2.2
-2.1	.002646	.001517	.000748	.000296	.000084	.000014	.000001	-2.1
-2.0	.004696	.003018	.001742	.000866	.000347	.000099	.000016	-2.0
-1.9	.007884	.005560	.003631	.002140	.001095	.000456	.000138	-1.9
-1.8	.012602	.009580	.006894	.004623	.002824	.001518	.000678	-1.8
-1.7	.019270	.015566	.012094	.008951	.006228	.003996	.002297	-1.7
-1.6	.028319	.024026	.019830	.015830	.012129	.008825	.006008	-1.6
-1.5	.040152	.035440	.030680	.025959	.021374	.017026	.013019	-1.5
-1.4	.055115	.050217	.045138	.039943	.034713	.029533	.024499	-1.4
-1.3	.073469	.068662	.063565	.058224	.052696	.047046	.041346	-1.3
-1.2	.095366	.090942	.086154	.081031	.075611	.069936	.064054	-1.2
-1.1	.120833	.117072	.112911	.108365	.103457	.098212	.092660	-1.1
-1.0	.149774	.146919	.143660	.140004	.135962	.131550	.126785	-1.0
-.9	.181969	.180206	.178059	.175530	.172625	.169354	.165726	-.9
-.8	.217090	.216535	.215630	.214376	.212776	.210836	.208564	-.8
-.7	.254718	.255412	.255798	.255873	.255641	.255104	.254270	-.7
-.6	.294367	.296281	.297924	.299296	.300398	.301231	.301802	-.6
-.5	.335510	.338548	.341350	.343912	.346235	.348320	.350171	-.5
-.4	.377599	.381616	.385423	.389014	.392388	.395546	.398490	-.4
-.3	.420094	.424908	.429529	.433950	.438169	.442185	.446000	-.3
-.2	.462478	.467885	.473109	.478143	.482985	.487632	.492084	-.2
-.1	.504275	.510062	.515674	.521104	.526346	.531398	.536259	-.1
.0	.545060	.551021	.556813	.562430	.567864	.573113	.578175	.0
.1	.584470	.590412	.596195	.601810	.607251	.612513	.617594	.1
.2	.622203	.627958	.633567	.639019	.644308	.649429	.654379	.2
.3	.658025	.663451	.668748	.673906	.678915	.683772	.688471	.3
.4	.691762	.696748	.701627	.706387	.711021	.715520	.719881	.4
.5	.723301	.727764	.732148	.736439	.740628	.744706	.748668	.5
.6	.752579	.756467	.760307	.764084	.767786	.771403	.774929	.6
.7	.779583	.782869	.786140	.789380	.792576	.795715	.798789	.7
.8	.804338	.807018	.809718	.812419	.815107	.817767	.820390	.8
.9	.826902	.828991	.831134	.833311	.835504	.837700	.839884	.9
1.0	.847359	.848887	.850501	.852180	.853905	.855659	.857427	1.0
1.1	.865812	.866821	.867945	.869162	.870452	.871795	.873176	1.1
1.2	.882381	.882919	.883598	.884395	.885287	.886256	.887284	1.2
1.3	.897190	.897312	.897596	.898018	.898555	.899188	.899898	1.3
1.4	.910373	.910134	.910074	.910167	.910392	.910728	.911157	1.4

Table A.2 *Continued*

z \ α_3	0.7	0.8	0.9	1.0	1.1	1.2	1.3	z
1.5	.922061	.921518	.921164	.920976	.920931	.921009	.921192	1.5
1.6	.932387	.931593	.930995	.930569	.930295	.930153	.930125	1.6
1.7	.941478	.940484	.939687	.939066	.938602	.938275	.938067	1.7
1.8	.949454	.948308	.947355	.946578	.945959	.945479	.945123	1.8
1.9	.956433	.955175	.954105	.953207	.952465	.951863	.951385	1.9
2.0	.962520	.961188	.960035	.959048	.958212	.957514	.956939	2.0
2.1	.967815	.966442	.965236	.964187	.963283	.962513	.961863	2.1
2.2	.972410	.971023	.969789	.968702	.967753	.966930	.966224	2.2
2.3	.976388	.975009	.973770	.972665	.971689	.970833	.970087	2.3
2.4	.979823	.978472	.977244	.976139	.975152	.974277	.973507	2.4
2.5	.982783	.981474	.980273	.979182	.978198	.977317	.976534	2.5
2.6	.985330	.984074	.982911	.981844	.980874	.979998	.979212	2.6
2.7	.987516	.986322	.985205	.984172	.983224	.982362	.981582	2.7
2.8	.989389	.988262	.987198	.986206	.985288	.984446	.983679	2.8
2.9	.990992	.989935	.988929	.987981	.987098	.986282	.985534	2.9
3.0	.992360	.991376	.990430	.989531	.988687	.987901	.987175	3.0
3.1	.993528	.992616	.991730	.990882	.990079	.989327	.988627	3.1
3.2	.994522	.993681	.992857	.992061	.991301	.990583	.989912	3.2
3.3	.995368	.994596	.993832	.993087	.992371	.991690	.991049	3.3
3.4	.996087	.995381	.994676	.993982	.993309	.992665	.992055	3.4
3.5	.996696	.996054	.995405	.994761	.994132	.993525	.992946	3.5
3.6	.997213	.996630	.996035	.995439	.994852	.994282	.993735	3.6
3.7	.997651	.997124	.996579	.996029	.995484	.994950	.994433	3.7
3.8	.998021	.997546	.997049	.996543	.996037	.995538	.995052	3.8
3.9	.998334	.997906	.997455	.996991	.996522	.996056	.995601	3.9
4.0	.998598	.998215	.997805	.997380	.996947	.996514	.996087	4.0
4.1	.998821	.998478	.998107	.997718	.997319	.996917	.996518	4.1
4.2	.999009	.998703	.998368	.998013	.997646	.997273	.996900	4.2
4.3	.999167	.998894	.998593	.998269	.997932	.997586	.997240	4.3
4.4	.999300	.999058	.998786	.998492	.998183	.997864	.997541	4.4
4.5	.999413	.999197	.998953	.998686	.998403	.998108	.997808	4.5
4.6	.999507	.999316	.999097	.998855	.998596	.998324	.998045	4.6
4.7	.999586	.999418	.999222	.999002	.998765	.998515	.998256	4.7
4.8	.999653	.999504	.999329	.999130	.998914	.998683	.998444	4.8
4.9	.999709	.999578	.999421	.999242	.999044	.998832	.998611	4.9
5.0	.999756	.999640	.999500	.999339	.999158	.998964	.998759	5.0
5.1	.999796	.999694	.999569	.999423	.999259	.999081	.998891	5.1
5.2	.999829	.999739	.999628	.999496	.999347	.999184	.999009	5.2
5.3	.999857	.999778	.999679	.999560	.999425	.999275	.999113	5.3
5.4	.999880	.999811	.999723	.999616	.999493	.999356	.999207	5.4
5.5	.999899	.999839	.999761	.999665	.999553	.999427	.999290	5.5
5.6	.999916	.999863	.999793	.999707	.999606	.999491	.999364	5.6
5.7	.999929	.999883	.999821	.999744	.999652	.999547	.999430	5.7
5.8	.999941	.999900	.999846	.999776	.999693	.999597	.999489	5.8
5.9	.999950	.999915	.999867	.999804	.999729	.999641	.999542	5.9

(Continued)

Table A.2 *Continued*

z \ α_3	1.4	1.5	1.6	1.7	1.8	1.9	2.0	z
-1.5	.009455	.006424	.003992	.002192	.001005	.000347	.000073	-1.5
-1.4	.019711	.015273	.011288	.007850	.005035	.002889	.001410	-1.4
-1.3	.035678	.030129	.024793	.019771	.015164	.011069	.007576	-1.3
-1.2	.058022	.051902	.045763	.039679	.033732	.028010	.022605	-1.2
-1.1	.086836	.080779	.074531	.068142	.061662	.055148	.048663	-1.1
-1.0	.121689	.116283	.110595	.104651	.098481	.092118	.085596	-1.0
-.9	.161757	.157458	.152847	.147940	.142755	.137310	.131623	-.9
-.8	.205968	.203060	.199849	.196349	.192571	.188528	.184232	-.8
-.7	.253146	.251740	.250061	.248120	.245926	.243491	.240823	-.7
-.6	.302116	.302182	.302007	.301600	.300971	.300130	.299084	-.6
-.5	.351793	.353194	.354379	.355359	.356141	.356733	.357146	-.5
-.4	.401225	.403755	.406089	.408233	.410194	.411982	.413603	-.4
-.3	.449617	.453041	.456276	.459330	.462210	.464922	.467475	-.3
-.2	.496344	.500415	.504302	.508011	.511548	.514920	.518133	-.2
-.1	.540932	.545418	.549721	.553848	.557803	.561594	.565225	-.1
.0	.583052	.587744	.592257	.596594	.600761	.604763	.608608	.0
.1	.622496	.627218	.631765	.636140	.640348	.644395	.648286	.1
.2	.659157	.663766	.668207	.672483	.676599	.680559	.684370	.2
.3	.693013	.697397	.701624	.705698	.709622	.713400	.717036	.3
.4	.724101	.728180	.732117	.735915	.739576	.743104	.746502	.4
.5	.752509	.756229	.759825	.763299	.766652	.769886	.773005	.5
.6	.778358	.781687	.784912	.788035	.791054	.793972	.796789	.6
.7	.801791	.804716	.807559	.810320	.812996	.815588	.818096	.7
.8	.822966	.825488	.827951	.830352	.832688	.834958	.837160	.8
.9	.842046	.844178	.846274	.848327	.850334	.852292	.854199	.9
1.0	.859197	.860960	.862706	.864430	.866126	.867789	.869417	1.0
1.1	.874582	.876001	.877422	.878840	.880246	.881635	.883004	1.1
1.2	.888356	.889460	.890585	.891721	.892862	.894000	.895131	1.2
1.3	.900670	.901489	.902345	.903228	.904127	.905038	.905953	1.3
1.4	.911662	.912229	.912845	.913500	.914184	.914890	.915612	1.4
1.5	.921463	.921807	.922212	.922666	.923160	.923685	.924233	1.5
1.6	.930194	.930345	.930565	.930843	.931170	.931535	.931931	1.6
1.7	.937964	.937949	.938011	.938137	.938317	.938543	.938807	1.7
1.8	.944874	.944720	.944646	.944642	.944697	.944803	.944951	1.8
1.9	.951017	.950745	.950558	.950443	.950392	.950394	.950443	1.9
2.0	.956474	.956107	.955825	.955618	.955476	.955392	.955356	2.0
2.1	.961321	.960876	.960517	.960234	.960018	.959860	.959752	2.1
2.2	.965624	.965119	.964698	.964353	.964075	.963856	.963690	2.2
2.3	.969444	.968892	.968424	.968030	.967702	.967434	.967217	2.3
2.4	.972834	.972249	.971744	.971312	.970946	.970637	.970381	2.4
2.5	.975842	.975235	.974704	.974244	.973847	.973507	.973219	2.5
2.6	.978512	.977891	.977344	.976864	.976445	.976081	.975767	2.6
2.7	.980882	.980255	.979698	.979205	.978771	.978389	.978056	2.7
2.8	.982985	.982359	.981799	.981299	.980855	.980462	.980115	2.8
2.9	.984851	.984233	.983674	.983173	.982724	.982323	.981967	2.9

Table A.2 *Continued*

z \ α_3	1.4	1.5	1.6	1.7	1.8	1.9	2.0	z
3.0	.986509	.985901	.985349	.984850	.984400	.983996	.983635	3.0
3.1	.987980	.987387	.986845	.986352	.985905	.985501	.985138	3.1
3.2	.989288	.988711	.988182	.987698	.987256	.986856	.986493	3.2
3.3	.990449	.989892	.989377	.988904	.988471	.988076	.987716	3.3
3.4	.991481	.990945	.990447	.989987	.989563	.989175	.988820	3.4
3.5	.992398	.991884	.991404	.990958	.990546	.990166	.989818	3.5
3.6	.993214	.992722	.992261	.991830	.991431	.991061	.990720	3.6
3.7	.993939	.993470	.993029	.992614	.992228	.991869	.991537	3.7
3.8	.994585	.994139	.993717	.993319	.992947	.992599	.992276	3.8
3.9	.995160	.994736	.994334	.993953	.993595	.993259	.992946	3.9
4.0	.995671	.995271	.994888	.994524	.994180	.993857	.993554	4.0
4.1	.996127	.995748	.995385	.995037	.994708	.994398	.994106	4.1
4.2	.996533	.996176	.995831	.995501	.995186	.994888	.994607	4.2
4.3	.996896	.996559	.996232	.995918	.995618	.995332	.995062	4.3
4.4	.997219	.996902	.996593	.996295	.996008	.995735	.995476	4.4
4.5	.997507	.997209	.996918	.996635	.996362	.996101	.995853	4.5
4.6	.997764	.997485	.997210	.996942	.996683	.996434	.996196	4.6
4.7	.997994	.997732	.997473	.997219	.996973	.996736	.996508	4.7
4.8	.998200	.997954	.997710	.997470	.997237	.997010	.996793	4.8
4.9	.998383	.998153	.997924	.997697	.997476	.997260	.997053	4.9
5.0	.998547	.998332	.998117	.997903	.997693	.997488	.997290	5.0
5.1	.998694	.998493	.998291	.998089	.997890	.997696	.997507	5.1
5.2	.998826	.998638	.998448	.998258	.998069	.997885	.997705	5.2
5.3	.998943	.998768	.998590	.998410	.998233	.998057	.997886	5.3
5.4	.999049	.998885	.998718	.998549	.998381	.998215	.998052	5.4
5.5	.999144	.998991	.998834	.998675	.998516	.998359	.998204	5.5
5.6	.999228	.999086	.998939	.998790	.998640	.998490	.998343	5.6
5.7	.999304	.999172	.999034	.998894	.998752	.998611	.998471	5.7
5.8	.999373	.999249	.999120	.998988	.998854	.998720	.998588	5.8
5.9	.999434	.999319	.999198	.999074	.998948	.998821	.998695	5.9
6.0	.999489	.999382	.999269	.999152	.999033	.998913	.998794	6.0
6.1	.999539	.999439	.999333	.999224	.999111	.998998	.998884	6.1
6.2	.999583	.999490	.999391	.999289	.999183	.999075	.998967	6.2
6.3	.999623	.999537	.999444	.999348	.999248	.999146	.999044	6.3
6.4	.999659	.999579	.999492	.999402	.999308	.999212	.999115	6.4
6.5	.999692	.999617	.999536	.999451	.999362	.999272	.999179	6.5
6.6	.999721	.999651	.999576	.999496	.999412	.999327	.999239	6.6
6.7	.999747	.999683	.999612	.999537	.999458	.999377	.999294	6.7
6.8	.999771	.999711	.999645	.999575	.999500	.999424	.999345	6.8
6.9	.999793	.999737	.999675	.999609	.999539	.999466	.999392	6.9
7.0	.999812	.999760	.999702	.999640	.999574	.999506	.999435	7.0
7.1	.999829	.999781	.999727	.999669	.999607	.999542	.999475	7.1
7.2	.999845	.999800	.999750	.999695	.999637	.999576	.999512	7.2
7.3	.999859	.999818	.999771	.999719	.999664	.999606	.999546	7.3
7.4	.999872	.999833	.999790	.999741	.999690	.999635	.999578	7.4

(Continued)

Table A.2 *Continued*

z\α₃	2.1	2.2	2.3	2.5	3.0	4.0	5.0	z
-1.5	.000006	.000000						-1.5
-1.4	.000536	.000133	.000014	.000000				-1.4
-1.3	.004754	.002642	.001225	.000088				-1.3
-1.2	.017611	.013123	.009232	.003520	.000000			-1.2
-1.1	.042274	.036054	.030080	.019209	.002164			-1.1
-1.0	.078952	.072227	.065465	.052022	.021633	.000000		-1.0
-.9	.125716	.119608	.113322	.100303	.066395	.009402		-.9
-.8	.179695	.174931	.169950	.159389	.130051	.064652	.009878	-.8
-.7	.237933	.234831	.231526	.224342	.203405	.151542	.091297	-.7
-.6	.297845	.296419	.294817	.291115	.279286	.246893	.205695	-.6
-.5	.357386	.357464	.357387	.356799	.353125	.338604	.317354	-.5
-.4	.415067	.416380	.417552	.419499	.422371	.421802	.415710	-.4
-.3	.469876	.472132	.474252	.478108	.485818	.495160	.499287	-.3
-.2	.521196	.524115	.526897	.532077	.543087	.558929	.569419	-.2
-.1	.568705	.572040	.575237	.581243	.594277	.613987	.628101	-.1
.0	.612300	.615847	.619256	.625682	.639747	.661396	.677278	.0
.1	.652028	.655626	.659087	.665623	.679975	.702206	.718639	.1
.2	.688036	.691564	.694959	.701374	.715480	.737373	.753582	.2
.3	.720537	.723906	.727150	.733283	.746777	.767735	.783248	.3
.4	.749775	.752927	.755963	.761705	.774353	.794012	.808558	.4
.5	.776011	.778909	.781702	.786989	.798654	.816814	.830258	.5
.6	.799508	.802132	.804664	.809465	.820080	.836657	.848950	.6
.7	.820522	.822867	.825132	.829436	.838987	.853973	.865123	.7
.8	.839295	.841363	.843366	.847181	.855690	.869129	.879177	.8
.9	.856054	.857857	.859607	.862953	.870463	.882430	.891438	.9
1.0	.871008	.872560	.874072	.876976	.883548	.894137	.902177	1.0
1.1	.884349	.885668	.886959	.889452	.895153	.904467	.911616	1.1
1.2	.896250	.897355	.898442	.900558	.905460	.913608	.919941	1.2
1.3	.906868	.907778	.908681	.910454	.914630	.921715	.927308	1.3
1.4	.916342	.917078	.917815	.919279	.922799	.928924	.933847	1.4
1.5	.924800	.925380	.925968	.927156	.930087	.935348	.939668	1.5
1.6	.932353	.932795	.933251	.934194	.936600	.941087	.944864	1.6
1.7	.939102	.939422	.939763	.940489	.942429	.946224	.949514	1.7
1.8	.945135	.945349	.945588	.946124	.947653	.950832	.953687	1.8
1.9	.950532	.950654	.950805	.951174	.952341	.954974	.957439	1.9
2.0	.955363	.955406	.955480	.955705	.956555	.958704	.960822	2.0
2.1	.959690	.959665	.959674	.959774	.960348	.962070	.963877	2.1
2.2	.963568	.963487	.963440	.963433	.963767	.965112	.966643	2.2
2.3	.967047	.966917	.966823	.966725	.966852	.967867	.969152	2.3
2.4	.970170	.970000	.969867	.969692	.969640	.970365	.971431	2.4
2.5	.972976	.972773	.972607	.972367	.972163	.972634	.973507	2.5
2.6	.975498	.975269	.975076	.974782	.974449	.974699	.975399	2.6
2.7	.977767	.977517	.977303	.976965	.976523	.976580	.977128	2.7
2.8	.979811	.979544	.979313	.978940	.978406	.978297	.978710	2.8
2.9	.981652	.981374	.981129	.980728	.980119	.979865	.980159	2.9

Table A.2 *Continued*

z \ α_3	2.1	2.2	2.3	2.5	3.0	4.0	5.0	z
3.0	.983313	.983026	.982771	.982348	.981677	.981301	.981489	3.0
3.1	.984811	.984519	.984258	.983819	.983098	.982616	.982711	3.1
3.2	.986165	.985870	.985605	.985154	.984394	.983822	.983836	3.2
3.3	.987389	.987093	.986826	.986368	.985578	.984930	.984872	3.3
3.4	.988496	.988201	.987934	.987472	.986660	.985949	.985828	3.4
3.5	.989498	.989206	.988940	.988478	.987650	.986887	.986712	3.5
3.6	.990406	.990119	.989855	.989394	.988558	.987752	.987528	3.6
3.7	.991230	.990947	.990687	.990230	.989390	.988550	.988285	3.7
3.8	.991977	.991700	.991445	.990994	.990153	.989287	.988986	3.8
3.9	.992655	.992385	.992135	.991691	.990855	.989968	.989637	3.9
4.0	.993272	.993009	.992765	.992329	.991500	.990598	.990241	4.0
4.1	.993832	.993577	.993339	.992913	.992094	.991183	.990803	4.1
4.2	.994343	.994095	.993864	.993448	.992641	.991724	.991326	4.2
4.3	.994807	.994568	.994344	.993938	.993145	.992227	.991814	4.3
4.4	.995231	.995000	.994782	.994388	.993610	.992694	.992268	4.4
4.5	.995617	.995394	.995184	.994801	.994039	.993120	.992692	4.5
4.6	.995969	.995754	.995552	.995180	.994436	.993532	.993089	4.6
4.7	.996291	.996084	.995889	.995529	.994804	.993908	.993459	4.7
4.8	.996585	.996386	.996198	.995850	.995143	.994259	.993806	4.8
4.9	.996854	.996663	.996482	.996146	.995458	.994586	.994131	4.9
5.0	.997100	.996917	.996742	.996418	.995750	.994892	.994435	5.0
5.1	.997325	.997150	.996982	.996670	.996020	.995177	.994720	5.1
5.2	.997531	.997363	.997202	.996901	.996272	.995443	.994988	5.2
5.3	.997720	.997559	.997405	.997115	.996505	.995693	.995240	5.3
5.4	.997894	.997740	.997592	.997313	.996722	.995927	.995476	5.4
5.5	.998053	.997906	.997764	.997496	.996924	.996145	.995699	5.5
5.6	.998199	.998058	.997922	.997665	.997111	.996350	.995908	5.6
5.7	.998333	.998199	.998069	.997821	.997286	.996543	.996106	5.7
5.8	.998457	.998329	.998204	.997966	.997449	.996723	.996291	5.8
5.9	.998570	.998448	.998329	.998100	.997601	.996893	.996467	5.9
6.0	.998675	.998558	.998444	.998225	.997743	.997052	.996632	6.0
6.1	.998771	.998660	.998551	.998341	.997875	.997202	.996788	6.1
6.2	.998860	.998754	.998649	.998448	.997999	.997343	.996936	6.2
6.3	.998942	.998840	.998741	.998547	.998114	.997476	.997075	6.3
6.4	.999017	.998921	.998825	.998640	.998222	.997601	.997207	6.4
6.5	.999087	.998995	.998904	.998726	.998323	.997719	.997332	6.5
6.6	.999151	.999063	.998976	.998806	.998418	.997830	.997450	6.6
6.7	.999211	.999127	.999044	.998881	.998506	.997935	.997562	6.7
6.8	.999266	.999186	.999106	.998950	.998589	.998034	.997668	6.8
6.9	.999316	.999240	.999164	.999015	.998667	.998127	.997769	6.9
7.0	.999363	.999291	.999218	.999075	.998740	.998216	.997864	7.0
7.1	.999407	.999338	.999269	.999131	.998809	.998299	.997955	7.1
7.2	.999447	.999381	.999315	.999184	.998873	.998378	.998041	7.2
7.3	.999485	.999422	.999359	.999233	.998934	.998453	.998123	7.3
7.4	.999519	.999459	.999399	.999278	.998990	.998523	.998200	7.4

Table A.3 Cumulative Distribution Function of the Standardized Inverse Gaussian Distribution $(\alpha_3; 0, 1)$

z \ α_3	0.0	0.1	0.2	0.3	0.4	0.5	0.6	z
-3.0	.001350	.000827	.000446	.000200	.000068	.000015	.000002	-3.0
-2.9	.001866	.001203	.000693	.000342	.000134	.000037	.000005	-2.9
-2.8	.002555	.001726	.001059	.000569	.000252	.000083	.000017	-2.8
-2.7	.003467	.002446	.001589	.000922	.000456	.000178	.000048	-2.7
-2.6	.004661	.003423	.002342	.001458	.000795	.000358	.000120	-2.6
-2.5	.006210	.004731	.003396	.002252	.001339	.000682	.000275	-2.5
-2.4	.008198	.006462	.004844	.003398	.002181	.001236	.000585	-2.4
-2.3	.010724	.008721	.006799	.005018	.003445	.002141	.001156	-2.3
-2.2	.013903	.011633	.009396	.007256	.005284	.003556	.002145	-2.2
-2.1	.017864	.015338	.012792	.010284	.007886	.005683	.003764	-2.1
-2.0	.022750	.019996	.017161	.014297	.011470	.008764	.006277	-2.0
-1.9	.028717	.025780	.022698	.019513	.016281	.013079	.010001	-1.9
-1.8	.035930	.032873	.029608	.026165	.022586	.018935	.015292	-1.8
-1.7	.044565	.041468	.038106	.034495	.030663	.026652	.022527	1.7
-1.6	.054799	.051760	.048409	.044747	.040786	.036550	.032079	-1.6
-1.5	.066807	.063939	.060724	.057150	.053216	.048925	.044297	-1.5
-1.4	.080757	.078184	.075243	.071914	.068182	.064037	.059476	-1.4
-1.3	.096800	.094652	.092131	.089212	.085871	.082085	.077838	-1.3
-1.2	.115070	.113475	.111520	.109174	.106410	.103199	.099513	-1.2
-1.1	.135666	.134746	.133492	.131873	.129860	.127421	.124526	-1.1
-1.0	.158655	.158516	.158081	.157321	.156207	.154708	.152794	-1.0
-.9	.184060	.184786	.185261	.185460	.185356	.184923	.184132	-.9
-.8	.211855	.213500	.214944	.216164	.217139	.217844	.218258	-.8
-.7	.241964	.244548	.246979	.249240	.251310	.253172	.254806	-.7
-.6	.274253	.277759	.281156	.284429	.287563	.290541	.293350	-.6
-.5	.308538	.312907	.317205	.321418	.325533	.329537	.333418	-.5
-.4	.344578	.349716	.354810	.359849	.364819	.369712	.374514	-.4
-.3	.382089	.387864	.393614	.399327	.404993	.410601	.416141	-.3
-.2	.420740	.426993	.433230	.439441	.445615	.451742	.457812	-.2
-.1	.460172	.466720	.473256	.479769	.486250	.492688	.499073	-.1
.0	.500000	.506647	.513283	.519898	.526479	.533019	.539507	.0
.1	.539828	.546376	.552914	.559430	.565914	.572357	.578748	.1
.2	.579260	.585517	.591767	.597999	.604203	.610369	.616487	.2
.3	.617911	.623701	.629493	.635277	.641041	.646776	.652471	.3
.4	.655422	.660591	.665781	.670978	.676171	.681350	.686504	.4
.5	.691462	.695890	.700362	.704867	.709390	.713920	.718447	.5
.6	.725747	.729345	.733021	.736759	.740547	.744369	.748213	.6
.7	.758036	.760755	.763589	.766523	.769540	.772625	.775762	.7
.8	.788145	.789969	.791953	.794075	.796318	.798663	.801094	.8
.9	.815940	.816891	.818045	.819378	.820869	.822498	.824247	.9
1.0	.841345	.841475	.841847	.842438	.843222	.844177	.845283	1.0
1.1	.864334	.863719	.863382	.863296	.863434	.863774	.864293	1.1
1.2	.884930	.883667	.882708	.882026	.881593	.881386	.881382	1.2
1.3	.903200	.901397	.899915	.898726	.897804	.897125	.896666	1.3
1.4	.919243	.917017	.915115	.913513	.912188	.911115	.910273	1.4

Source: From Cohen and Whitten (1988), Table A.3.3, pp. 308–315, by courtesy of Marcel Dekker, Inc.

Table A.3 *Continued*

z \ α_3	0.0	0.1	0.2	0.3	0.4	0.5	0.6	z
1.5	.933193	.930659	.928441	.926520	.924875	.923486	.922332	1.5
1.6	.945201	.942469	.940035	.937885	.936003	.934372	.932973	1.6
1.7	.955435	.952607	.950049	.947753	.945709	.943904	.942325	1.7
1.8	.964070	.961236	.958636	.956268	.954130	.952215	.950510	1.8
1.9	.971283	.968519	.965946	.963571	.961399	.959427	.957650	1.9
2.0	.977250	.974615	.972126	.969798	.967642	.965661	.963853	2.0
2.1	.982136	.979675	.977314	.975077	.972978	.971026	.969226	2.1
2.2	.986097	.983842	.981642	.979527	.977517	.975627	.973864	2.2
2.3	.989276	.987244	.985227	.983257	.981361	.979556	.977855	2.3
2.4	.991802	.990001	.988177	.986369	.984603	.982901	.981279	2.4
2.5	.993790	.992217	.990591	.988950	.987324	.985737	.984208	2.5
2.6	.995339	.993983	.992552	.991080	.989599	.988135	.986707	2.6
2.7	.996533	.995381	.994135	.992829	.991494	.990155	.988833	2.7
2.8	.997445	.996479	.995407	.994259	.993066	.991852	.990638	2.8
2.9	.998134	.997334	.996421	.995423	.994365	.993272	.992166	2.9
3.0	.998650	.997995	.997226	.996366	.995434	.994458	.993457	3.0
3.1	.999032	.998503	.997851	.997124	.996312	.995446	.994544	3.1
3.2	.999313	.998889	.998359	.997733	.997029	.996265	.995459	3.2
3.3	.999517	.999182	.998747	.998220	.997614	.996944	.996227	3.3
3.4	.999663	.999401	.999048	.998607	.998089	.997505	.996870	3.4
3.5	.999767	.999565	.999280	.998914	.998473	.997967	.997407	3.5
3.6	.999841	.999686	.999458	.999157	.998784	.998347	.997855	3.6
3.7	.999892	.999774	.999594	.999347	.999033	.998658	.998229	3.7
3.8	.999928	.999839	.999698	.999496	.999234	.998913	.998539	3.8
3.9	.999952	.999886	.999776	.999613	.999394	.999121	.998797	3.9
4.0	.999968	.999920	.999834	.999703	.999522	.999290	.999010	4.0
4.1	.999979	.999944	.999878	.999773	.999624	.999428	.999187	4.1
4.2	.999987	.999961	.999911	.999827	.999705	.999540	.999333	4.2
4.3	.999991	.999973	.999935	.999869	.999769	.999630	.999453	4.3
4.4	.999995	.999982	.999953	.999901	.999819	.999704	.999553	4.4
4.5	.999997	.999987	.999966	.999925	.999859	.999763	.999634	4.5
4.6	.999998	.999991	.999975	.999944	.999890	.999810	.999701	4.6
4.7	.999999	.999994	.999982	.999958	.999915	.999849	.999756	4.7
4.8	.999999	.999996	.999987	.999968	.999934	.999879	.999801	4.8
4.9	1.000000	.999997	.999991	.999976	.999949	.999904	.999838	4.9
5.0	1.000000	.999998	.999994	.999982	.999960	.999924	.999869	5.0
5.1	1.000000	.999999	.999996	.999987	.999969	.999939	.999893	5.1
5.2	1.000000	.999999	.999997	.999990	.999977	.999952	.999913	5.2
5.3	1.000000	1.000000	.999998	.999993	.999982	.999962	.999930	5.3
5.4	1.000000	1.000000	.999998	.999995	.999986	.999970	.999943	5.4
5.5	1.000000	1.000000	.999999	.999996	.999989	.999976	.999954	5.5
5.6	1.000000	1.000000	.999999	.999997	.999992	.999981	.999963	5.6
5.7	1.000000	1.000000	.999999	.999998	.999994	.999985	.999970	5.7
5.8	1.000000	1.000000	1.000000	.999998	.999995	.999988	.999976	5.8
5.9	1.000000	1.000000	1.000000	.999999	.999996	.999991	.999980	5.9

(Continued)

Table A.3 *Continued*

z \ α_3	0.7	0.8	0.9	1.0	1.1	1.2	1.3	z
-3.0	.000000	.000000	.000000					-3.0
-2.9	.000000	.000000	.000000	.000000				-2.9
-2.8	.000001	.000000	.000000	.000000				-2.8
-2.7	.000007	.000000	.000000	.000000	.000000			-2.7
-2.6	.000025	.000002	.000000	.000000	.000000			-2.6
-2.5	.000077	.000011	.000000	.000000	.000000	.000000		-2.5
-2.4	.000209	.000047	.000005	.000000	.000000	.000000		-2.4
-2.3	.000506	.000159	.000028	.000002	.000000	.000000	.000000	-2.3
-2.2	.001105	.000448	.000122	.000016	.000000	.000000	.000000	-2.2
-2.1	.002215	.001095	.000413	.000098	.000010	.000000	.000000	-2.1
-2.0	.004115	.002379	.001141	.000406	.000086	.000007	.000000	-2.0
-1.9	.007161	.004684	.002691	.001271	.000438	.000086	.000005	-1.9
-1.8	.011765	.008482	.005590	.003238	.001541	.000532	.000103	-1.8
-1.7	.018373	.014306	.010470	.007033	.004175	.002055	.000744	-1.7
-1.6	.027434	.022702	.018001	.013484	.009341	.005784	.003024	-1.6
-1.5	.039362	.034176	.028819	.023402	.018081	.013053	.008560	-1.5
-1.4	.054507	.049151	.043448	.037464	.031296	.025085	.019019	-1.4
-1.3	.073118	.067919	.062248	.056128	.049601	.042738	.035647	-1.3
-1.2	.095328	.090622	.085379	.079588	.073252	.066385	.059023	-1.2
-1.1	.121142	.117240	.112791	.107768	.102147	.095909	.089045	-1.1
-1.0	.150434	.147594	.144243	.140347	.135873	.130789	.125063	-1.0
-.9	.182955	.181363	.179324	.176808	.173783	.170215	.166069	-.9
-.8	.218354	.218110	.217500	.216499	.215078	.213212	.210872	-.8
-.7	.256193	.257314	.258150	.258681	.258887	.258750	.258247	-.7
-.6	.295973	.298397	.300608	.302590	.304331	.305818	.307036	-.6
-.5	.337163	.340762	.344203	.347478	.350575	.353488	.356206	-.5
-.4	.379218	.383812	.388290	.392642	.396863	.400945	.404883	-.4
-.3	.421604	.426982	.432267	.437454	.442535	.447505	.452362	-.3
-.2	.463818	.469750	.475602	.481367	.487041	.492618	.498095	-.2
-.1	.505397	.511653	.517832	.523930	.529940	.535859	.541681	-.1
.0	.545934	.552293	.558577	.564779	.570895	.576919	.582849	.0
.1	.585080	.591344	.597533	.603641	.609663	.615594	.621430	.1
.2	.622549	.628546	.634471	.640319	.646084	.651761	.657347	.2
.3	.658118	.663708	.669235	.674691	.680072	.685373	.690588	.3
.4	.691623	.696700	.701726	.706695	.711600	.716436	.721199	.4
.5	.722959	.727448	.731903	.736319	.740688	.745004	.749262	.5
.6	.752068	.755924	.759769	.763595	.767395	.771162	.774889	.6
.7	.778939	.782143	.785363	.788588	.791810	.795020	.798212	.7
.8	.803596	.806154	.808755	.811387	.814040	.816705	.819372	.8
.9	.826096	.828031	.830037	.832100	.834209	.836352	.838520	.9
1.0	.846521	.847872	.849320	.850849	.852448	.854103	.855804	1.0
1.1	.864970	.865785	.866721	.867762	.868893	.870100	.871372	1.1
1.2	.881557	.881893	.882370	.882970	.883680	.884483	.885368	1.2
1.3	.896405	.896320	.896393	.896606	.896943	.897388	.897929	1.3
1.4	.909641	.909197	.908922	.908799	.908812	.908945	.909185	1.4

Table A.3 *Continued*

z \ α_3	0.7	0.8	0.9	1.0	1.1	1.2	1.3	z
1.5	.921393	.920649	.920082	.919675	.919412	.919277	.919257	1.5
1.6	.931790	.930803	.929997	.929354	.928859	.928499	.928259	1.6
1.7	.940954	.939778	.938781	.937948	.937265	.936718	.936294	1.7
1.8	.949006	.947690	.946547	.945565	.944732	.944034	.943460	1.8
1.9	.956058	.954644	.953396	.952303	.951354	.950538	.949845	1.9
2.0	.962216	.960742	.959424	.958253	.957220	.956315	.955528	2.0
2.1	.967577	.966077	.964719	.963499	.962408	.961439	.960584	2.1
2.2	.972232	.970732	.969362	.968117	.966992	.965981	.965078	2.2
2.3	.976264	.974787	.973425	.972176	.971037	.970003	.969070	2.3
2.4	.979747	.978312	.976976	.975740	.974603	.973562	.972614	2.4
2.5	.982749	.981369	.980073	.978864	.977743	.976709	.975759	2.5
2.6	.985331	.984016	.982772	.981601	.980507	.979489	.978548	2.6
2.7	.987546	.986305	.985119	.983995	.982936	.981945	.981021	2.7
2.8	.989443	.988280	.987159	.986087	.985071	.984112	.983212	2.8
2.9	.991064	.989981	.988928	.987914	.986944	.986023	.985154	2.9
3.0	.992447	.991445	.990462	.989507	.988588	.987709	.986873	3.0
3.1	.993625	.992703	.991790	.990896	.990029	.989194	.988395	3.1
3.2	.994627	.993782	.992939	.992106	.991291	.990502	.989742	3.2
3.3	.995476	.994707	.993931	.993158	.992397	.991654	.990934	3.3
3.4	.996196	.994498	.994787	.994073	.993364	.992667	.991988	3.4
3.5	.996806	.996175	.995526	.994868	.994210	.993559	.992921	3.5
3.6	.997320	.996752	.996162	.995559	.994950	.994344	.993745	3.6
3.7	.997754	.997245	.996710	.996158	.995597	.995034	.994474	3.7
3.8	.998120	.997665	.997181	.996678	.996162	.995640	.995119	3.8
3.9	.998428	.998022	.997586	.997128	.996655	.996173	.995688	3.9
4.0	.998687	.998326	.997934	.997519	.997086	.996642	.996192	4.0
4.1	.998904	.998584	.998233	.997857	.997462	.997053	.996637	4.1
4.2	.999086	.998803	.998489	.998150	.997790	.997415	.997030	4.2
4.3	.999239	.998989	.998709	.998403	.998075	.997732	.997377	4.3
4.4	.999366	.999147	.998897	.998622	.998325	.998011	.997683	4.4
4.5	.999473	.999280	.999058	.998811	.998542	.998255	.997954	4.5
4.6	.999562	.999393	.999196	.998975	.998731	.998470	.998194	4.6
4.7	.999636	.999489	.999314	.999116	.998896	.998658	.998405	4.7
4.8	.999698	.999569	.999415	.999238	.999040	.998823	.998592	4.8
4.9	.999750	.999637	.999502	.999343	.999165	.998968	.998756	4.9
5.0	.999793	.999695	.999575	.999434	.999274	.999095	.998902	5.0
5.1	.999828	.999744	.999638	.999513	.999368	.999207	.999030	5.1
5.2	.999858	.999784	.999692	.999580	.999451	.999305	.999144	5.2
5.3	.999883	.999819	.999738	.999639	.999523	.999391	.999244	5.3
5.4	.999903	.999848	.999777	.999689	.999585	.999466	.999333	5.4
5.5	.999920	.999872	.999810	.999732	.999639	.999532	.999411	5.5
5.6	.999934	.999893	.999838	.999770	.999687	.999590	.999480	5.6
5.7	.999945	.999910	.999863	.999802	.999728	.999640	.999541	5.7
5.8	.999955	.999925	.999883	.999830	.999763	.999685	.999594	5.8
5.9	.999963	.999937	.999901	.999853	.999794	.999724	.999642	5.9

(Continued)

Table A.3 *Continued*

z\α_3	1.4	1.5	1.6	1.7	1.8	1.9	2.0	z
-1.5	.004865	.002204	.000674	.000094	.000002	.000000		-1.5
-1.4	.013347	.008372	.004418	.001749	.000406	.000029	.000000	-1.4
-1.3	.028488	.021482	.014926	.009188	.004673	.001713	.000331	-1.3
-1.2	.051233	.043119	.034844	.026640	.018835	.011857	.006210	-1.2
-1.1	.081556	.073460	.064802	.055664	.046179	.036556	.027105	-1.1
-1.0	.118667	.111575	.103769	.095242	.086004	.076091	.065577	-1.0
-.9	.161309	.155899	.149802	.142979	.135396	.127020	.117823	-.9
-.8	.208025	.204642	.200686	.196123	.190912	.185015	.178385	-.8
-.7	.257358	.256059	.254328	.252138	.249463	.246274	.242537	-.7
-.6	.307972	.308612	.308943	.308950	.308617	.307928	.306865	-.6
-.5	.358722	.361028	.363116	.364980	.366612	.368004	.369149	-.5
-.4	.408673	.412309	.415788	.419106	.422260	.425248	.428065	-.4
-.3	.457100	.461717	.466211	.470579	.474821	.478935	.482920	-.3
-.2	.503468	.508735	.513894	.518944	.523884	.528713	.533432	-.2
-.1	.547405	.553028	.558548	.563964	.569276	.574482	.579583	-.1
.0	.588680	.594411	.600039	.605564	.610985	.616302	.621514	.0
.1	.627169	.632808	.638345	.643779	.649110	.654337	.659459	.1
.2	.662838	.668231	.673526	.678721	.683815	.688808	.693700	.2
.3	.695716	.700754	.705699	.710550	.715307	.719968	.724534	.3
.4	.725885	.730491	.735015	.739455	.743810	.748079	.752261	.4
.5	.753458	.757588	.761650	.765640	.769558	.773402	.777171	.5
.6	.778572	.782206	.785787	.789313	.792781	.796189	.799536	.6
.7	.801379	.804515	.807617	.810680	.813701	.816678	.819608	.7
.8	.822036	.824689	.827325	.829940	.832530	.835090	.837619	.8
.9	.840704	.842896	.845091	.847281	.849463	.851631	.853782	.9
1.0	.857540	.859303	.861086	.862882	.864684	.866487	.868287	1.0
1.1	.872698	.874067	.875472	.876905	.878360	.879829	.881308	1.1
1.2	.886323	.887336	.888400	.889505	.890644	.891811	.892999	1.2
1.3	.898553	.899250	.900008	.900820	.901677	.902572	.903499	1.3
1.4	.909519	.909936	.910425	.910977	.911584	.912238	.912933	1.4
1.5	.919340	.919514	.919768	.920093	.920480	.920923	.921413	1.5
1.6	.928127	.928092	.928144	.928272	.928469	.928727	.929038	1.6
1.7	.935982	.935771	.935650	.935610	.935643	.935741	.935897	1.7
1.8	.942999	.942641	.942375	.942193	.942086	.942047	.942070	1.8
1.9	.949263	.948784	.948398	.948097	.947873	.947719	.947628	1.9
2.0	.954852	.954276	.953793	.953394	.953072	.952821	.952634	2.0
2.1	.959834	.959183	.958622	.958145	.957744	.957413	.957146	2.1
2.2	.964275	.963567	.962946	.962407	.961942	.961546	.961213	2.2
2.3	.968231	.967483	.966817	.966230	.965715	.965267	.964881	2.3
2.4	.971754	.970979	.970282	.969660	.969108	.968619	.968191	2.4
2.5	.974890	.974100	.973384	.972738	.972158	.971640	.971179	2.5
2.6	.977681	.976886	.976160	.975500	.974902	.974362	.973878	2.6
2.7	.980164	.979373	.978645	.977979	.977370	.976817	.976315	2.7
2.8	.982373	.981592	.980870	.980203	.979591	.979030	.978519	2.8
2.9	.984337	.983573	.982861	.982201	.981590	.981027	.980511	2.9

Table A.3 *Continued*

z \ α_3	1.4	1.5	1.6	1.7	1.8	1.9	2.0	z
3.0	.986083	.985340	.984643	.983994	.983390	.982830	.982313	3.0
3.1	.987635	.986916	.986239	.985604	.985010	.984457	.983943	3.1
3.2	.989015	.988323	.987668	.987050	.986470	.985926	.985419	3.2
3.3	.990241	.989578	.988947	.988349	.987785	.987253	.986755	3.3
3.4	.991331	.990698	.990093	.989516	.988969	.988452	.987965	3.4
3.5	.992299	.991697	.991119	.990565	.990037	.989536	.989062	3.5
3.6	.993159	.992589	.992038	.991507	.991000	.990516	.990056	3.6
3.7	.993923	.993384	.992860	.992354	.991868	.991402	.990958	3.7
3.8	.994602	.994094	.993598	.993116	.992651	.992204	.991775	3.8
3.9	.995205	.994727	.994258	.993801	.993357	.992929	.992517	3.9
4.0	.995740	.995292	.994850	.994416	.993994	.993585	.993190	4.0
4.1	.996216	.995796	.995380	.994970	.994569	.994179	.993801	4.1
4.2	.996639	.996246	.995855	.995469	.995089	.994717	.994356	4.2
4.3	.997014	.996648	.996281	.995917	.995557	.995205	.994860	4.3
4.4	.997347	.997006	.996663	.996320	.995981	.995646	.995318	4.4
4.5	.997644	.997326	.997005	.996683	.996363	.996046	.995735	4.5
4.6	.997906	.997612	.997312	.997010	.996708	.996409	.996113	4.6
4.7	.998140	.997866	.997587	.997304	.997020	.996737	.996457	4.7
4.8	.998348	.998094	.997834	.997569	.997302	.997036	.996770	4.8
4.9	.998532	.998297	.998055	.997808	.997557	.997306	.997055	4.9
5.0	.998695	.998479	.998254	.998023	.997788	.997551	.997314	5.0
5.1	.998841	.998640	.998432	.998216	.997996	.997774	.997550	5.1
5.2	.998970	.998785	.998591	.998391	.998185	.997976	.997765	5.2
5.3	.999085	.998914	.998735	.998548	.998355	.998159	.997960	5.3
5.4	.999187	.999030	.998863	.998690	.998510	.998325	.998138	5.4
5.5	.999277	.999133	.998979	.998817	.998649	.998477	.998301	5.5
5.6	.999358	.999225	.999082	.998932	.998776	.998614	.998448	5.6
5.7	.999429	.999307	.999176	.999036	.998890	.998739	.998583	5.7
5.8	.999492	.999380	.999259	.999130	.998994	.998852	.998706	5.8
5.9	.999549	.999446	.999334	.999214	.999087	.998955	.998818	5.9
6.0	.999599	.999505	.999401	.999290	.999172	.999049	.998920	6.0
6.1	.999644	.999557	.999462	.999359	.999249	.999134	.999013	6.1
6.2	.999683	.999604	.999516	.999421	.999319	.999211	.999098	6.2
6.3	.999718	.999646	.999565	.999477	.999382	.999281	.999176	6.3
6.4	.999749	.999683	.999609	.999527	.999439	.999345	.999246	6.4
6.5	.999777	.999717	.999648	.999573	.999491	.999404	.999311	6.5
6.6	.999802	.999746	.999684	.999614	.999538	.999457	.999370	6.6
6.7	.999824	.999773	.999715	.999651	.999581	.999505	.999424	6.7
6.8	.999843	.999797	.999744	.999685	.999619	.999549	.999473	6.8
6.9	.999861	.999818	.999770	.999715	.999654	.999589	.999518	6.9
7.0	.999876	.999837	.999793	.999742	.999686	.999625	.999559	7.0
7.1	.999890	.999855	.999814	.999767	.999715	.999658	.999596	7.1
7.2	.999902	.999870	.999832	.999789	.999741	.999688	.999631	7.2
7.3	.999913	.999884	.999849	.999809	.999765	.999716	.999662	7.3
7.4	.999922	.999896	.999864	.999828	.999786	.999741	.999691	7.4

(Continued)

Table A.3 *Continued*

z\α₃	2.1	2.2	2.3	2.5	3.0	4.0	5.0	z
-1.5								-1.5
-1.4	.000000							-1.4
-1.3	.000014	.000000	.000000					-1.3
-1.2	.002366	.000480	.000021					-1.2
-1.1	.018263	.010608	.004799	.000129				-1.1
-1.0	.054597	.043373	.032248	.012543	.000000			-1.0
-.9	.107792	.096930	.085272	.059978	.004076			-.9
-.8	.170980	.162750	.153649	.132662	.063754			-.8
-.7	.238220	.233284	.227688	.214331	.165727	.006350		-.7
-.6	.305411	.303545	.301245	.295244	.270614	.157885	.000000	-.6
-.5	.370038	.370664	.371018	.370872	.364976	.321474	.199098	-.5
-.4	.430711	.433183	.435478	.439527	.446384	.444418	.414582	-.4
-.3	.486777	.490506	.494105	.500920	.515738	.536137	.544306	-.3
-.2	.538041	.542540	.546931	.555390	.574724	.606335	.629925	-.2
-.1	.584579	.589472	.594262	.603537	.625023	.661482	.690804	-.1
.0	.626622	.631627	.636530	.646033	.668102	.705781	.736403	.0
.1	.664479	.669395	.674210	.683538	.705178	.742030	.771865	.1
.2	.698490	.703181	.707772	.716662	.737246	.772151	.800234	.2
.3	.729005	.733381	.737663	.745950	.765114	.797504	.823429	.3
.4	.756357	.760367	.764291	.771885	.789442	.819076	.842730	.4
.5	.780865	.784483	.788026	.794887	.810768	.837602	.859020	.5
.6	.802820	.806040	.809198	.815321	.829535	.853641	.872933	.6
.7	.822489	.825320	.828100	.833505	.846108	.867623	.884935	.7
.8	.840114	.842572	.844992	.849713	.860793	.879889	.895377	.8
.9	.855913	.853020	.860103	.864183	.873844	.890707	.904531	.9
1.0	.870081	.871864	.873633	.877122	.885475	.900294	.912606	1.0
1.1	.882792	.884278	.885761	.888709	.895869	.908828	.919771	1.1
1.2	.894203	.895419	.896643	.899100	.905179	.916454	.926160	1.2
1.3	.904452	.905426	.906417	.908433	.913536	.923293	.931883	1.3
1.4	.913662	.914420	.915203	.916824	.921055	.929446	.937030	1.4
1.5	.921944	.922511	.923110	.924380	.927832	.934998	.941676	1.5
1.6	.929396	.929795	.930231	.931192	.933952	.940022	.945885	1.6
1.7	.936104	.936357	.936650	.937340	.939487	.944579	.949708	1.7
1.8	.942147	.942273	.942442	.942894	.944502	.948721	.953191	1.8
1.9	.947593	.947610	.947673	.947918	.949053	.952496	.956372	1.9
2.0	.952505	.952428	.952399	.952466	.953188	.955941	.959284	2.0
2.1	.956937	.956781	.956674	.956588	.956950	.959093	.961956	2.1
2.2	.960938	.960717	.960544	.960327	.960378	.961980	.964412	2.2
2.3	.964552	.964276	.964048	.963721	.963504	.964629	.966675	2.3
2.4	.967819	.967498	.967225	.966806	.966358	.967064	.968762	2.4
2.5	.970773	.970416	.970106	.969610	.968968	.969305	.970692	2.5
2.6	.973445	.973060	.972720	.972163	.971355	.971370	.972478	2.6
2.7	.975863	.975457	.975095	.974488	.973542	.973276	.974133	2.7
2.8	.978054	.977632	.977252	.976606	.975547	.975037	.975670	2.8
2.9	.980038	.979606	.979214	.978539	.977387	.976665	.977098	2.9

Table A.3 *Continued*

z \ α_3	2.1	2.2	2.3	2.5	3.0	4.0	5.0	z
3.0	.981836	.981399	.980999	.980302	.979076	.978172	.978427	3.0
3.1	.983467	.983028	.982623	.981912	.980629	.979570	.979665	3.1
3.2	.984946	.984508	.984102	.983384	.982057	.980866	.980820	3.2
3.3	.986289	.985854	.985450	.984729	.983372	.982069	.981898	3.3
3.4	.987508	.987079	.986679	.985959	.984583	.983188	.982905	3.4
3.5	.988615	.988194	.987800	.987086	.985699	.984228	.983847	3.5
3.6	.989621	.989209	.988822	.988117	.986728	.985197	.984729	3.6
3.7	.990535	.990134	.989755	.989062	.987678	.986099	.985556	3.7
3.8	.991366	.990977	.990608	.989928	.988555	.986940	.986331	3.8
3.9	.992122	.991745	.991386	.990723	.989366	.987725	.987058	3.9
4.0	.992810	.992446	.992098	.991452	.990115	.988457	.987741	4.0
4.1	.993436	.993085	.992749	.992121	.990808	.989142	.988383	4.1
4.2	.994006	.993669	.993344	.992735	.991450	.989781	.988986	4.2
4.3	.994526	.994202	.993889	.993300	.992044	.990379	.989554	4.3
4.4	.994999	.994688	.994387	.993818	.992594	.990939	.990089	4.4
4.5	.995430	.995133	.994844	.994295	.993104	.991463	.990593	4.5
4.6	.995823	.995539	.995262	.994734	.993576	.991954	.991068	4.6
4.7	.996181	.995910	.995645	.995138	.994015	.992414	.991516	4.7
4.8	.996508	.996249	.995996	.995509	.994422	.992846	.991939	4.8
4.9	.996806	.996560	.996318	.995851	.994799	.993250	.992339	4.9
5.0	.997078	.996844	.996614	.996166	.995150	.993630	.992716	5.0
5.1	.997326	.997104	.996885	.996456	.995476	.993987	.993072	5.1
5.2	.997553	.997342	.997133	.996724	.995779	.994322	.993409	5.2
5.3	.997760	.997560	.997361	.996970	.996060	.994637	.993728	5.3
5.4	.997949	.997760	.997571	.997198	.996322	.994933	.994030	5.4
5.5	.998122	.997943	.997763	.997407	.996566	.995212	.994316	5.5
5.6	.998280	.998110	.997940	.997601	.996793	.995474	.994586	5.6
5.7	.998425	.998264	.998102	.997780	.997004	.995720	.994843	5.7
5.8	.998557	.998405	.998252	.997944	.997200	.995952	.995086	5.8
5.9	.998677	.998534	.998389	.998097	.997383	.996171	.995316	5.9
6.0	.998788	.998652	.998515	.998237	.997554	.996377	.995535	6.0
6.1	.998889	.998761	.998631	.998367	.997713	.996571	.995742	6.1
6.2	.998981	.998861	.998738	.998487	.997861	.996754	.995939	6.2
6.3	.999066	.998952	.998836	.998598	.997999	.996927	.996126	6.3
6.4	.999143	.999036	.998927	.998701	.998128	.997090	.996304	6.4
6.5	.999214	.999113	.999010	.998796	.998248	.997243	.996472	6.5
6.6	.999279	.999184	.999086	.998884	.998361	.997388	.996633	6.6
6.7	.999338	.999249	.999157	.998965	.998465	.997525	.996785	6.7
6.8	.999393	.999309	.999222	.999040	.998563	.997655	.996930	6.8
6.9	.999443	.999364	.999282	.999110	.998655	.997777	.997068	6.9
7.0	.999489	.999414	.999337	.999174	.998740	.997892	.997199	7.0
7.1	.999531	.999461	.999388	.999234	.998820	.998001	.997324	7.1
7.2	.999569	.999503	.999435	.999289	.998894	.998104	.997442	7.2
7.3	.999604	.999543	.999478	.999340	.998964	.998202	.997555	7.3
7.4	.999636	.999579	.999518	.999387	.999029	.998294	.997663	7.4

Table A.4 Cumulative Distribution Function of the Standardized Gamma Distribution $(\alpha_3; 0, 1)$

z \ α_3	0.0	0.1	0.2	0.3	0.4	0.5	0.6	z
-3.0	.001350	.000824	.000430	.000177	.000047	.000005	.000000	-3.0
-2.9	.001866	.001199	.000673	.000309	.000099	.000016	.000000	-2.9
-2.8	.002555	.001721	.001034	.000523	.000199	.000043	.000002	-2.8
-2.7	.003467	.002440	.001557	.000862	.000378	.000108	.000011	-2.7
-2.6	.004661	.003416	.002305	.001382	.000686	.000244	.000042	-2.6
-2.5	.006210	.004724	.003352	.002157	.001192	.000509	.000130	-2.5
-2.4	.008198	.006454	.004794	.003286	.001994	.000992	.000340	-2.4
-2.3	.010724	.008713	.006745	.004890	.003217	.001816	.000783	-2.3
-2.2	.013903	.011624	.009340	.007116	.005020	.003149	.001623	-2.2
-2.1	.017864	.015330	.012736	.010136	.007594	.005202	.003088	-2.1
-2.0	.022750	.019989	.017108	.014150	.011165	.008231	.005468	-2.0
-1.9	.028717	.025774	.022652	.019377	.015982	.012525	.009103	-1.9
-1.8	.035930	.032869	.029572	.026050	.022315	.018402	.014372	-1.8
-1.7	.044565	.041467	.038086	.034414	.030445	.026188	.021666	-1.7
-1.6	.054799	.051762	.048407	.044709	.040646	.036202	.031368	-1.6
-1.5	.066807	.063945	.060744	.057165	.053176	.048740	.043821	-1.5
-1.4	.080757	.078193	.075288	.071990	.068260	.064051	.059306	-1.4
-1.3	.096800	.094665	.092202	.089353	.086077	.082321	.078021	-1.3
-1.2	.115070	.113492	.111616	.109379	.106746	.103663	.100068	-1.2
-1.1	.135666	.134766	.133612	.132139	.130319	.128104	.125440	-1.1
-1.0	.158655	.158539	.158221	.157639	.156773	.155584	.154029	-1.0
-.9	.184060	.184811	.185416	.185819	.186007	.185953	.185624	-.9
-.8	.211855	.213527	.215109	.216549	.217845	.218978	.219926	-.8
-.7	.241964	.244575	.247147	.249634	.252039	.254353	.256561	-.7
-.6	.274253	.277785	.281319	.284815	.288281	.291711	.295099	-.6
-.5	.308538	.312931	.317357	.321779	.326207	.330640	.335076	-.5
-.4	.344578	.349737	.354944	.360167	.365419	.370699	.376006	-.4
-.3	.382089	.387881	.393723	.399591	.405494	.411432	.417406	-.3
-.2	.420740	.427005	.433311	.439640	.445999	.452389	.458809	-.2
-.1	.460172	.466727	.473304	.479896	.486505	.493132	.499776	-.1
.0	.500000	.506647	.513299	.519949	.526602	.533255	.539910	.0
.1	.539828	.546371	.552896	.559408	.565906	.572390	.578861	.1
.2	.579260	.585506	.591719	.597909	.604073	.610214	.616330	.2
.3	.617911	.623686	.629417	.635125	.640803	.646453	.652075	.3
.4	.655422	.660573	.665682	.670774	.675842	.680887	.685909	.4
.5	.691462	.695869	.700245	.704622	.708990	.713347	.717694	.5
.6	.725747	.729322	.732891	.736487	.740096	.743716	.747346	.6
.7	.758036	.760731	.763453	.766235	.769060	.771923	.774822	.7
.8	.788145	.789945	.791815	.793782	.795826	.797941	.800121	.8
.9	.815940	.816868	.817911	.819091	.820384	.821782	.823273	.9
1.0	.841345	.841454	.841721	.842165	.842758	.843487	.844340	1.0
1.1	.864334	.863701	.863268	.863045	.863004	.863128	.863403	1.1
1.2	.884930	.883651	.882608	.881802	.881205	.880797	.880562	1.2
1.3	.903200	.901384	.899830	.898533	.897464	.896602	.895931	1.3
1.4	.919243	.917007	.915046	.913353	.911899	.910664	.909629	1.4

Source: From Cohen and Whitten (1988), Table A.3.4, pp. 316–323, by courtesy of Marcel Dekker, Inc.

Table A.4 *Continued*

z \ α_3	0.0	0.1	0.2	0.3	0.4	0.5	0.6	z
1.5	.933193	.930652	.928388	.926393	.924639	.923108	.921782	1.5
1.6	.945201	.942465	.939998	.937790	.935819	.934066	.932518	1.6
1.7	.955435	.952606	.950027	.947688	.945574	.943669	.941960	1.7
1.8	.964070	.961237	.958626	.956230	.954040	.952044	.950232	1.8
1.9	.971283	.968521	.965947	.963557	.961348	.959315	.957449	1.9
2.0	.977250	.974619	.972136	.969804	.967626	.965600	.963723	2.0
2.1	.982136	.979680	.977332	.975100	.972991	.971010	.969158	2.1
2.2	.986097	.983847	.981665	.979563	.977554	.975648	.973848	2.2
2.3	.989276	.987250	.985254	.983303	.981417	.979608	.977883	2.3
2.4	.991802	.990007	.988207	.986421	.984672	.982976	.981343	2.4
2.5	.993790	.992222	.990621	.989006	.987403	.985830	.984301	2.5
2.6	.995339	.993989	.992582	.991138	.989685	.988240	.986821	2.6
2.7	.996533	.995386	.994165	.992888	.991582	.990268	.988962	2.7
2.8	.997445	.996484	.995435	.994316	.993154	.991969	.990777	2.8
2.9	.998134	.997338	.996447	.995477	.994451	.993390	.992310	2.9
3.0	.998650	.998001	.997250	.996415	.995517	.994574	.993602	3.0
3.1	.999032	.998508	.997882	.997170	.996390	.995558	.994600	3.1
3.2	.999313	.998894	.998377	.997775	.997102	.996372	.995599	3.2
3.3	.999517	.999185	.998763	.998258	.997681	.997045	.996360	3.3
3.4	.999663	.999404	.999062	.998641	.998150	.997598	.996996	3.4
3.5	.999767	.999567	.999292	.998944	.998528	.998053	.997526	3.5
3.6	.999841	.999688	.999468	.999183	.998833	.998425	.997965	3.6
3.7	.999892	.999776	.999603	.999369	.999077	.998728	.998330	3.7
3.8	.999928	.999840	.999705	.999515	.999272	.998976	.998632	3.8
3.9	.999952	.999887	.999781	.999629	.999427	.999177	.998881	3.9
4.0	.999968	.999921	.999839	.999717	.999551	.999340	.999086	4.0
4.1	.999979	.999945	.999882	.999785	.999649	.999472	.999255	4.1
4.2	.999987	.999962	.999914	.999837	.999726	.999578	.999394	4.2
4.3	.999991	.999973	.999937	.999877	.999787	.999664	.999507	4.3
4.4	.999995	.999982	.999955	.999907	.999834	.999733	.999600	4.4
4.5	.999997	.999988	.999967	.999930	.999872	.999788	.999676	4.5
4.6	.999998	.999992	.999977	.999948	.999901	.999832	.999738	4.6
4.7	.999999	.999994	.999983	.999961	.999924	.999867	.999788	4.7
4.8	.999999	.999996	.999988	.999971	.999941	.999895	.999829	4.8
4.9	1.000000	.999998	.999992	.999979	.999955	.999917	.999862	4.9
5.0	1.000000	.999998	.999994	.999984	.999965	.999935	.999889	5.0
5.1	1.000000	.999999	.999996	.999988	.999974	.999949	.999911	5.1
5.2	1.000000	.999999	.999997	.999991	.999980	.999960	.999928	5.2
5.3	1.000000	1.000000	.999998	.999994	.999985	.999968	.999943	5.3
5.4	1.000000	1.000000	.999999	.999995	.999988	.999975	.999954	5.4
5.5	1.000000	1.000000	.999999	.999997	.999991	.999981	.999963	5.5
5.6	1.000000	1.000000	.999999	.999998	.999993	.999985	.999971	5.6
5.7	1.000000	1.000000	1.000000	.999998	.999995	.999988	.999977	5.7
5.8	1.000000	1.000000	1.000000	.999999	.999996	.999991	.999981	5.8
5.9	1.000000	1.000000	1.000000	.999999	.999997	.999993	.999985	5.9

(Continued)

Table A.4 *Continued*

z \ α_3	0.7	0.8	0.9	1.0	1.1	1.2	1.3	z
-3.0								-3.0
-2.9								-2.9
-2.8	.000000							-2.8
-2.7	.000000							-2.7
-2.6	.000001							-2.6
-2.5	.000008	.000000						-2.5
-2.4	.000049	.000000						-2.4
-2.3	.000191	.000007						-2.3
-2.2	.000571	.000075	.000000					-2.2
-2.1	.001417	.000368	.000012					-2.1
-2.0	.003047	.001203	.000189	.000000				-2.0
-1.9	.005866	.003047	.000986	.000057				-1.9
-1.8	.010335	.006480	.003126	.000776	.000001			-1.8
-1.7	.016941	.012130	.007461	.003358	.000589			-1.7
-1.6	.026153	.020602	.014827	.009080	.003899	.000451		-1.6
-1.5	.038387	.032418	.025926	.018988	.011839	.005089	.000414	-1.5
-1.4	.053967	.047970	.041252	.033769	.025523	.016655	.007706	-1.4
-1.3	.073105	.067486	.061064	.053725	.045348	.035820	.025095	-1.3
-1.2	.095885	.091026	.085380	.078813	.071164	.062229	.051767	-1.2
-1.1	.122259	.118480	.114005	.108708	.102438	.094997	.086132	-1.1
-1.0	.152053	.149592	.146565	.142877	.138403	.132993	.126446	-1.0
-.9	.184981	.183978	.182556	.180648	.178168	.175012	.171048	-.9
-.8	.220664	.221160	.221379	.221277	.220800	.219885	.218454	-.8
-.7	.258649	.260599	.262391	.263998	.265392	.266536	.267388	-.7
-.6	.298439	.301720	.304932	.308063	.311095	.314012	.316792	-.6
-.5	.339511	.343942	.348363	.352768	.357150	.361500	.365807	-.5
-.4	.381340	.386698	.392079	.397480	.402897	.408324	.413755	-.4
-.3	.423415	.429459	.435535	.441643	.447779	.453939	.460119	-.3
-.2	.465260	.471740	.478249	.484784	.491343	.497922	.504517	-.2
-.1	.506438	.513116	.519809	.526515	.533232	.539956	.546683	-.1
.0	.546566	.553222	.559878	.566530	.573177	.579816	.586444	.0
.1	.585317	.591759	.598186	.604597	.610988	.617359	.623706	.1
.2	.622422	.628490	.634534	.640552	.646544	.652506	.658437	.2
.3	.657671	.663239	.668780	.674294	.679778	.685232	.690654	.3
.4	.690908	.695885	.700839	.705770	.710677	.715558	.720411	.4
.5	.722031	.726358	.730672	.734974	.739262	.743535	.747790	.5
.6	.750985	.754630	.758281	.761935	.765590	.769244	.772895	.6
.7	.777754	.780714	.783700	.786709	.789738	.792783	.795841	.7
.8	.802359	.804651	.806992	.809378	.811803	.814263	.816755	.8
.9	.824852	.826510	.828240	.830037	.831894	.833806	.835767	.9
1.0	.845306	.846377	.847543	.848796	.850129	.851535	.853008	1.0
1.1	.863815	.864354	.865009	.865771	.866630	.867578	.868609	1.1
1.2	.880486	.880554	.880756	.881081	.881519	.882060	.882697	1.2
1.3	.895433	.895095	.894904	.894849	.894919	.895103	.895394	1.3
1.4	.908778	.908098	.907574	.907194	.906949	.906827	.906819	1.4

Table A.4 *Continued*

z \ α_3	0.7	0.8	0.9	1.0	1.1	1.2	1.3	z
1.5	.920646	.919684	.918884	.918235	.917724	.917343	.917081	1.5
1.6	.931158	.929973	.928952	.928083	.927355	.926758	.926284	1.6
1.7	.940435	.939082	.937889	.936847	.935945	.935175	.934527	1.7
1.8	.948594	.947121	.945802	.944629	.943593	.942685	.941899	1.8
1.9	.955745	.954194	.952790	.951523	.950389	.949378	.948484	1.9
2.0	.961992	.960402	.958946	.957620	.956417	.955332	.954360	2.0
2.1	.967433	.965834	.964358	.963000	.961757	.960623	.959595	2.1
2.2	.972158	.970577	.969105	.967740	.966479	.965319	.964257	2.2
2.3	.976249	.974708	.973260	.971907	.970648	.969480	.968402	2.3
2.4	.979782	.978296	.976890	.975566	.974324	.973164	.972085	2.4
2.5	.982824	.981407	.980056	.978774	.977562	.976422	.975355	2.5
2.6	.985437	.984098	.982811	.981580	.980409	.979300	.978255	2.6
2.7	.987676	.986422	.985205	.984033	.982911	.981840	.980825	2.7
2.8	.989591	.988423	.987282	.986174	.985105	.984080	.983100	2.8
2.9	.991224	.990144	.989080	.988040	.987029	.986052	.985114	2.9
3.0	.992614	.991622	.990636	.989664	.988713	.987788	.986895	3.0
3.1	.993794	.992888	.991979	.991076	.990186	.989315	.988468	3.1
3.2	.994794	.993970	.993137	.992302	.991473	.990657	.989857	3.2
3.3	.995640	.994895	.994134	.993365	.992597	.991834	.991083	3.3
3.4	.996355	.995683	.994991	.994287	.993577	.992867	.992165	3.4
3.5	.996957	.996354	.995728	.995084	.994431	.993773	.993118	3.5
3.6	.997463	.996925	.996359	.995774	.995174	.994567	.993958	3.6
3.7	.997888	.997409	.996901	.996369	.995821	.995262	.994697	3.7
3.8	.998244	.997820	.997364	.996883	.996383	.995870	.995348	3.8
3.9	.998543	.998167	.997760	.997326	.996872	.996402	.995920	3.9
4.0	.998792	.998461	.998098	.997708	.997296	.996866	.996424	4.0
4.1	.999000	.998709	.998387	.998037	.997664	.997272	.996866	4.1
4.2	.999173	.998918	.998632	.998319	.997983	.997627	.997254	4.2
4.3	.999317	.999094	.998842	.998562	.998259	.997936	.997596	4.3
4.4	.999436	.999242	.999020	.998771	.998498	.998205	.997895	4.4
4.5	.999536	.999367	.999171	.998950	.998705	.998440	.998158	4.5
4.6	.999618	.999472	.999300	.999103	.998884	.998645	.998388	4.6
4.7	.999686	.999559	.999408	.999235	.999039	.998823	.998590	4.7
4.8	.999742	.999633	.999501	.999347	.999173	.998979	.998767	4.8
4.9	.999788	.999694	.999579	.999443	.999288	.999114	.998922	4.9
5.0	.999827	.999746	.999645	.999526	.999387	.999231	.999058	5.0
5.1	.999858	.999788	.999701	.999596	.999473	.999333	.999177	5.1
5.2	.999884	.999824	.999749	.999656	.999547	.999422	.999281	5.2
5.3	.999905	.999854	.999789	.999708	.999611	.999499	.999372	5.3
5.4	.999923	.999879	.999822	.999751	.999666	.999566	.999452	5.4
5.5	.999937	.999900	.999851	.999789	.999713	.999624	.999522	5.5
5.6	.999949	.999917	.999875	.999820	.999754	.999675	.999583	5.6
5.7	.999958	.999931	.999895	.999848	.999789	.999718	.999636	5.7
5.8	.999966	.999943	.999912	.999871	.999819	.999756	.999682	5.8
5.9	.999972	.999953	.999926	.999890	.999845	.999789	.999723	5.9

(Continued)

Table A.4 *Continued*

z \ α_3	1.4	1.5	1.6	1.7	1.8	1.9	2.0	z
-1.5								-1.5
-1.4	.000685							-1.4
-1.3	.013409	.002331						-1.3
-1.2	.039499	.025184	.009098					-1.2
-1.1	.075506	.062662	.046967	.027549	.003895			-1.1
-1.0	.118505	.108815	.096882	.081970	.062896	.037476	.000000	-1.0
-.9	.166109	.159975	.152353	.142839	.130847	.115471	.095163	-.9
-.8	.216408	.213628	.209961	.205208	.199108	.191299	.181269	-.8
-.7	.267897	.268002	.267630	.266688	.265067	.262625	.259182	-.7
-.6	.319410	.321836	.324036	.325972	.327596	.328853	.329680	-.6
-.5	.370058	.374240	.378336	.382328	.386197	.389919	.393469	-.5
-.4	.419184	.424602	.430001	.435370	.440699	.445976	.451188	-.4
-.3	.466314	.472518	.478724	.484924	.491112	.497278	.503415	-.3
-.2	.511124	.517738	.524351	.530958	.537552	.544125	.550671	-.2
-.1	.553409	.560129	.566837	.573528	.580196	.586831	.593430	-.1
.0	.593056	.599648	.606215	.612751	.619251	.625710	.632121	.0
.1	.630026	.636314	.642567	.648780	.654948	.661066	.667129	.1
.2	.664335	.670195	.676015	.681790	.687516	.693190	.698806	.2
.3	.696042	.701392	.706702	.711968	.717187	.722355	.727468	.3
.4	.725235	.730027	.734784	.739504	.744182	.748817	.753403	.4
.5	.752026	.756239	.760428	.764588	.768718	.772813	.776870	.5
.6	.776539	.780175	.783798	.787406	.790995	.794562	.798103	.6
.7	.798909	.801983	.805060	.808135	.811206	.814267	.817316	.7
.8	.819273	.821814	.824373	.826945	.829527	.832113	.834701	.8
.9	.837771	.839814	.841891	.843995	.846124	.848270	.850431	.9
1.0	.854540	.856126	.857760	.859437	.861150	.862894	.864665	1.0
1.1	.869714	.870886	.872120	.873408	.874746	.876126	.877544	1.1
1.2	.883420	.884224	.885100	.886041	.887042	.888096	.889197	1.2
1.3	.895783	.896260	.896820	.897455	.898157	.898922	.899741	1.3
1.4	.906916	.907111	.907395	.907760	.908201	.908711	.909282	1.4
1.5	.916929	.916881	.916927	.917060	.917273	.917561	.917915	1.5
1.6	.925924	.925669	.925513	.925447	.925465	.925560	.925726	1.6
1.7	.933994	.933568	.933241	.933007	.932858	.932790	.932794	1.7
1.8	.941226	.940660	.940192	.939818	.939530	.939323	.939190	1.8
1.9	.947701	.947022	.946441	.945952	.945548	.945225	.944977	1.9
2.0	.953493	.952727	.952055	.951474	.950976	.950557	.950213	2.0
2.1	.958668	.957837	.957096	.956442	.955870	.955374	.954951	2.1
2.2	.963289	.962411	.961620	.960911	.960281	.959724	.959238	2.2
2.3	.967411	.966504	.965678	.964930	.964256	.963653	.963117	2.3
2.4	.971086	.970164	.969317	.968542	.967838	.967200	.966627	2.4
2.5	.974359	.973434	.972577	.971788	.971064	.970403	.969803	2.5
2.6	.977273	.976354	.975498	.974704	.973971	.973295	.972676	2.6
2.7	.979865	.978961	.978114	.977323	.976588	.975906	.975276	2.7
2.8	.982169	.981287	.980456	.979675	.978944	.978262	.977629	2.8
2.9	.984216	.983362	.982551	.981786	.981065	.980389	.979758	2.9

Table A.4 *Continued*

z \ α_3	1.4	1.5	1.6	1.7	1.8	1.9	2.0	z
3.0	.986034	.985211	.984426	.983680	.982974	.982309	.981684	3.0
3.1	.987648	.986859	.986102	.985379	.984692	.984042	.983427	3.1
3.2	.989079	.988326	.987600	.986904	.986239	.985605	.985004	3.2
3.3	.990348	.989633	.988940	.988271	.987630	.987016	.986431	3.3
3.4	.991473	.990795	.990136	.989497	.988881	.988289	.987723	3.4
3.5	.992469	.991830	.991205	.990597	.990007	.989438	.988891	3.5
3.6	.993351	.992750	.992159	.991582	.991019	.990474	.989948	3.6
3.7	.994131	.993568	.993012	.992465	.991930	.991409	.990905	3.7
3.8	.994821	.994295	.993772	.993256	.992749	.992253	.991770	3.8
3.9	.995432	.994941	.994451	.993965	.993485	.993014	.992553	3.9
4.0	.995972	.995515	.995057	.994600	.994147	.993700	.993262	4.0
4.1	.996449	.996025	.995597	.995168	.994742	.994319	.993903	4.1
4.2	.996870	.996477	.996078	.995677	.995276	.994878	.994483	4.2
4.3	.997242	.996879	.996508	.996133	.995757	.995381	.995008	4.3
4.4	.997571	.997235	.996891	.996542	.996189	.995836	.995483	4.4
4.5	.997860	.997551	.997232	.996907	.996577	.996245	.995913	4.5
4.6	.998116	.997831	.997537	.997234	.996926	.996615	.996302	4.6
4.7	.998342	.998080	.997808	.997527	.997239	.996948	.996654	4.7
4.8	.998540	.998300	.998049	.997789	.997521	.997248	.996972	4.8
4.9	.998716	.998496	.998264	.998023	.997774	.997519	.997261	4.9
5.0	.998870	.998669	.998456	.998233	.998001	.997764	.997521	5.0
5.1	.999006	.998822	.998626	.998420	.998206	.997984	.997757	5.1
5.2	.999126	.998958	.998778	.998588	.998389	.998183	.997971	5.2
5.3	.999232	.999078	.998913	.998738	.998554	.998362	.998164	5.3
5.4	.999325	.999185	.999034	.998872	.998702	.998523	.998338	5.4
5.5	.999407	.999279	.999141	.998992	.998834	.998669	.998497	5.5
5.6	.999479	.999363	.999236	.999099	.998954	.998800	.998640	5.6
5.7	.999542	.999437	.999321	.999195	.999061	.998918	.998769	5.7
5.8	.999598	.999502	.999396	.999281	.999157	.999025	.998886	5.8
5.9	.999647	.999560	.999463	.999358	.999243	.999121	.998992	5.9
6.0	.999690	.999611	.999523	.999426	.999321	.999208	.999088	6.0
6.1	.999728	.999656	.999576	.999487	.999391	.999286	.999175	6.1
6.2	.999761	.999696	.999623	.999542	.999453	.999357	.999253	6.2
6.3	.999790	.999732	.999665	.999591	.999509	.999420	.999324	6.3
6.4	.999816	.999763	.999703	.999635	.999560	.999477	.999389	6.4
6.5	.999838	.999791	.999736	.999674	.999605	.999529	.999447	6.5
6.6	.999858	.999815	.999765	.999709	.999645	.999576	.999500	6.6
6.7	.999876	.999837	.999792	.999740	.999682	.999617	.999547	6.7
6.8	.999891	.999856	.999815	.999768	.999714	.999655	.999590	6.8
6.9	.999904	.999873	.999836	.999793	.999744	.999689	.999629	6.9
7.0	.999916	.999888	.999854	.999815	.999770	.999720	.999665	7.0
7.1	.999926	.999901	.999870	.999835	.999794	.999748	.999696	7.1
7.2	.999935	.999913	.999885	.999852	.999815	.999773	.999725	7.2
7.3	.999943	.999923	.999898	.999868	.999834	.999795	.999751	7.3
7.4	.999950	.999932	.999909	.999882	.999851	.999815	.999775	7.4

(Continued)

Table A.4 *Continued*

z \ α_3	2.1	2.2	2.3	2.5	3.0	4.0	5.0	z
-1.5								-1.5
-1.4								-1.4
-1.3								-1.3
-1.2								-1.2
-1.1								-1.1
-1.0								-1.0
-.9	.066775	.020163						-.9
-.8	.168238	.150925	.126921	.000000				-.8
-.7	.254500	.248258	.239997	.214268				-.7
-.6	.329998	.329714	.328711	.323915	.279143			-.6
-.5	.396821	.399943	.402800	.407558	.411126	.000000		-.5
-.4	.456324	.461371	.466313	.475830	.496757	.516555	.000000	-.4
-.3	.509513	.515565	.521561	.533355	.561337	.608339	.639089	-.3
-.2	.557182	.563651	.570070	.582735	.613116	.666828	.710210	-.2
-.1	.599986	.606491	.612941	.625647	.656061	.709851	.753804	-.1
.0	.638478	.644776	.651010	.663264	.692458	.743678	.785204	.0
.1	.673132	.679071	.684941	.696455	.723770	.771331	.809575	.1
.2	.704361	.709850	.715271	.725889	.751006	.794523	.829335	.2
.3	.732524	.737518	.742446	.752097	.774898	.814332	.845823	.3
.4	.757939	.762420	.766844	.775509	.795998	.831482	.859865	.4
.5	.780886	.784858	.788783	.796481	.814732	.846486	.872005	.5
.6	.801616	.805096	.808540	.815311	.831442	.859725	.882627	.6
.7	.820349	.823363	.826353	.832252	.846402	.871487	.892008	.7
.8	.837286	.839863	.842430	.847519	.859839	.881996	.900358	.8
.9	.852602	.854778	.856955	.861298	.871942	.891431	.907836	.9
1.0	.866457	.868266	.870087	.873751	.882869	.899937	.914571	1.0
1.1	.878993	.880470	.881970	.885018	.892756	.907633	.920665	1.1
1.2	.890339	.891518	.892728	.895224	.901718	.914618	.926200	1.2
1.3	.900610	.901523	.902476	.904477	.909857	.920975	.931246	1.3
1.4	.909910	.910588	.911311	.912874	.917258	.926775	.935860	1.4
1.5	.918331	.918802	.919324	.920498	.923998	.932079	.940091	1.5
1.6	.925958	.926249	.926594	.927428	.930144	.936938	.943980	1.6
1.7	.932867	.933002	.933194	.933729	.935754	.941398	.947563	1.7
1.8	.939126	.939127	.939186	.939463	.940881	.945498	.950870	1.8
1.9	.944798	.944684	.944630	.944683	.945570	.949273	.953928	1.9
2.0	.949938	.949727	.949576	.949437	.949863	.952753	.956761	2.0
2.1	.954596	.954304	.954072	.953770	.953795	.955966	.959390	2.1
2.2	.958818	.958460	.958160	.957721	.957402	.958935	.961831	2.2
2.3	.962645	.962233	.961878	.961324	.960710	.961682	.964103	2.3
2.4	.966114	.965659	.965260	.964612	.963748	.964226	.966218	2.4
2.5	.969260	.968772	.968337	.967613	.966539	.966584	.968191	2.5
2.6	.972112	.971600	.971138	.970354	.969105	.968771	.970032	2.6
2.7	.974698	.974169	.973687	.972857	.971464	.970802	.971752	2.7
2.8	.977043	.976504	.976008	.975143	.973636	.972689	.973360	2.8
2.9	.979170	.978625	.978122	.977234	.975635	.974444	.974865	2.9

Table A.4 *Continued*

z \ α_3	2.1	2.2	2.3	2.5	3.0	4.0	5.0	z
3.0	.981100	.980554	.980047	.979144	.977476	.976077	.976275	3.0
3.1	.982849	.982307	.981801	.980891	.979172	.977597	.977596	3.1
3.2	.984436	.983901	.983399	.982489	.980736	.979014	.978836	3.2
3.3	.985876	.985351	.984855	.983951	.982178	.980334	.979999	3.3
3.4	.987182	.986669	.986182	.985289	.983508	.981565	.981092	3.4
3.5	.988367	.987867	.987391	.986513	.984736	.982714	.982119	3.5
3.6	.989442	.988957	.988494	.987634	.985869	.983787	.983085	3.6
3.7	.990418	.989949	.989499	.988660	.986916	.984788	.983993	3.7
3.8	.991303	.990851	.990416	.989600	.987882	.985724	.984848	3.8
3.9	.992106	.991671	.991252	.990461	.988775	.986599	.985653	3.9
4.0	.992834	.992418	.992014	.991249	.989601	.987417	.986412	4.0
4.1	.993495	.993097	.992710	.991971	.990364	.988183	.987128	4.1
4.2	.994095	.993715	.993344	.992633	.991069	.988899	.987802	4.2
4.3	.994640	.994278	.993923	.993240	.991722	.989570	.988438	4.3
4.4	.995134	.994790	.994451	.993796	.992325	.990198	.989039	4.4
4.5	.995583	.995256	.994933	.994306	.992884	.990788	.989606	4.5
4.6	.995990	.995680	.995373	.994774	.993401	.991338	.990141	4.6
4.7	.996359	.996066	.995774	.995202	.993879	.991854	.990647	4.7
4.8	.996695	.996417	.996140	.995595	.994322	.992339	.991125	4.8
4.9	.996999	.996737	.996475	.995956	.994732	.992793	.991577	4.9
5.0	.997276	.997028	.996780	.996287	.995112	.993219	.992005	5.0
5.1	.997526	.997293	.997058	.996590	.995465	.993619	.992409	5.1
5.2	.997754	.997534	.997313	.996868	.995791	.993995	.992792	5.2
5.3	.997961	.997754	.997545	.997124	.996093	.994347	.993154	5.3
5.4	.998148	.997954	.997757	.997358	.996373	.994678	.993497	5.4
5.5	.998319	.998136	.997951	.997573	.996633	.994989	.993822	5.5
5.6	.998473	.998302	.998128	.997771	.996874	.995281	.994129	5.6
5.7	.998614	.998453	.998289	.997952	.997097	.995555	.994421	5.7
5.8	.998741	.998591	.998437	.998118	.997304	.995813	.994697	5.8
5.9	.998857	.998716	.998571	.998271	.997496	.996056	.994958	5.9
6.0	.998962	.998831	.998694	.998411	.997674	.996284	.995207	6.0
6.1	.999057	.998935	.998807	.998540	.997839	.996498	.995442	6.1
6.2	.999144	.999029	.998910	.998659	.997993	.996699	.995665	6.2
6.3	.999223	.999115	.999003	.998767	.998135	.996889	.995877	6.3
6.4	.999294	.999194	.999089	.998867	.998267	.997067	.996078	6.4
6.5	.999359	.999266	.999167	.998959	.998389	.997235	.996268	6.5
6.6	.999418	.999331	.999239	.999043	.998503	.997393	.996449	6.6
6.7	.999471	.999390	.999304	.999120	.998609	.997541	.996621	6.7
6.8	.999520	.999444	.999364	.999191	.998707	.997681	.996784	6.8
6.9	.999564	.999494	.999419	.999256	.998798	.997813	.996939	6.9
7.0	.999604	.999539	.999469	.999316	.998883	.997937	.997086	7.0
7.1	.999640	.999579	.999514	.999371	.998961	.998054	.997226	7.1
7.2	.999673	.999617	.999556	.999422	.999034	.998164	.997358	7.2
7.3	.999703	.999651	.999594	.999469	.999102	.998268	.997485	7.3
7.4	.999731	.999682	.999629	.999511	.999165	.998365	.997604	7.4

Glossary

Glossary

THE GREEK ALPHABET

A	α	alpha
B	β	beta
Γ	γ	gamma
Δ	δ	delta
E	ε	epsilon
Z	ζ	zeta
H	η	eta
Θ	θ	theta
I	ι	iota
K	κ	kappa
Λ	λ	lambda
M	μ	mu
N	ν	nu
Ξ	ξ	xi
O	o	omicron
Π	π	pi
P	ρ	rho
Σ	σ	sigma
T	τ	tau
Υ	υ	upsilon
Φ	φ	phi
X	χ	chi
Ψ	ψ	psi
Ω	ω	omega

A, B, C, D, E, etc.	Arbitrary constants or functions—definitions vary from chapter to chapter.
$\alpha_3 = E\{[X - E(X)]/\sigma\}^3$	Population third standard moment; a measure of skewness.
$\bar{\alpha}_k$	kth standardized moment of a truncated distribution; used in Chapter 2.
$a_3 = \left[\sum_1^n (x_i - \bar{x})^3/n\right] \Big/ \left[\sum_1^n (x_i - \bar{x})^2/n\right]^{3/2}$	Sample third standard moment.
$\alpha_4 = E\{[X - E(X)]/\sigma\}^4$	Population fourth standard moment; a measure of kurtosis.
$a_4 = \left[\sum_1^n (x_i - \bar{x})^4/n\right] \Big/ \left[\sum_1^n (x_i - \bar{x})^2/n\right]^2$	Sample fourth standard moment.
α	A parameter of the Pareto distribution; also a parameter of the extreme value distribution, and elsewhere.
$\alpha(\xi) = \{1 - Q(\xi)[Q(\xi) - \xi]\}/[Q(\xi) - \xi]^2$	A function of the standardized terminus in singly truncated samples from the normal distribution.
$\alpha(h, \xi) = [1 - \Omega(\Omega - \xi)]/(\Omega - \xi)^2$	A function of the standardized terminus and the proportion of censored observations in singly censored samples from the normal distribution, $h = c/N$.

Arranged in alphabetical order with Greek characters inserted according to their English pronunciation.

Glossary

$\hat{\alpha} = \alpha(\hat{\xi}) = s^2/(\bar{x} - T)^2$	Maximum likelihood estimates of α in singly truncated and singly censored samples from a normal distribution.
β	A scale parameter in various distributions.
β_1 and β_2	Pearson's betas; $\beta_1 = \alpha_3^2$ and $\beta_2 = \alpha_4$.
c	Number of censored observations in a censored sample; n_0 is sometimes used as an alternate symbol for c.
c_j	Number of observations censored at point (time) T_j; $c = \sum_1^k c_j$.
cdf	Abbreviation for cumulative distribution function.
CV	Coefficient of variation (sometimes abbreviated as v); $CV = \sigma/[E(X) - \gamma]$.
δ	The Weibull shape parameter.
$e = 2.7182818\ldots$	Base of natural logarithms.
$E(\ )$	The expectation symbol.
$f(\ )$, $g(\ )$, and $h(\ )$	Symbols for probability density functions.
$F(\ )$, $G(\ )$, and $H(\ )$	Symbols for cumulative density functions.
γ	A threshold parameter in various skewed distributions.
$\Gamma(z) = \int_0^\infty x^{z-1} e^{-x}\, dx$	The gamma function.
$\Gamma_k = \Gamma(1 + k/\delta)$	A function of the Weibull shape parameter.
Γ'	Derivative of the gamma function.

h	Proportion of censored observations in censored samples; $h = c/N$ except for the Rayleigh distributions where $h = n_0/n$ (i.e., $h = c/n$) (cf. Chapter 9).
$h(x) = f(x)/[1 - F(x)]$	The hazard function.
$H(x) = \int_0^x h(t)\, dt$	The cumulative hazard function.
$H_1(\xi_1, \xi_2)$ and $H_2(\xi_1, \xi_2)$	Estimating functions for doubly truncated samples from the normal distribution.
$H_3(z; h)$	Rayleigh distribution censored sample estimating function ($h = c/n$).
$J(n, \sigma)$	The lognormal estimating function.
$J_2(z)$	Rayleigh distribution truncated sample estimating function, $p = 2$.
$J_3(z)$	Rayleigh distribution truncated sample estimating function, $p = 3$.
$J_1(\xi_1, \xi_2)$ and $J_2(\xi_1, \xi_2)$	Estimating functions for centrally truncated samples from the normal distribution.
κ_r	Symbol for the rth order cumulant.
λ	The Poisson parameter; also used as a symbol for various functions.
$\lambda(\alpha, h) = \lambda(\xi, h) = \Omega(\xi, h)/[\Omega(\xi, h) - \xi]$	Normal distribution estimating function for singly censored samples.
$L(\ )$	Symbol for the likelihood function.
$\ln L(\ )$	Symbol for the loglikelihood function.
$\ln(\ )$	Symbol for the natural logarithm (base e).
$\log(\ )$	Symbol for common logarithm (base 10).
$Mo(X)$	Mode of X.

Glossary

Me(X)	Median of X.
$M_x(t) = E(e^{tX})$	Moment generating function.
μ	Symbol for a distribution mean; in the three-parameter inverse Gaussian distribution, mean $= \gamma + \mu$.
μ_k	Symbol for the kth central moment.
μ'_k	Symbol for the kth noncentral moment.
$\bar{\mu}_k$	Symbol for kth central truncated moment.
$\bar{\mu}'_k$	Symbol for kth noncentral truncated moment.
μ_{ij}	Symbol for variance–covariance factors used with truncated and censored normal samples.
n	Number of complete observations in a sample.
N	Total number of observations in a sample; in censored samples $N = n + c$; in truncated samples and in complete samples $N = n$.
v_1	First moment of truncated or censored sample about left terminus, $v_1 = \bar{x} - T_1$.
ω	An alternate shape parameter for the lognormal distribution where $\omega = \exp(\sigma^2)$.
$\Omega(h, \xi) = [h/(1 - h)]Q(-\xi)$	An estimating function involved in singly censored samples from the normal distribution.
$\Omega_1(a_1, \xi_1) = a_1 Q(-\xi_1)$ and $\Omega_2(a_2, \xi_2) = a_2 Q(\xi_2)$	Estimating functions involved in doubly censored samples from the normal distribution, $a_1 = c_1/n$ and $a_2 = c_2/n$.

p	A dimension symbol in the multidimensional Rayleigh distribution; also a symbol for probability.
pdf	Abbreviation for probability density function.
$\phi(z) = (1/\sqrt{2\pi}) \exp(-z^2/2)$	pdf of the standard normal distribution.
$\Phi(z) = \int_{-\infty}^{z} \phi(t)\, dt$	cdf of the standard normal distribution.
ϕ_{ij}	Symbol for variance–covariance factors used with complete samples.
π	Symbol for pi = 3.14159 . . .
$\psi(x) = \partial \ln \Gamma(x)/\partial x = \Gamma'(x)/\Gamma(x)$	The digamma function.
$\psi'(x) = \partial \psi(x)/\partial x$	The trigamma function.
$\psi(t)$	Used in Chapter 3 as a symbol for the cumulant generating function.
$Q(\xi) = \phi(\xi)/[1 - \Phi(\xi)]$	Involved in singly truncated normal distribution samples.
$\bar{Q}_1(\xi_1, \xi_2) = \phi(\xi_1)/[\Phi(\xi_2) - \Phi(\xi_1)]$	Involved in doubly truncated normal distribution samples.
$\bar{Q}_2(\xi_1, \xi_2) = \phi(\xi_2)/[\Phi(\xi_2) - \Phi(\xi_1)]$	Involved in doubly truncated normal distribution samples.
ρ	Shape parameter of the gamma distribution.
σ	Symbol for distribution standard deviation; a shape parameter in the lognormal distribution.
$s = \left[\sum_{1}^{n} (x_i - \bar{x})/(n-1) \right]^{1/2}$	Sample standard deviation.
T, T_j	Points or times at which truncated or censoring occurs.
θ	An alternate scale parameter sometimes used in lieu of β; also an arbitrary parameter.

Glossary

$\theta(\alpha) = \theta(\xi) = Q(\xi)/[Q(\xi) - \xi]$	Estimating function in singly truncated normal distribution samples.
$U_i = \dfrac{c}{n}\left[\dfrac{\phi_i}{\Phi_1 + 1 - \Phi_2}\right], i = 1, 2$	Functions involved in doubly censored samples from the normal distribution when the total number of censored observations is known, but the number in each tail separately is unknown.
$V(\ )$	Symbol for variance, $V(X) = \sigma_x^2$.
$W(n, \delta)$	Weibull distribution estimating function.
$w = T_2 - T_1$	Sample range in a doubly truncated sample.
X, Y, and sometimes other capital letters	Symbols for random variables.
X_1 or $X_{1:N}$	First-order statistic in a random sample of size N.
x_i, y_i, etc.	Sample observations.
$\bar{x}, \bar{y}$, etc.	Sample means, $\bar{x} = \sum_1^n x_i/n$.
$\xi_j = [T_j - E(X)]/\sqrt{V(X)}$	Standardized points of truncation or censoring.
$Z = [X - E(X)]/\sqrt{V(X)}$	Standardized random variable.
$Z_{1:N}$	Standardized first-order statistic in a random sample of size N.
$z_1 = (x_1 - \bar{x})/s$	Observed value of standardized first-order statistic.

Bibliography

Aitchison, J., and Brown, J. A. C. (1957) *The Lognormal Distribution*. Cambridge University Press, London.

Aitken, A. C. (1934) Note on selection from a multivariate population. *Proc. Edinburgh Math. Soc., 4,* 106–110.

Anscombe, F. J. (1948) The transformation of Poisson, binomial and negative binomial data. *Biometrika, 35,* 246–254.

Anscombe, F. J. (1949) The statistical analysis of insect counts based on the negative binomial distribution. *Biometrics, 5,* 165–173.

Anscombe, F. J. (1950) Sampling theory of the negative binomial and logarithmic series distributions. *Biometrika, 37,* 358–382.

Arnold, B. C. (1983) *Pareto Distributions*. International Co-op Publishing Co., Fairland, Md.

Balakrishnan, N. (1985) Order statistics from the half logistic distribution. *J. Statist. Comp. Simul., 20,* 287–309.

Balakrishnan, N., and Kocherlakota, S. (1985) On the double Weibull distribution: Order statistics and estimation. *Sankhya B, 47,* 161–178.

Balakrishnan, N., and Malik, H. J. (1985) Some general identities involving order statistics. *Comm. Statist. Theor. Meth.,* 14(2), 333–339.

Barnett, V. D. (1966) Order statistic estimators of the location of the Cauchy distribution. *J. Amer. Statist. Assoc., 61,* 1205–1218.

Baten, W. D. (1938) *Mathematical Statistics*. Wiley, New York.

Beal, G. (1940) The fit and significance of contageous distributions when applied to observations on larval insects. *Ecology, 21,* 460–470.
Beal, G., and Rescia, R. R. (1953) A generalization of Neyman's contageous distributions. *Biometrics, 9,* 354–386.
Birnbaum, Z. W. (1950) Effect of linear truncation on a multi-normal population. *Ann. Math. Statist., 21,* 272–279.
Birnbaum, Z. W., Paulsen, E., and Andrews, F. C. (1950) On the effect of selection performed on some coordinates of a multi-dimensional population. *Psychometrika, 15,* 191–204.
Bliss, C. I. (1948) Estimation of the mean and its error from incomplete Poisson distributions. *Conn. Agricultural Exp. Station Bul. 513,* 12pp.
Bliss, C. I. (1953) Fitting the negative binomial distribution to biological data. *Biometrics, 9,* 176–196.
Bliss, C. I., and Fisher, R. A. (1953) Fitting the negative binomial distribution to biological data, with a note on the efficient fitting of the negative binomial. *Biometrics, 9,* 176–200.
Blom, G. (1954) Transformations of the binomial, negative binomial, Poisson and chi-square distributions. *Biometrika, 41,* 302–316 (corrections in *Biometrika, 43,* 235).
Blom, G. (1956) On linear estimates with nearly minimum variances. *Arkiv. Matematik, 3,* 365–369.
Boltzmann, L. E. (1877) Bemerkungen uber einige probleme der mechanishen warmetheorie. Wein. Ber. 75, 62, or *Wiss. Abh., 2,* 122–148.
Bortkiewicz, L. von (1898) *Das Gesatz der Kleinen Zahlen.* Teubner, Leipzig.
Bortkiewicz, L. von (1922) Variationsbreite und mittlerer Fehler. *Sitzungsberichte der Berliner Mathematischen Gellschaft, 21,* 3–11.
Bowman, K. O., and Shenton, L. R. (1965) Asymptotic covariance for the Maximum Likelihood Estimators of the Parameters of a Negative Binomial Distribution. T.R. K-1643, Union Carbide Corporation, Oak Ridge.
Bowman, K. O., and Shenton, L. R. (1966) Biases of Estimators for the Negative Binomial Distribution. T.R. ORNL-4005, Union Carbide Corporation, Oak Ridge.
Bowman, K. O., and Shenton, L. R. (1987) *Properties of Estimators for the Gamma Distribution.* Marcel Dekker, New York.
Brass, W. (1958) Simplified methods of fitting the truncated negative binomial distribution. *Biometrika, 45,* 59–68.
Calitz, F. (1973) Maximum likelihood estimation of the parameters of the lognormal distribution—a reconsideration. *Austral. J. Statist., 3,* 185–190.
Campbell, F. L. (1945) A Study of Truncated Bivariate Normal Distributions. Ph.D. dissertation, University of Michigan.
Chan, M. (1982) Modified Moment and Maximum Likelihood Estimators for Parameters of the Three-Parameter Inverse Gaussian Distribution. Ph.D. dissertation, University of Georgia, Athens.

Bibliography

Chan, M., Cohen, A. C., and Whitten, B. J. (1983) The standardized inverse Gaussian distribution: tables of the cumulative probability function. *Comm. Statist. Simul. Comp. B.*, *12*, 423–442.

Chan, M., Cohen, A. C., and Whitten, B. J. (1984) Modified maximum likelihood and modified moment estimators for the three-parameter inverse Gaussian distribution. *Comm. Statist. Simul. Comp. B*, *13*, 47–68.

Chapman, D. G. (1951) Some properties of the hypergeometric distribution with applications to zoological sample censuses. *Univ. Calif. Publ. Statist.*, *1*, 131–159.

Chapman, D. G. (1952) Inverse, multiple and sequential sample censuses. *Biometrics*, *8*, 286–306.

Cheng, R. C. H., and Amin, N. A. K. (1981) Maximum likelihood estimation of parameters in the inverse Gaussian distribution with unknown origin. *Technometrics*, *23*, 257–263.

Chhikara, R. S., and Folks, J. L. (1974) Estimation of the inverse Gaussian distribution function. *J. Amer. Statist. Assoc.*, *69*, 250–254.

Chhikara, R. S., and Folks, J. L. (1988) *The Inverse Gaussian Distribution: Theory, Methodology, and Applications.* Marcel Dekker, New York.

Cirillo, R. (1979) *The Economics of Vilfredo Pareto.* Frank Case, London.

Cochran, W. G. (1946) Use of IBM equipment in an investigation of the truncated normal problem. *Proc. Res. Forum*, IBM Corporation, 40–43.

Cohen, A. C. (1941) Estimation of Parameters in Truncated Pearson Frequency Distributions. Ph.D. dissertation, University of Michigan, Ann Arbor.

Cohen, A. C. (1949) On estimating the mean and standard deviation of truncated normal distributions. *J. Amer. Statist. Assoc.*, *44*, 518–525.

Cohen, A. C. (1950) Estimating the mean and variance of normal populations from singly and doubly truncated samples. *Ann. Math. Statist.*, *21*, 557–569.

Cohen, A. C. (1951a) Estimating parameters of logarithmic-normal distributions by maximum likelihood. *J. Amer. Statist. Assoc.*, *46*, 206–212.

Cohen, A. C. (1951b) On estimating the mean and variance of singly truncated normal frequency distributions from the first three sample moments. *Ann. Inst. Statist. Math.*, *3*, 37–44.

Cohen, A. C. (1953) Estimating parameters in truncated Pearson frequency distributions without resort to higher moments. *Biometrika*, *40*, 50–57.

Cohen, A. C. (1955a) Restriction and selection in samples from bivariate normal distributions. *J. Amer. Statist. Assoc.*, *50*, 884–893.

Cohen, A. C. (1955b) Maximum likelihood estimation of the dispersion parameter of a chi-distributed radial error from truncated and censored samples with applications to target analysis. *J. Amer. Statist. Assoc.*, *50*, 1122–1135.

Cohen, A. C. (1957) On the solution of estimating equations for truncated and censored samples from normal populations. *Biometrika*, *44*, 225–236.

Cohen, A. C. (1959) Simplified estimators for the normal distribution when samples are singly censored or truncated. *Technometrics*, *1*, 217–237.

Cohen, A. C. (1961) Tables for maximum likelihood estimates: singly truncated and singly censored samples. *Technometrics, 3,* 535–541.

Cohen, A. C. (1963) Progressively censored samples in life testing. *Technometrics, 5,* 237–339.

Cohen, A. C. (1965a) Maximum likelihood estimation in the Weibull distribution based on complete and censored samples. *Technometrics, 7,* 579–588.

Cohen, A. C. (1965b) Estimation in the Negative Binomial Distribution, T.R. 14, Department of Statistics, University of Georgia, Athens.

Cohen, A. C. (1966) Life testing and early failure. *Technometrics, 8,* 539–549.

Cohen, A. C. (1969) A Generalization of the Weibull Distribution. NASA Contractor T.R. 61293 (Contract NAS 8-11175), Marshall Space Flight Center, Alabama.

Cohen, A. C. (1970a) Curtailed attribute sampling. *Technometrics, 12,* 295–298.

Cohen, A. C. (1970b) Estimating the Bernoulli parameter and the average sample size from truncated samples. In *Random Counts in Scientific Work.* Patil, G. P., ed., Pennsylvania State, University Park, pp. 127–134.

Cohen, A. C. (1973) The reflected Weibull distribution. *Technometrics, 15,* 867–873.

Cohen, A. C. (1975) Multi-censored sampling in the three-parameter Weibull distribution. *Technometrics, 17,* 347–351.

Cohen, A. C. (1976) Progressively censored sampling in the three-parameter lognormal distribution. *Technometrics, 18,* 99–103.

Cohen, A. C. (1988) In *Lognormal Distributions: Theory and Applications.* Crow, E. L., and Shimizu, K, eds. Marcel Dekker, New York, pp. 113–172.

Cohen, A. C., and Helm, R. (1973) Estimation in the exponential distribution. *Technometrics, 14,* 841–846.

Cohen, A. C., and Norgaard, N. J. (1977) Progressively censored sampling in the three-parameter gamma distribution. *Technometrics, 19,* 333–340.

Cohen, A. C., and Whitten, B. J. (1980) Estimation in the three-parameter lognormal distribution. *J. Amer. Statist. Assoc., 75,* 399–404.

Cohen, A. C., and Whitten, B. J. (1981) Estimation of lognormal distributions. *Amer. J. Math. Manage. Sci., 1,* 139–153.

Cohen, A. C., and Whitten, B. J. (1982) Modified moment and maximum likelihood estimators for parameters of the three-parameter gamma distribution. *Comm. Statist. Simul. Comp., 11*(2), 197–216.

Cohen, A. C., and Whitten, B. J. (1985) Modified moment estimation for the three-parameter inverse Gaussian distribution. *J. Qual. Tech., 17,* 147–154.

Cohen, A. C., and Whitten, B. J. (1986) Modified moment estimation for the three-parameter gamma distribution. *J. Qual. Tech., 18,* 53–62.

Cohen, A. C., and Whitten, B. J. (1988) *Parameter Estimation in Reliability and Life Span Models.* Marcel Dekker, New York.

Bibliography

Cohen, A. C., Whitten, B. J., and Ding Y. (1984) Modified moment estimation for the three-parameter Weibull distribution. *J. Qual. Tech., 16*, 159–167.

Cohen, A. C., Whitten, B. J., and Ding, Y. (1985) Modified moment estimation for the three-parameter lognormal distribution. *J. Qual. Tech., 17*, 92–99.

Cooley, C. G., and Cohen, A. C. (1970) Tables of maximum likelihood estimating functions for singly truncated and singly censored samples from the normal distribution. NASA Contractor Report NASA CR-61330, George C. Marshall Space Flight Center, Marshall Space Flight Center, Alabama.

Craig, C. C. (1936) A new exposition and chart for the Pearson system of frequency curves. *Ann. Math. Statist., 7*, 16–28.

Craig, C. C. (1968a) The ASN for single and double attribute sampling plans. *ASQC Conv. Trans.*, 63–67.

Craig, C. C. (1968b) The average sample number for truncated single and double attribute sampling plans. *Technometrics, 10*, 685–692.

Cramér, H. (1946) *Mathematical Methods of Statistics*. Princeton University Press, Princeton, N.J.

Crow, E. L., and Shimizu, K. (1988) *Lognormal Distributions: Theory and Applications*. Marcel Dekker, New York.

David, F. N., and Johnson, N. L. (1952) The truncated Poisson. *Biometrics, 8*, 275–285.

David, F. N., and Johnson, N. L. (1954) Statistical treatment of censored data I. Fundamental formulae. *Biometrika, 41*, 228–240.

David, H. A. (1962) Order statistics in short-cut tests. In *Contributions to Order Statistics*. Sarhan, A. E., and Greenberg, B. G., eds. Wiley, New York, pp. 94–128.

David, H. A. (1970) *Order Statistics* (2nd ed., 1981). Wiley, New York.

David, H. A., Kennedy, W. J., and Knight, R. D. (1977) Means, variances, and covariances of normal order statistics in the presence of an outlier. *Selected Tables in Mathematical Statistics*, Vol. 5, American Mathematical Society, Providence, pp. 75–204.

Des Raj (1952) On estimating the parameters of normal populations from singly truncated samples. *Ganita, 3*, 41–57.

Des Raj (1953a) Estimation of the parameters of type III populations from truncated samples. *J. Amer. Statist. Assoc., 48*, 336–349.

Des Raj (1953b) On estimating the parameters of bivariate normal populations from doubly and singly linearly truncated samples. *Sankhya, 12*, 277–290.

Dubey, S. D. (1963) On some statistical inferences for Weibull laws. (abstract). *J. Amer. Statist. Assoc., 58*, 549.

Dubey, S. D. (1967) On some permissible estimators of the location parameter of Weibull and certain other distributions. *Technometrics, 9*, 293–307.

Elandt-Johnson, R. C., and Johnson, N. L. (1980) *Survival Models and Data Analysis*. Wiley, New York.

Elderton, W. P. (1938) *Frequency Curves and Correlation*, 3rd ed. Cambridge University Press, New York.

Elderton, W. P., and Johnson, N. L. (1969) *Systems of Frequency Curves.* Cambridge University Press, New York.

Epstein, B. (1960) Estimation from life test data. *Technometrics, 2,* 447–454.

Epstein, B., and Sobel, M. (1953) Life testing. *J. Amer. Statist. Assoc., 48,* 485–502.

Esary, J. D., and Proschan, F. (1963) Relationship between system failure rate and component failure rates. *Technometrics, 5,* 183–189.

Finney, D. J. (1941) On the distribution of a variate whose logarithm is normally distributed. *J. Roy. Statist. Soc. Ser. B., 7,* 155–161.

Finney, D. J., and Varley, G. C. (1955) An example of the truncated Poisson distribution. *Biometrics, 2,* 387–394.

Fisher, R. A. (1931) Properties and applications of Hh functions. *Introduction to British A.A.S. Math. Tables, 1,* xxvi–xxxv.

Fisher, R. A. (1941) The negative binomial distribution. *Ann. of Eugenics, London, 11,* 182–187.

Fisher, R. A. (1953) Note on the efficient fitting of the negative binomial. Appended to paper by Bliss, C. I. (1953), *Biometrics, 9,* 197–200.

Folks, J. L., and Chhikara, R. S. (1978) The inverse Gaussian distribution and its statistical application—a review. *J. Roy. Statist. Soc. Ser. B., 40,* 268–289.

Francis, V. J. (1946) On the distribution of the sum of n sample values drawn from a truncated normal population. *J. Roy. Statist. Soc. (Suppl.), 8,* 223–232.

Freeman, P. R. (1973) An algorithm for computing individual hypergeometric terms. *Appl. Statist., 22,* 130–133.

Galton, F. (1897) An examination into the registered speeds of American trotting horses with remarks on their value as hereditary data. *Proc. Roy. Soc. Lond., 62,* 310–314.

Giesbrecht, F., and Kempthorne, O. (1976) Maximum likelihood estimation in the three-parameter lognormal distribution. *J. Roy. Statist. Soc., Ser. B., 38,* 257–264.

Gnedenko, B. (1943) Sur le distribution limite du terme maximum d'une serie aleatoire. *Ann. Math., 44,* 423–453.

Goedicke, V. (1953) *Introduction to the Theory of Statistics.* Harper and Brothers, New York.

Govindarajulu, Z. (1966a) Characterization of the exponential and power distribution. *Skand. Aktuarietidskr., 49,* 132–136.

Govindarajulu, Z. (1966b) Best linear estimates under symmetric censoring of the parameters of a double exponential population. *J. Amer. Statist. Assoc., 61,* 248–258.

Bibliography

Govindarajulu, Z. (1968)   Certain general properties of unbiased estimates of location and scale parameters based on ordered observations. *Siam. J. Appl. Math.*, *16*, 533–551.

Greenwood, M., and Yule, G. U. (1920)   An enquiry into the nature of frequency distributions of multiple happenings with particular reference to the occurrence of multiple attacks of disease or repeated accidents. *J. Roy. Statist. Soc., Ser. A.*, *83*, 255–279.

Grundy, P. M. (1952)   The fitting of grouped truncated and grouped censored normal distributions. *Biometrika*, *39*, 252–259.

Guenther, W. C. (1975)   A review of the inverse hypergeometric distribution. *Statist. Neerlandica 29*, 129–144.

Guenther, W. C. (1977)   *Sampling Inspection in Statistical Quality Control*. Charles Griffin, London.

Guenther, W. C. (1983)   Hypergeometric distributions. In *Encyclopedia of Statistical Sciences*, Vol. 3. Kotz, S., Johnson, N. L., and Read, C. B., eds. pp. 707–712.

Gumbel, E. J. (1958)   *Statistics of Extremes*. Columbia University Press, New York.

Gupta, A. K. (1952)   Estimation of the mean and standard deviation of a normal population from a censored sample. *Biometrika*, *39*, 260–273.

Gupta, S. S., Qureishi, A. S., and Shah, B. K. (1967)   Best linear unbiased estimators of parameters of the logistic distribution using order statistics. *Technometrics*, *9*, 43–56.

Hagstroem, K. G. (1960)   Early characterization of the Pareto distribution. *Skand. Aktuarietidskr.*, *430*, 65–88.

Hald, A. (1949)   Maximum likelihood estimation of the parameters of a normal distribution which is truncated at a known point. *Skand. Aktuarietidskr.*, *8*, 65–88.

Haldane, J. B. S. (1941)   The fitting of binomial distributions. *Ann. Eugen. Lond.*, *11*, 179.

Halperin, M. (1952)   Maximum-likelihood estimation in truncated samples. *Ann. Math. Statist.*, *23*, 226–238.

Harris, C. M. (1968)   The Pareto distribution as a queue service discipline. *Oper. Res.*, *16*, 307–316.

Harter, H. L. (1961)   Expected values of normal order statistics. *Biometrika*, *48*, 151–166.

Harter, H. L. and Moore, A. H. (1965)   Maximum likelihood estimation of the parameters of the gamma and Weibull populations from censored samples. *Technometrics*, *7*, 639–643.

Harter, H. L. and Moore, A. H. (1966)   Local-maximum-likelihood estimation of three-parameter lognormal populations from complete and censored samples. *J. Amer. Statist. Assoc.*, *61*, 842–851.

Harter, H. L. and Moore, A. H. (1967) Asymptotic variances and covariances of maximum likelihood estimators of parameters of Weibull and gamma populations from censored samples. *Ann. Math. Statist., 38,* 557–571.

Hartley, H. O. (1958) Maximum likelihood estimation from incomplete data. *Biometrics, 14,* 174–194.

Herd, G. R. (1956) Estimation of the parameters of a population from a multicensored sample. Ph.D. dissertation, Iowa State College, Cedar Falls.

Herd, G. R. (1957) Estimation of reliability functions. *Proc. Third Nat. Symp. on Reliability and Quality Control.*

Herd, G. R. (1960) Estimation of reliability from incomplete data. *Proc. Sixth Nat. Symp. on Reliability and Quality Control.*

Heyde, C. C. (1963) On a property of the lognormal distribution. *J. Roy. Statist. Soc., Ser. B, 25,* 392–393.

Hill, B. M. (1963) The three-parameter lognormal distribution and Baysian analysis of a point-source epidemic. *J. Amer. Statist. Assoc., 58,* 72–84.

Hirano, K. (1986) Rayleigh distribution. In *Encyclopedia of Statistical Science,* Vol. 7. Kotz, S., Johnson, N. L., and Read, C. B., eds. Wiley, New York, pp. 647–649.

Hotelling, H. (1948) Fitting generalized truncated normal distributions. Abstracts of Madison Meeting, *Ann. Math Statist., 19,* 596.

Ipsen, J., Jr. (1949) A practical method of estimating the mean and standard deviation of truncated normal distributions. *Hum. Biol., 21,* 1–16.

Irwin, J. O. (1959) On the estimation of the mean of a Poisson distribution from a sample with the zero class missing. *Biometrics, 15,* 342–326.

Jaech, J. L. (1964) Estimation of the Weibull distribution shape parameter when no more than two failures occur per lot. *Technometrics, 6,* 415–422.

Johnson, N. L., and Kotz S. (1969) *Distributions in Statistics: Discrete Distributions.* Wiley, New York.

Johnson, N. L., and Kotz, S. (1970) *Continuous Univariate Distributions,* Vol. I. Houghton Mifflin, Boston.

Kane, V. E. (1978) Some topics in multivariate classifications. Presentation SREB Summer Research Conference in Statistics, DeGray State Park Lodge, Arkadelphia, Ark.

Kane, V. E. (1982) Standard and goodness-of-fit parameter estimation methods for the three-parameter lognormal distribution. *Comm. Statist. A,11*(17), 1935–1957.

Kao, J. H. K. (1958) Computer methods for estimating Weibull parameters in reliability studies. *IRE Trans. Reliab. Qual. Control, 13,* 15–22.

Kao, J. H. K. (1959) A graphical estimation of mixed Weibull parameters in life testing of electron tubes. *Technometrics, 1,* 389–407.

Keyfitz, N. (1938) Graduation by a truncated normal. *Ann. Math. Statist., 9,* 66–67.

Bibliography

Koniger, W. (1981) Die anwendung der extremal-3-verteilung bei der regenauswertung und der niedrigwasseranalyse. *GWF-Wasser/Abwasser, 122*, 460–466.

Lawless, J. F. (1982) *Statistical Models and Methods for Lifetime Data.* Wiley, New York.

Lawley, D. N. (1943) A note on Karl Pearson's selection formulae. *Proceedings of the Royal Society, Edinburgh, A., 66*, 28–30.

Lee, Alice (1915) Table of Gaussian 'tail' functions when the 'tail' is larger than the body. *Biometrika, 10*, 208–215.

Lehman, E. H. (1963) Shapes, moments and estimators of the Weibull distribution. *IEEE Trans. Reliab., R-12, 3*, 32–38.

Lemon, G. H. (1975) Maximum likelihood estimation for the three parameter Weibull distribution based on censored samples. *Technometrics, 17*, 247–254.

Leone, F., Rutenberg, Y. H., and Topp, C. W. (1960) Order statistics and estimators for the Weibull distribution. *Case Statist. Lab. Pub. No. 1026* (AFOSR Rep. TN 60–389, Case Institute of Technology), Cleveland.

Lieberman, G. J., and Owen, D. B. (1961) *Tables of the Hypergeometric Probability Distribution.* Stanford University Press, Stanford, Calif.

Lloyd, E. H. (1952) Least-squares estimation of location and scale parameters using order statistics. *Biometrika, 39*, 88–95.

Lloyd, D. K., and Lipow, M. (1962) *Reliability: Management, Methods and Mathematics.* Prentice-Hall, Englewood Cliffs, N.J.

Lund, R. E. (1980) An algorithm for computing cumulative hypergeometric sums. *Appl. Statist. 29*, 221–223.

Mahalanobis, P. C. (1934) Tables of random samples from a normal population. *Sankhya, 1*, 289–328.

Malik, H. J. (1966) Exact moments of order statistics from the Pareto distribution. *Skand. Aktuarietidskr., 49*, 144–157.

Malik, H. J. (1970) Products of Pareto variables. *Metrika, 15*, 19–22.

Mandelbrodt, B. (1960) The Pareto–Levy law and the distribution on income. *Int. Econ. Rev., 1*, 79–106.

Mann, N. R. (1968) Estimation procedures for the two-parameter Weibull and extreme-value distributions. *Technometrics, 10*, 231–256.

Mann, N. R., and Fertig, K. W. (1973) Tables for obtaining Weibull confidence bounds and tolerance bounds based on best linear invariant estimates of parameters of the extreme value distribution. *Technometrics, 15*, 87–102.

Mann, N. R., and Fertig, K. W. (1975a) A goodness-of-fit test for the two-parameter vs. three-parameter Weibull: confidence bounds for threshold. *Technometrics, 17*, 237–245.

Mann, N. R., and Fertig, K. W. (1975b) Simplified efficient point and interval estimators for Weibull parameters. *Technometrics, 17*, 361–368.

Mann, N. R., Schafer, R. E., and Singpurwalla, N. D. (1974) *Methods for Statistical Analysis of Reliability and Life Data.* Wiley, New York.
Mantel, N, 1951) Evaluation of a class of diagnostic tests, *Biometrics, 3,* 240–246.
Maxwell, J. C. (1860) Illustrations of the dynamical theory of gases. *Phil. Mag., 19,* 19–32; and *20,* 21–37.
Mendenhall, W. (1958) A bibliography on life testing and related topics. *Biometrika, 45,* 521–543.
Menon, M. V. (1963) Estimation of the shape and scale parameters of the Weibull distribution. *Technometrics, 5,* 175–182.
Molina, E. C. (1942) *Poisson's Exponential Binomial Limit,* Van Nostrand, New York.
Mood, A. M., Graybill, F. A., and Boes, D. C. (1974) *Introduction to the Theory of Statistics.* McGraw-Hill, New York.
Moore, P. G. (1954) A note on truncated Poisson distributions. Biometrics, 10, 402–406.
Morris, K. W. (1963) A note on direct and inverse binomial sampling. *Biometrika, 50,* 544–545.
Muench, H. (1938) Discrete frequency distributions arising from mixtures of several probability values. *J. Amer. Statist. Assoc., 33,* 390–398.
Muniruzzaman, A. N. M. (1950) On some distributions in connection with Pareto's law. *Proc. 1st. Pakistan Statist. Conf.,* pp. 90–93.
Muniruzzaman, A. N. M. (1957) On measures of location and dispersion and tests of hypotheses on a Pareto population. *Bull. Calcutta Statist. Assoc., 7,* 115–123.
Munro, A. H., and Wixley, R. A. J. (1970) Estimators based on order statistics of small samples from a three-parameter lognormal distribution. *J. Amer. Statist. Assoc., 65,* 212–225.
Nelson, W. (1968) A method for statistical hazard plotting of incomplete failure data that are arbitrarily censored. *TIS Report 68-C-007,* General Electric Research & Development Center, Schenectady, N.Y.
Nelson, W. (1969) Hazard plotting for incomplete failure data. *J. Qual. Tech., 1,* 27–52.
Nelson, W. (1972) Theory and applications of hazard plotting for censored failure data. *Technometrics, 14,* 945–966.
Nelson, W. (1982) *Applied Life Data Analysis.* Wiley, New York.
Nelson, W., and Schmee, J. (1979) Inference for (log)normal life distributions from small singly censored samples and BLUEs. *Technometrics, 21,* 43–54.
Ogawa, J. (1951) Contributions to the theory of systematic statistics, I. *Osaka Math J., 3,* 175–213.
Ogawa, J. (1952) Contributions to the theory of systematic statistics, II. *Osaka Math. J., 4,* 41–61.

Bibliography

Ogawa, J. (1957) A further contribution to the theory of systematic statistics (abstract). *Ann. Math. Statist.*, 28, 525–526.

Ogawa, J. (1962) Distribution and moments of order statistics. In *Contributions to Order Statistics*, Sarhan, A. E., and Greenberg, B. G., Wiley, New York, chap. 2.

Padgett, W. J., and Wei, L. J. (1979) Estimation for the three-parameter inverse Gaussian distribution. *Comm. Statist. Theor. Meth.*, 8, 129–137.

Pareto, V. (1897) *Cours d'Economie Politique*. Rouge and Cie, Lausanne.

Patel, J. K., and Read, C. B. (1982) *Handbook of the Normal Distribution*. Marcel Dekker, New York, chap. 7.

Patil, G. P. (1960) On the evaluation of the negative binomial distribution with samples. *Technometrics*, 2, 501–505.

Patil, G. P. (1963a) Note on the equivalence of the binomial and inverse binomial acceptance sampling plans and an acknowledgment. *Technometrics*, 5, 119–121.

Patil, G. P. (1963b) A characterization of the exponential-type distribution. *Biometrika*, 50, 205–207.

Patil, G. P., and Joshi, S. W. (1968) *A Dictionary and Bibliography of Discrete Distributions*. Hafner, New York.

Pearson, Karl (1902) On the systematic fitting of frequency curves. *Biometrika*, 2, 2–7.

Pearson, Karl, and Lee, Alice (1908) On the generalized probable error in multiple normal correlation. *Biometrika*, 6, 59–69.

Phatak, A. G., and Bhatt, M. M. (1967) Estimation of fraction defective in curtailed sampling plans. *Technometrics*, 9, 219–228.

Pittman, E. J. G. (1936) Sufficient statistics and intrinsic accuracy. *Proc. Cambridge Philos. Soc.*, 32, 567.

Plackett, R. L. (1953) The truncated Poisson distribution. *Biometrics*, 9, 485–488.

Plackett, R. L. (1958) Linear estimation from censored data. *Ann. Math. Statist.*, 29, 131–142.

Poisson, S. D. (1837) *Resserches sur la Probabilite des Jugements en Matiere Criminelle et en Matiere Civile, Precedees des Regles Generales du Calcul des Probabilities*. Bacheli, Imprimeur-Libraire pour les Mathematiques, la Physique, etc., Paris.

Prescott, P. (1970) Estimation of the standard deviation of a normal population from doubly censored samples using normal scores. *Biometrika*, 57, 409–419.

Prescott, P. (1974) Variances and covariances of order statistics from the gamma distribution. *Biometrika*, 61, 607–613.

Procassini, A. A., and Romano, A. (1961) Transistor reliability estimates improved with the Weibull distribution function. *Motorola Eng. Bull.*, *9*(2), 16–18.

Proschan, F. (1963) Theoretical explanation of observed decreasing failure rate. *Technometrics, 5,* 375–383.

Quant, R. E. (1966) Old and new methods of estimation and the Pareto distribution. *Metrika, 10,* 55–82.

Quensel, C. E. (1945) Studies of the logarithmic normal curve. *Skand. Aktuarietidskr., 28,* 141–153.

Rayleigh, J. W. S. (1919) On the problem of random vibrations, and of random flights in one, two, or three dimensions. *Philos. Mag., 37,* 321–347.

Rider, P. R. (1953) Truncated Poisson distributions. *J. Amer. Statist. Assoc., 48,* 826–830.

Rider, P. R. (1955) Truncated binomial and negative binomial distributions. *J. Amer. Statist. Assoc., 50,* 877–883.

Ringer, L. J., and Sprinkle, E. E. (1972) Estimation of the parameters of the Weibull distribution from multi-censored samples. *IEEE Trans. Reliab., R-21,* 46–51.

Roberts, H. R. (1962a) Some results in life testing based on hypercensored samples from an exponential distribution. Ph.D. dissertation, George Washington University, Washington, D.C.

Roberts, H. R. (1962b) Life test experiments with hypercensored samples. *Proc. 18th. Annual Quality Control Conf.*, Rochester Society for Quality Control, Rochester, N.Y.

Rockette, H., Antle, C., and Klimko, L. A. (1974) Maximum likelihood estimation with the Weibull model. *J. Amer. Statist. Assoc., 69,* 246–249.

Rukhin, A. L. (1984) Improved estimation in lognormal models. *Tech. Rep. 84-38,* Department of Statistics, Purdue University, West Lafayette, Ind.

Rutherford, E., and Geiger, H. (1910) The probability variations in the distribution of alpha particles. *Phil. Mag., Ser. 6, 20,* 698.

Salvosa, L. R. (1936) Tables of Pearson's type III function. *Ann. Math. Statist., 1,* 191–198; appendix, 1–125.

Sampford, M. R. (1952) The estimation of response-time distributions, Part II. *Biometrics, 9,* 307–369.

Sampford, M. R. (1955) The truncated negative binomial distribution. *Biometrika, 42,* 58–69.

Sarhan, A. E. (1954) Estimation of the mean and standard deviation by order statistics. *Ann. Math. Statist., 25,* 317–328.

Sarhan, A. E. (1955a) Estimation of the mean and standard deviation by order statistics, Part II. *Ann. Math. Statist., 26,* 505–511.

Sarhan, A. E. (1955b) Estimation of the mean and standard deviation by order statistics, Part III. *Ann. Math. Statist., 26,* 576–592.

Sarhan, A. E., and Greenberg, B. G. (1956) Estimation of location and scale parameters by order statistics from singly and doubly censored samples. Part

Bibliography

I. The normal distribution up to size 10. *Ann. Math. Statist.*, *27*, 427–451 (correction, *40*, 325).

Sarhan, A. E., and Greenberg, B. G. (1957) Tables for best linear estimates by order statistics of the parameters of single exponential distributions from singly and doubly censored samples. *J. Amer. Statist. Assoc.*, *52*, 58–87.

Sarhan, A. E., and Greenberg, B. G. (1958) Estimation of location and scale parameters by order statistics from singly and doubly censored samples. Part II. Tables for the normal distribution for samples of size 11 to 15. *Ann. Math. Statist.*, *29*, 79–105.

Sarhan, A. E., and Greenberg, B. G., eds. (1962) *Contributions to Order Statistics*. Wiley, New York.

Saw, J. G. (1958) Moments of sample moments of censored samples from a normal population. *Biometrika*, *45*, 211–221.

Saw, J. G. (1959) Estimation of the normal population parameters given a singly censored sample. *Biometrika*, *46*, 150–159.

Saw, J. G. (1961a) Estimation of the normal population parameters given a type I censored sample. *Biometrika*, *48*, 367–377.

Saw, J. G. (1961b) The bias of the maximum likelihood estimates of the location and scale parameters given a type II censored normal sample. *Biometrika*, *48*, 448–451.

Scarborough, J. B. (1930) *Numerical Analysis*. Johns Hopkins Press, Baltimore, pp. 191–193.

Schmee, J., and Nelson, W. B. (1977) Estimates and approximate confidence limits for (log)normal life distributions from singly censored samples by maximum likelihood. *Tech Rept. 76CRD250*, General Electric Company, Schenectady, N.Y.

Schmee, J., Gladstein, D., and Nelson, W. (1985) Confidence limits of a normal distribution from singly censored samples using maximum likelihood. *Technometrics*, *27*, 119–128.

Schneider, H. (1986) *Truncated and Censored Samples from Normal Populations*. Marcel Dekker, New York.

Schroedinger, E. (1915) Zur theorie der fall—und steigversuche an teilchen mit Brownscher bewegung. *Phys. Z.*, *16*, 289–295.

Shook, B. L. (1930) Synopsis of elementary mathematical statistics. *Ann. Math. Statist.*, *1*, 14–41.

Singh, S. N. (1963) A note on inflated Poisson distributions. *J. Indian Statist. Assoc.*, *1*, 140–144.

Singh, S. N. (1962–1963) Inflated Poisson distributions. *J. Sci. Res. Banaras Hindu Univ.*, *13*, 317–326.

Smoluchowsky, M. W. (1915) Notiz uber die berechnung der Browschen molekularbewegung bei der Ehrenhaft-Millikanschen versuchanordnung. *Phys. Z.*, *16*, 318–321.

Srivastava, M. S. (1965) A characterization of Pareto's distribution (abstract). *Ann. Math. Statist., 36,* 361–362.

Stedinger, J. R. (1980) Fitting lognormal distributions to hydrologic data. *Water Resources Res., 16,* 481–490.

Stevens, W. L. (1937) The truncated normal distribution. (Appendix to paper by C. I. Bliss, The calculation of the time mortality curve.) *Ann. Appl. Biol., 24,* 815–852.

Sundaraiyer, V. H. (1986) Estimation of parameters of the inverse Gaussian distribution from censored samples. Ph.D. dissertation, Department of Statistics, University of Georgia, Athens.

Teichroew, D. (1956) Tables of expected values of order statistics for samples of size twenty and less from the normal distribution. *Ann. Math. Statist., 27,* 410–426.

Thompson, G. W., Friedman, M., and Garellis, E. (1954) Estimation of mean and variance of normal populations from doubly truncated samples. *Tech. Rep. Ethyl Corp. Res. Lab.,* Detroit.

Tietjen, G. L., Kahaner, D. K., and Beckman, R. J. (1977) Variances and covariances of the normal order statistics for sample sizes 2 to 50. *Selected Tables in Mathematical Statistics,* Vol. 5, American Mathematics Society, Providence, pp. 1–73.

Tiku, M. L. (1967) Estimating the mean and standard deviation from a censored normal sample. *Biometrika, 54,* 64–74.

Tiku, M. L. (1968a) Estimating the parameters of log-normal distributions from censored samples. *J Amer. Statist. Assoc., 63,* 134–140.

Tiku, M. L. (1968b) Estimating the parameters of normal and logistic distributions from censored samples. *Austral. J. Statist., 10,* 64–74.

Tiku, M. L. and Kumra, S. (1981) Expected values, variances and covariances of order statistics for a family of symmetric distributions (Student's t). In *Selected Tables in Mathematical Statistics, 8.* 141–270, Amer. Math. Soc.

Tiku, M. L., Tan, W. Y., and Balakrishnan, N. (1986) *Robust Inference.* Marcel Dekker, New York.

Tippett, L. H. C. (1925) On the extreme individuals and the range of samples taken from a normal population. *Biometrika, 17,* 364–387.

Tippett, L. H. C. (1932) A modified method of counting particles. *Proc. Roy. Soc. Lond., Ser. A, 137,* 434–436.

Tischendorf, J. A. (1955) Linear Estimation Techniques Using Order Statistics. Ph.D. dissertation, Purdue University, Lafayette, Ind.

Tukey, J. W. (1949) Sufficiency, truncation and selection. *Ann. Math. Statist., 20,* 309–311.

Tweedie, M. C. K. (1956) Some statistical properties of the inverse Gaussian distribution. *Va. J. Sci., 7,* 160–165.

Tweedie, M. C. K. (1957a) Statistical properties of the inverse Gaussian distribution I. *Ann. Math. Statist., 28,* 362–377.

Bibliography

Tweedie, M. C. K. (1957b) Statistical properties of the inverse Gaussian distribution II. *Ann. Math. Statist., 28,* 695–705.

Votaw, D. F., Rafferty, J. A., and Deemer, W. L. (1950) Estimation of parameters in a truncated trivariate normal distribution. *Psychometrika, 15,* 339–347.

Wald, A. (1944) On cumulative sums of random variables. *Ann. Math. Statist., 15,* 283–296.

Wallace Year Book (1892–1896) American Trotting Association, Vols. 8–12.

Wasan, M. T. (1968) On an inverse Gaussian process. *Skand. Aktuarietidskr., 51,* 69–96.

Wasan, M. T., and Roy, L. K. (1969) Tables of inverse Gaussian percentage points. *Technometrics, 11,* 590–603.

Whitten, B. J., Cohen, A. C., and Sundaraiyer, V. (1988) A pseudo-complete sample technique for estimation from censored samples. *Comm. Statist. Theor. Meth., 17*(7), 2239–2258.

Wilks, S. S. (1932) On estimates from fragmentary data. *Ann. Math. Statist., 3,* 163–196.

Wingo, D. R. (1973) Solution of the three-parameter Weibull equations by constrained modified quasi-linearization (progressively censored samples). *IEEE Trans. Reliab., R-22, 2,* 96–102.

Wingo, D. R. (1975) The use of interior penalty functions to overcome lognormal estimation anomalies. *J. Statist. Comp. Simul., 4,* 49–61.

Wingo, D. R. (1976) Moving truncations barrier-function methods for estimation in three-parameter lognormal models. *Comm. Statist. B, 5,* 65–80.

Wycoff, J., Bain, L., and Engelhardt, M. (1980) Some complete and censored sampling results for the three-parameter Weibull distribution. *J. Statist. Comp. Simul., 11,* 139–151.

Yuan, P. T. (1933) On the logarithmic frequency distributions and the semilogarithmic correlation surface. *Ann. Math. Statist., 4,* 30–74.

Zanakis, S. H. (1977) Computational experience with some nonlinear optimization algorithms in deriving MLE for the three-parameter Weibull distribution. In *Algorithmic Methods in Probability: TIMS Studies in Management Sciences,* vol. 7. Neuts, M. F., ed. North Holland, Amsterdam.

Zanakis, S. H. (1979a) Monte Carlo study of some simple estimators of the Weibull distribution. *J. Statist. Comp. Simul., 9,* 101–116.

Zanakis, S. H. (1979b) Extended pattern search with transformations for the three-parameter Weibull MLE problem. *Manage. Sci., 25,* 1149–1161.

Zanakis, S. H., and Mann, N. R. (1981) A good simple percentile estimator of the Weibull shape parameter for use when all three parameters are unknown. Unpublished manuscript.

Zipf, G. K. (1949) *Human Behavior and the Principle of Least Effort.* Addison-Wesley, Reading, Mass.

Index

Aitchison, J., 96
Aitkin, A. C., 186
Amin, N. A. K., 106, 107
Andrews, F. C., 3
Anscombe, F. J., 208, 209
Arnold, B. C., 165

Bain, L., 85
Balakrishnan, N., 70
Barnett, V. D., 70
Baten, W. D., 195
Beal, G., 240, 241
Bhatt, M. M., 220, 224
Binomial distribution, 212–215
    asymptotic variance of estimates, 214
    moments of, 212
    probability function of, 212
    truncated with zero class missing, 213
        Mantel's estimator, 214
        mle, 213–214
Birnbaum, Z. W., 3, 186

Bivariate normal distribution, 186–193
    pdf, of, 186
    truncated samples, 187–189
        reliability of estimates, 190–191
Bliss, C. I., 3, 200, 208
Blom, G., 66, 81
Blom's nearly best linear estimators, 81
Boltzmann, L. E., 150
Bortkiewicz, L. von, 139, 200, 231
Bowman, K. O., 208
Brass, W., 208, 210, 211, 240
Brown, J. A. C., 96
Brownian motion (*see* Inverse Gaussian distribution)

Calitz, F., 98
Campbell, F. L., 3, 186
Carver, H. C., 195
Censored samples, 3, 4, 6 (*see also* specific distributions)

305

[Censored samples, Continued]
  pseudo-complete sample technique for parameter estimation, 121–127
    first approximations, 122
    iterative procedures, 122
    stopping rules, 123
    where applicable, 121
Chan, M., 106, 107
Chapman, D. G., 187
Cheng, R. C. H., 106, 107
Chhikara, R. S., 106
Circular normal distribution (see Rayleigh distribution)
Cirillo, R., 165
Cochran, W. S., 3
Cohen, A. C., 3, 5, 8, 12, 15, 16, 17, 18, 21, 27, 30, 31, 33, 35, 41, 42, 43, 44, 45, 49, 50, 52, 56, 58, 60, 85, 86, 88, 89, 92, 94, 96, 98, 100, 104, 105, 106, 123, 124, 125, 138, 157, 158, 159, 160, 163, 165, 167, 174, 183, 186, 189, 194, 200, 208, 220, 230, 235, 236, 241
Complete samples, 121
Cooley, C. G., 3
Craig, C. C., 3, 180, 220, 224
Cramér, H., 182
Crow, E. L., 96
Curtailed attribute samples, 220–224
  parameter estimation, from, 222–224
  probability function of sample size, 221–222
  variance of estimates, 223

David, F. N., 200, 208, 210, 236
David, H. A., 3, 66, 70

Des Raj, 3, 186, 189
Ding, Y., 92
Discrete distributions, 199–219
  (see also specific distributions)
Double exponential distribution
  (see Extreme value distribution)
Dubey, S. D., 85

Elandt-Johnson, R. C., 139
Elderton, W. P., 180
Engelhardt, W. P., 85
Epstein, B., 3
Esary, J. D., 85
Exponential distribution, 128–138
  characteristics of, 129
  cdf of, 128
  first order statistic, distribution of, 130
  hazard function and cumulative function, 129
  one parameter special case, estimation in, 133
    asymptotic variance of estimates, 133–134
  pdf of, 128
  progressively censored sample, 129–131
    hazard plot estimates of, 134–138
    loglikelihood function, 129
    mle of, 130
    modified mle of, 131–132
  singly right truncated sample, 132–133
    loglikelihood function, 132
    mle of, 132–133
    modified mle of, 133
Extreme value distributions, 139–145
  characteristics of, 140–141

Index

[Extreme value distributions, Continued]
  cdfs types I, II, and III, 139
  parameter estimation
  pdfs, 140
  progressively censored samples, 144–145
    loglikelihood function of, 144
    mle of, 144
  singly right censored samples, 141–144
    loglikelihood function of, 143
    mle of, 143
  standardized distributions, pdfs of, 141

Fertig, K. W., 85, 145
Finney, D. J., 3
Fisher, R. A., 2, 208, 209
Folded normal distribution, 147–148
Folks, J. L., 106
Francis, V. J., 3
Freeman, P. R., 219
Friedman, M., 33

Galton, F., 2
Gamma distribution, 113–127
  maximum likelihood estimators, 115–118
    for censored samples, 115–116
    for truncated samples, 117–118, 120–121
  modified maximum likelihood estimators for censored samples, 116–117
  moments of, 114
  pdf of, 114
  progressively censored samples from, 120
  standardized pdf, 114

[Gamma distribution, Continued]
  two-parameter distribution, 118–119
Garellis, E., 33
Geiger, H., 207
Giesbrecht, F., 96, 100, 101
Gnedenko, B., 140
Goedicke, V., 191
Govindarajulu, Z., 70
Greenberg, B. G., 66, 67, 70, 71, 79, 82, 83
Greenwood, M., 208
Grundy, P. M., 3
Guenther, W. C., 219
Gumbel, E. J., 139
Gumbel distribution (see Extreme value distribution)
Gupta, A. K., 3, 26, 28, 66, 70, 81
Gupta, S. S., 70
Gupta's alternate linear estimators, 70, 81

Hagstroem, K. G., 165
Hald, A., 1, 3
Haldane, J. B. S., 208–210
Half normal distribution (see Folded normal distribution)
Halperin, M., 3
Harris, C. M., 165
Harter, H. L., 85, 98, 101
Hartley, H. O., 208, 236
Hazard function and cumulative hazard function, 129, (see also specific distributions)
Herd, G. R., 50
Higher moment estimation (see Method of higher moments)
Hill, B. M., 96, 98
Hotelling, H., 187
Hypergeometric distribution, 215–219

[Hypergeometric distribution, Continued]
  probability function, 215
  truncated distribution, 216
  parameter estimation, 217

Inflated zero distributions, 236–243, (*see also* specific distributions)
Instantaneous failure rate (*see* Hazard function)
Inverse Gaussian distribution, 106–113
  limiting normal distribution, 108
  mle for progressively censored samples, 108–111
    loglikelihood function, 109
  mle for truncated samples, 112–113
  modified maximum likelihood estimators for censored samples, 111
  moments of, 107
  pdf of, and standardized pdf of, 107
  standardized cdf of, 108

Jaech, J. L., 85
Johnson, N. L., 3, 96, 106, 113, 139, 165, 166, 171, 180, 200, 208, 210, 214, 219, 236
Joshi, S. W., 219

Kane, V. E., 96
Kao, J. H. K., 85
Kempthoprne, O., 96, 100, 101
Kerrich, J. E., 1
Keyfitz, N., 3
Kocherlakota, S., 70
Koniger, W., 85

Kotz, S., 96, 106, 113, 139, 165, 166, 171, 214, 219
Kumra, S., 70

Lawless, J. F., 139, 145
Lawley, D. N., 186
Lee, A., 2
Lehman, E. H., 85
Lemon, G. H., 85
Leone, F., 85
Lieberman, G. J., 219
Linear estimators, 66–84
  alternative estimators, 70, 83
  calculation of, 67
  derivations, 67–70
Lipow, M., 85
Lloyd, D. K., 85
Lloyd, E. H., 66
Lognormal distribution, 96–105
  characteristics of, 97
  global maximum likelihood estimation, 97–98
  local maximum likelihood estimation, 98–100
    in progressively censored samples, 98–100
  modified maximum likelihood estimators, in censored samples, 100–104
  pdf and cdf, of, 97
  relation to normal distribution, 96–97
Log Weibull distribution (*see* extreme value distribution)
Lund, R. E., 219

Mahalanobis, P. C., 16
Malik, H. J., 70, 165
Mandelbrodt, B., 165
Mann, N. R., 85, 139, 145
Mantel, N., 214

Index

Maximum likelihood estimation (*see* specific distributions)
Maxwell, J. C., 150
Maxwell-Boltzmann distribution, 150–151
   characteristics of, 150
   parameter estimation (*see* Rayleigh distribution)
   pdf of, 150
Mendenhall, W., 3
Menon, M. V., 85
Method of higher moments, 174
Misclassified inspection data, 224–235
   from binomial distribution, 232–235
      parameter estimates, 233
      sampling errors, 234–235
   from Poisson distribution, 227–228
      parameter estimates, 227–228
      sampling errors, 228–229
Modified maximum likelihood estimators (*see* specific distributions)
Molina, E. C., 202, 206, 207
Mood, A. M., 219
Moore, A. H., 85, 98
Moore, P. G., 3, 200, 207
Morris, K. W., 222
Muench, H., 236, 241
Multivariate distributions, 185
Multivariate normal distribution, 185–198
   pdf of p-dimensional normal, 186
   restricted samples from, 187–189
      mle of, 187–188
      reliability of estimates, 194–195

Muniruzzaman, A. N. M., 165
Munro, A. M., 96

Negative binomial distribution, 208–212
   estimate variances, 211
   inflated zero distribution, 239–243
   probability function of, 208
   truncated, zero class missing, 209
      Brass estimators, 211
      mle of, 210
Nelson, W. B., 5, 29, 50, 56, 85, 134, 136, 137, 138
Norgaard, N. J., 3
Normal distribution, 8–84
   cdf and pdf of, 9
   centrally censored samples, 62–64
   centrally truncated samples, 64–65
   doubly censored samples, 46–47
      computational procedures, 48–49
      sampling errors of estimates, 47–48
      total number censored observations known, but not number in each tail separately, 59–62
   doubly truncated samples, 31–45
      computational procedures, 41–43
      mle, derivations, 32–33
      sampling errors of estimates, 40–41
   progressively censored samples, 50–59
      cumulative hazard plot estimates, 56
      errors of estimates, 54

[Normal distribution, Continued]
  mle, derivations, 50–54
  probit estimates, 54–56
  special cases, 53
  singly left censored samples from, 18–30, 47
    mle of, 18–26, derivations, 18–20
    sampling errors of estimates, 25–26
  singly left truncated normal distribution, 9–18
    equivalence of mle and moment estimators, 8
    moment estimators of, 10–12
    moments of, 9–10
    pdf of, 9
    three-moment estimators of, 13–16
Norton, H. W., 43

Ogawa, J., 66
Owen, D. B., 219

Padgett, W. J., 106
Pareto, V., 165
Pareto distribution, 165–173
  cdf of, 166
  estimate reliability, 171–172
  first order statistic, 167–168
    characteristics of, 168
    distribution of, 167
  hazard function of, 166
  moments of, 166
  parameter estimation, 168–171
  pdf of, 165
  singly right censored samples, 170–171
    loglikelihood function, 170
    mle of, 170
    modified mle of, 170–171

[Pareto distribution, Continued]
  singly right truncated samples, 168–170
    loglikelihood function, 169
    mle of, 169
Patel, J. K., 219
Patil, G. P., 219, 222
Pearson, K., 2, 113, 174, 186
Pearson distributions, 174–184
  distribution types, 180
  doubly truncated samples estimating equations, 178
  genesis, (differential equation), 174
  method of higher moments, 178
  moments of, 175–176
  normal distribution as special case, 181
  singly truncated samples, 180–181
  truncated distributions, 177
    moments of, 177
    recursion formula for moments, 177
  type III (gamma) distribution as special case, 181–182
Phatak, A. G., 220, 224
Pittman, E. J. G., 187
Plackett, R. L., 3, 200
Poisson, S. D., 200
Poisson distribution, 199–208
  asymptotic variance of estimates, 205–206
  cumulative probability function of, 201
    derivative of, 201
  inflated zero distribution, 236–239
  limiting function of binomial distribution, 200
  mle, 202–207

Index

[Poisson distribution, Continued]
  doubly censored samples, 204–205
  doubly truncated samples, 203–204
  singly censored samples, 203
  singly truncated samples, 202–203
  moments of, 200–201
  probability function of, 200
Prescott, P., 70
Procasinni, A. A., 85
Proschan, F., 85

Quant, R. E., 171
Quensel, C. E., 3
Qureishi, A. S., 70

Rayleigh, J. W. S., 146
Rayleigh distributions, 146–164
  cdf of, 147
  folded normal (a special case), 147–148
  moments of, 147
  parameter estimation, 151–156
    asymptotic variances of mle for censored and truncated samples, 156
    mle for singly right censored samples, 151–155
    mle for singly right truncated samples, 155–156
  pdf of (dimension p), 146
  three-dimensional Rayleigh (*see* Maxwell-Boltzmann distribution)
  two-dimensional Rayleigh, (circular normal), 148–150
    cdf and pdf of, 149
    characteristics of, 149
    relation to Weibull distribution, 151

Read, C. B., 219
Rescia, R. R., 240, 241
Rider, P. R., 3, 200, 207, 208
Ringer, L. J., 85
Roberts, H. R., 50
Rockette, H., 85
Romano, A., 85
Roy, L. K., 106
Rukhin, A. L., 96
Rutherford, E., 207

Sample types, definitions of, 4–5
  censored, type I, 4
  censored type II, 4
  truncated, 4
Sampford, M. R., 3, 50, 208, 210, 236
Sarhan, A. E., 66, 67, 70, 71, 79, 82, 83
Saw, J. G., 3, 81
Saw's nonlinear estimators, 81
Scarborough, J. B., 43
Schneider, H., 3
Schroedinger, E., 106
Scott, J. A., 241
Shah, B. K., 70
Shenton, L. R., 208
Shimizu, K., 96
Shook, B. L., 182
Singh, S. N., 236
Smoluchowsky, M. W., 106
Sobel, M., 3
Sprinkle, E. E., 85
Srivastava, M. S., 165
Stedinger, J. R., 96
Stevens, W. L., 3
Sundaraiyer, V. H., 3, 5, 126

Teicheroew, D., 66
Thompson, G. W., 33
Tietjen, G. L., 70
Tiku, M. L., 70

Tippett, L. H. C., 200
Truncated distributions (*see* Truncated samples)
Truncated samples, 2, (*see also* specific distributions)
Tukey, J. W., 187
Tweedie, M. C. K., 106

Votaw, D. F., 186

Wald, A., 106
Wallace year book, 2
Wasan, M. T., 106
Wei, L. J., 106
Weibull distribution, 85–90
　characteristics, 86
　pdf of, 85
　progressively censored samples, 88–90
　　hazard plot estimates, 135–136
　　relation to extreme value distribution, 135

[Weilbull distribution, Continued]
　singly censored samples, 86–88
　three-parameter distribution, 90–92
　　errors of estimates, 93–94
　　mle of, 91
　　modified mle of, 92–93
Whitten, B. J., 3, 5, 12, 21, 33, 35, 85, 92, 94, 98, 104, 106, 121, 123, 124, 125, 126, 127, 138, 157, 158, 160, 163, 165, 167, 179
Wilks, S. S., 186
Wingo, D. R., 85, 96
Wixley, R. A. J., 96
Wycoff, J., 85

Yuan, P. T., 96
Yule, G. U., 208

Zanakis, S. H., 85
Zipf, G. K., 165